Chemistry
for CBSE Class **IX**

Atul Kumar Singhal

Ph.D. (Organometallic Chemistry)

Academic Head, Aakash Institute

Modipuram, Meerut, Uttar Pradesh

Universities Press

CHEMISTRY FOR CBSE CLASS IX

UNIVERSITIES PRESS (INDIA) PRIVATE LIMITED

Registered Office
3-6-747/1/A & 3-6-754/1, Himayatnagar, Hyderabad 500 029, Telangana, India
info@universitiespress.com; www.universitiespress.com

Distributed by
Orient Blackswan Private Limited

Registered Office
3-6-752 Himayatnagar, Hyderabad 500 029 Telangana, India

Other Offices
Bengaluru / Chennai / Guwahati / Hyderabad / Kolkata
Mumbai / New Delhi / Noida / Patna / Visakhapatnam

© Universities Press (India) Private Limited 2024
First published 2024

Cover and book design:
© Universities Press (India) Private Limited 2024

ISBN 978-93-93330-62-8

Typeset in Minion Pro 9.5 points *by*
Cameo Corporate Services Limited, Coimbatore

Printed in India by
Shiva Printech Pvt Ltd, Tronica city (U.P.)

Published by
Universities Press (India) Private Limited
3-6-747/1/A & 3-6-754/1, Himayatnagar, Hyderabad 500 029, Telangana, India

501965

Care has been taken to confirm the accuracy of the information presented in this book. The author and the publisher, however, cannot accept any responsibility for errors or omissions or for consequences from application of the information in this book, and make no warranty, express or implied, with respect to its contents.

Preface

Chemistry for CBSE Class IX is an excellent preparatory book for students studying in Class IX of the CBSE Board. This book provides class-tested material and practice problems that help students understand theories and concepts and, as a consequence, develop problem-solving skills to attempt exams with full confidence.

The book is written in lucid language with lots of solved examples, and aims to assist students in understanding concepts even without the help of an instructor, as solutions are provided for most of the questions.

Salient features

- Structured according to the latest syllabi and exam pattern of the CBSE Board
- Includes MCQs, Very Short, Short, Long, HOTS and Case-study/Application-based questions
- Includes NCERT exercises and Exemplar questions along with solutions
- The app accompanying this book supports more chapter-wise MCQs for competitive exam practice

The purpose of this book is two-fold: to strengthen the students' foundation in Chemistry in order to excel in school exams and to prepare students in Chemistry for competitive exams.

I welcome suggestions and constructive criticism from the readers. Students and teachers can share their feedback at singhal.atul1974@gmail.com.

Atul Kumar Singhal

Acknowledgements

The contentment that accompanies the successful completion of my work would remain essentially incomplete if I fail to mention the people whose constant support has encouraged me.

I am grateful to all my revered teachers, especially the late J.K. Mishra, Dr D.K. Rastogi, the late A.K. Rastogi and my honorable guide, Dr. S.K. Agarwala. Their knowledge and wisdom assisted me in no small measure in presenting this work.

I express my immense gratitude to my colleagues for collaborating so patiently and constructively during the various stages of this project. I am extremely thankful to Ronit Singh Kushwaha for typing out the manuscript.

I am indebted to my father, B.K. Singhal, mother, Usha Singhal, and brothers Amit Singhal and Katar Singh, who have been my motivators at every step. Their never-ending affection has provided me with moral support and encouragement while writing this book. Last but not the least, I express my deepest gratitude to my wife, Urmila, my daughters Khushi and Shanvi, and my little but witty-beyond-years son Shashwat, who always supported me during my work.

I would like to express my heartfelt gratitude to Madhavi Sethupathi for her brilliant editing. I would also like to show my appreciation to Kallol Das, Thomas Mathew Rajesh and Madhu Reddy for providing all the necessary help and assistance.

Atul Kumar Singhal

Contents

Preface   *iii*
Acknowledgements   *iv*

Chapter 1   **Matter in Our Surroundings**   **1.1**
Introduction | Matter | Classification of Matter | Characteristic Features of Matter |
Properties of Matter | Particle Nature of Matter | States of Matter | Change of State of Matter

Chapter 2   **Is the Matter Around Us Pure?**   **2.1**
Introduction | Pure Substances | Elements | Compounds | Mixtures | Types of Mixtures |
Properties of Mixtures | Physical Changes | Chemical Changes | Solutions | Suspensions |
Colloids | Separation of Mixtures | Separation of a Mixture of Two Solids | Separation of a
Mixture of a Solid and a Liquid | Separation of a Mixture of Two or More Liquids

Chapter 3   **Atoms and Molecules**   **3.1**
Introduction | Laws of Chemical Combination | Law of Conservation of Mass | Law of
Constant Proportions | Dalton's Atomic Theory | Features of Dalton's Atomic Theory |
Significance of Dalton's Atomic Theory | Drawbacks of Dalton's Atomic Theory | Atoms
| Symbols of Elements | Atomic Mass | How Do Atoms Exist? | Ions | Simple Ions and
Compound Ions | Valency | Valency of Ions | Variable Valency | Chemical Formulae |
Formulae of Elements | Formulae of Ionic Compounds | Formulae of Molecular Compounds
| Writing the Formula of a Compound | Gram Atomic Mass | Gram Molecular Mass |
Formula Unit Mass | Mole

Chapter 4   **Structure of the Atom**   **4.1**
Introduction | Charged Particles in Matter | Features of Cathode Rays | Electron | Features
of Anode Rays | Proton | Neutron | Features of a Neutron | Structure of the Atom and
Development of the Atomic Model | Thomson's Atomic Model | Rutherford's Atomic Model
| Bohr's Atomic Model | Composition of the Nucleus | Electron Distribution in Orbits |
Electronic Configurations of the First Twenty Elements | Valency | Isotopes | Isobars

To
My parents and teachers

CHAPTER 1

Matter in Our Surroundings

After studying this unit, you will be able to:

- Define matter, its types and its properties
- Explain the properties of solids, liquids and gases
- Describe the changes in the states of matter
- Explain the effect of temperature and pressure on the states of matter
- Define melting, freezing, vaporisation and condensation
- List the types of latent heat
- Explain sublimation and liquification

1.1 INTRODUCTION

When we look at our surroundings, we see a vast variety of things possessing different shapes, sizes, textures, and so on. Everything in our universe is made up of material which is called **matter**. The air we breathe, the food we eat, clouds, stars, plants and animals, even a small drop of water or a particle of sand – everything is matter. When we look around, we observe that all these things occupy space, that is, they have volume and mass.

Early Indian philosophers classified matter into five basic elements – the *panch tatva* – air, earth, fire, sky and water (Fig. 1.1). According to them, everything, living or non-living, was made up of these five elements. Ancient Greek philosophers also arrived at a similar classification of matter.

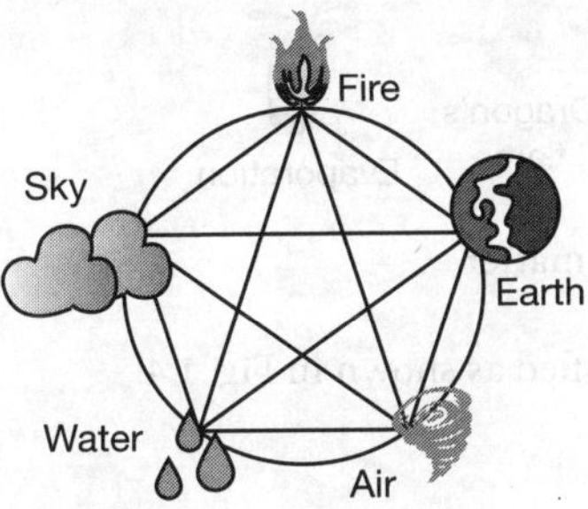

Fig. 1.1 Panch tatva

Modern-day scientists have classified matter on the basis of their physical properties and their chemical nature. In this chapter, we will learn about matter based on its physical properties. The chemical aspects of matter will be taken up in subsequent chapters.

1.2 MATTER

Anything that occupies space and has mass is known as matter. Everything around us is made up of tiny pieces or particles. These particles are called atoms or molecules. For example, air or gases like ammonia are considered as matter because they have mass and occupy space; although we cannot see them, their presence can be felt by their characteristic smell. Vacuum cannot be considered as matter as it does not occupy space nor does it have mass. Heat, light, humidity, electricity, sound and magnetism cannot be matter as they do not have mass and do not occupy space. Some more examples of matter are food, water, air, clothes, table, chair, plants and trees (Fig. 1.2).

Thus, anything that occupies space, possesses mass and the presence of which can be felt by using one or more of our five senses is matter.

1.2.1 Classification of Matter

Matter can be classified into several categories depending on its physical and chemical nature:

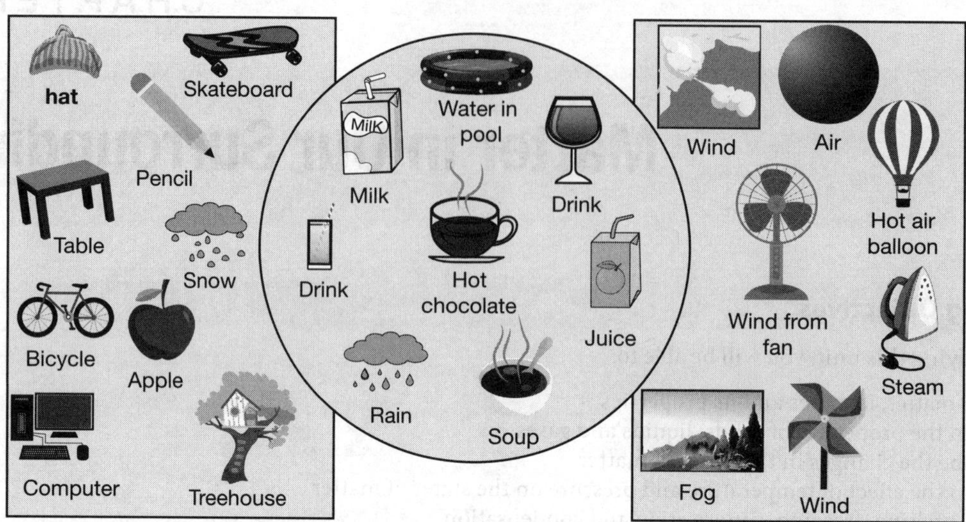

Fig. 1.2 Examples of matter

- **Physical classification**: On the basis of their physical states, all matter can be classified into four groups: solids, liquids, gases and plasma (Fig. 1.3).

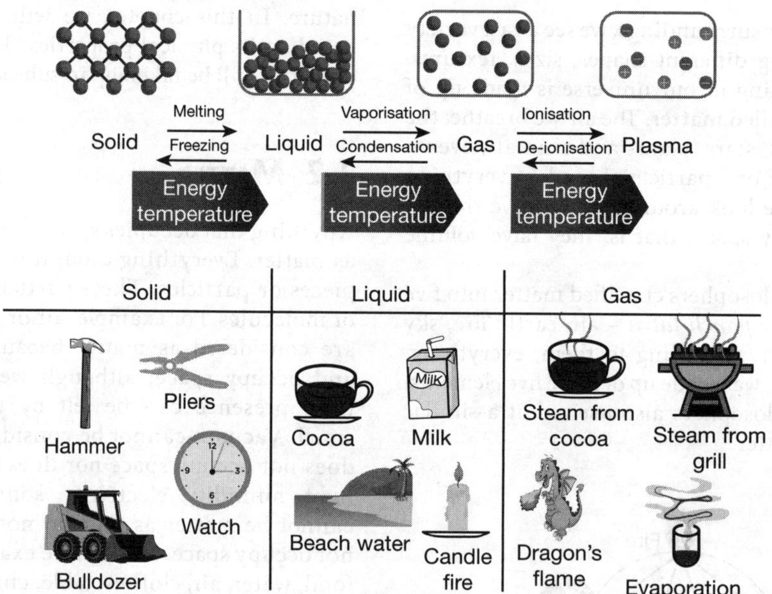

Fig. 1.3 Physical classification of matter

- **Chemical classification**: On the basis of purity, matter can be classified as shown in Fig. 1.4.

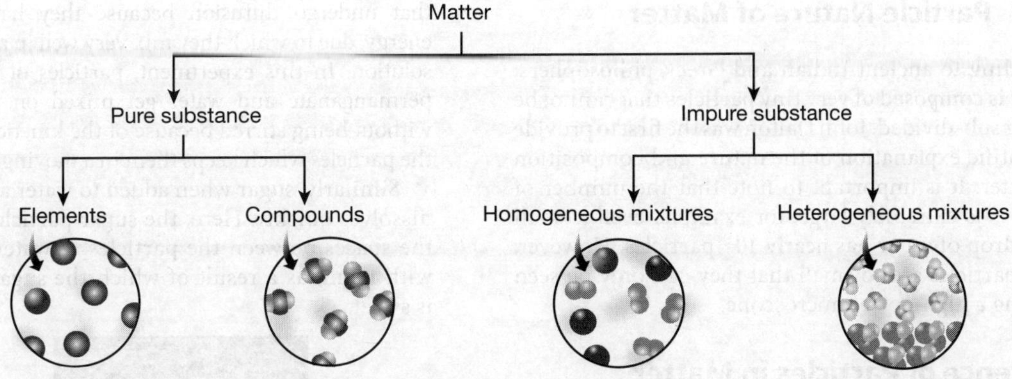

Fig. 1.4 Chemical classification of matter

1.2.2 Characteristic Features of Matter

- **Mass:** Any form of matter has mass. Mass is the amount of matter present in any object. For example, even though air is invisible, it has mass.
- **Space:** Any form of matter also occupies space. For example, a chair in a room, a book on a bookshelf, a ring in a box.
- **Property of inertia:** Matter has the property of inertia. For example, a football can move only when it is pushed by an external force such as by a player; a book lying on a table will remain as it is unless some external force disturbs its position, by picking it up or pushing it.
- **Influence of gravity:** When anything is thrown upwards with force, it comes to the ground automatically due to the force of attraction exerted by the earth; this force is called gravity. Anything composed of matter is under the influence of gravity. For example, fruits always fall down from the tree; water always flows from a higher level to a lower level.
- **Matter cannot be destroyed or created:** During all physical and chemical changes, the total mass of the matter before and after the change remains constant (Lavoisier's Law of Conservation of Mass).

1.2.3 Properties of Matter

Every substance in nature has a special set of properties that allows us to recognise and distinguish it from other substances. Such properties of matter can be categorised as physical or chemical. The measurement of physical properties is possible without changing the identity and composition of the substance. Such properties are colour, hardness, odour, boiling point, melting point, density, and so on (Fig. 1.5).

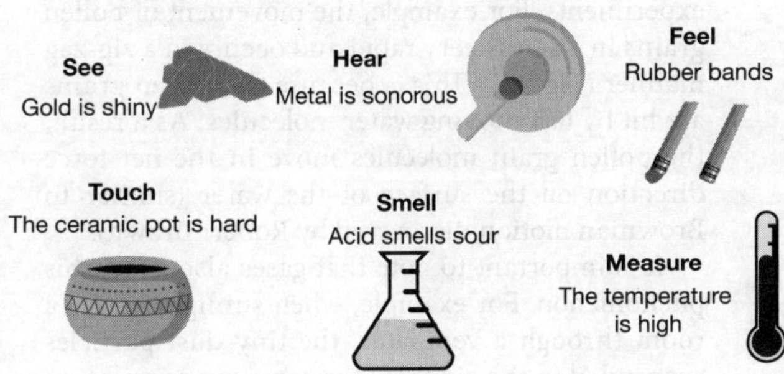

Fig. 1.5 Physical properties of matter

Chemical properties represent the way by which a substance can change or react to give other substances (Fig. 1.6). Flammability is such a property and it is the ability of a substance to burn in air or in the presence of oxygen (O_2).

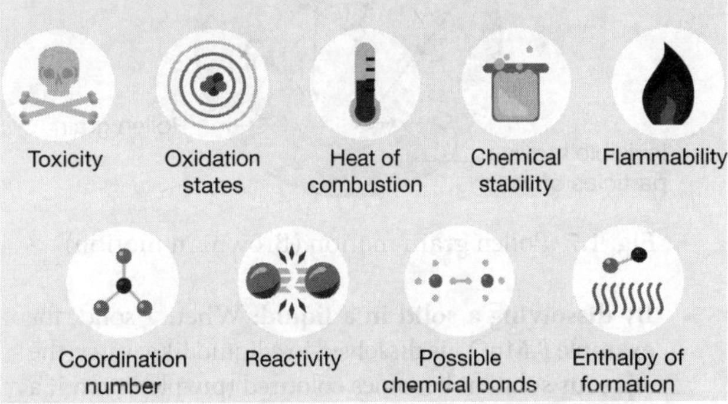

Fig. 1.6 Chemical properties of matter

1.2.4 Particle Nature of Matter

According to ancient Indian and Greek philosophers, matter is composed of very tiny particles that cannot be further sub-divided. John Dalton was the first to provide a scientific explanation of the nature and composition of matter. It is important to note that the number of particles is extremely large; for example, even a small 1-mL drop of water has nearly 10^{21} particles. However, these particles are so small that they can only be seen by using a high-power microscope.

Existence of Particles in Matter

The existence of particles in matter can be proved by the following observations:

- **Brownian motion:** The existence of particles in matter and their motion can be proved based on the results of diffusion and Brownian motion experiments. For example, the movement of pollen grains in water is very rapid and occurs in a zig-zag manner (Fig. 1.7). This is because the pollen grains are hit by fast-moving water molecules. As a result, the pollen grain molecules move in the net force direction on the surface of the water (similar to Brownian motion discovered by Robert Brown).

    It is important to note that gases also show this phenomenon. For example, when sunlight enters a room through a ventilator, the tiny dust particles suspended in the air show the same zig-zag motion as they are constantly hit by the fast-moving particles of air. This means that from Brownian motion, both the particle nature and the movement of particles is proved.

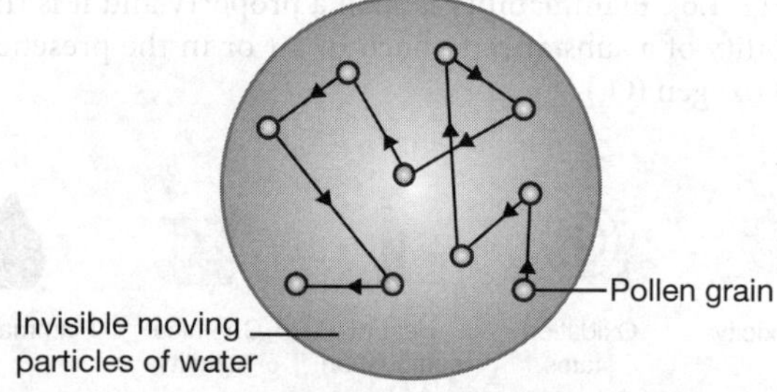

Fig. 1.7 Pollen grain motion (Brownian motion)

- **By dissolving a solid in a liquid:** When a solid, for example $KMnO_4$, is dissolved in a liquid like water, the aqueous solution becomes coloured (purple), even if a very small amount or just a crystal of $KMnO_4$ is added to the water (Fig. 1.8). This means that each crystal of $KMnO_4$ (matter) is composed of tiny particles

that undergo diffusion because they have kinetic energy, due to which they mix very well in an aqueous solution. In this experiment, particles of potassium permanganate and water get mixed on their own without being stirred because of the kinetic energy of the particles which keeps them in a moving state.

Similarly, sugar when added to water and stirred dissolves quickly. Here, the sugar particles get into the spaces between the particles of water and mix with them, as a result of which the sugar solution is sweet.

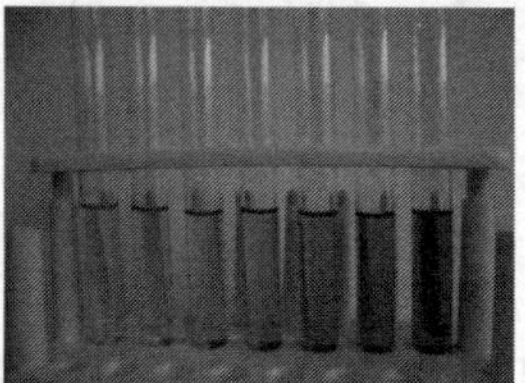

Fig. 1.8 Dissolution of $KMnO_4$

- **Mixing of two or more gases:** Air is a colourless gas or a mixture of gases. A gas vessel or jar may seem empty, but it is actually filled with air but we cannot see it as it is colourless. The presence of air in a jar can be confirmed by adding a brown-red liquid form of bromine that is easily vaporised. When a jar having air is placed over the jar containing the bromine vapours, after some time, both the jars become red-brown in colour (Fig. 1.9). This means that the particles in the air and bromine vapours undergo collision so they mix uniformly (diffusion). Although bromine is heavier than air, its vapours move upwards due to their high kinetic energy.

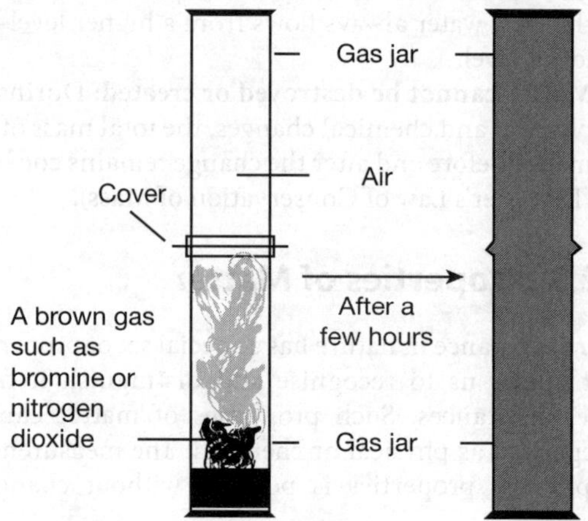

Fig. 1.9 Mixing of bromine vapours with air

Characteristics of Particles of Matter

The following are the important characteristics of particles of matter:

- **Particles of matter are very, very small**: Matter is composed of very tiny particles. This can be demonstrated when we add a crystal of KMnO4 or copper sulfate (blue vitriol) to water. The aqueous solution becomes purple or blue and the size of the crystal becomes smaller and smaller. Finally, the crystal decomposes into a number of tiny particles as the colour becomes uniform throughout. In Fig. 1.10, we can see that on dilution, the potassium permanganate solution become lighter and lighter in colour. This is because $KMnO_4$ is made up of millions of tiny particles which spread continuously and impart colour on dilution (addition of water). The same is seen in the case of copper sulfate solution, where the dilution colour changes from dark blue to light blue.

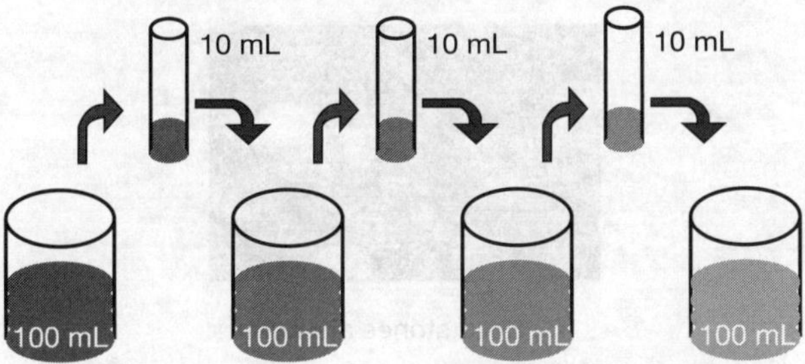

Fig. 1.10 Dilution of potassium permanganate

- **Particles of matter have spaces between them**: This can be easily proved by dissolving salt or sugar in water. When we make tea, coffee or lemonade, particles of one type of matter get into the spaces between the particles of the other. This shows that there is space between the particles of matter. It is interesting to note that in a sugar solution, the volume of water does not change as the sugar molecules only occupy the space between the particles of water (Fig. 1.11).

Fig. 1.11 Sugar solution

- **Particles of matter move continuously**: Particles of matter possess kinetic energy. As the temperature rises, the particles move faster. So, we can say that with an increase in temperature, the kinetic energy of the particles also increases. The movement of particles can be easily confirmed by diffusion. When we burn an incense stick in a room, the fragrances spread all around in a very short period of time as the gaseous particles move or diffuse in all possible directions. Another example is when we spray a perfume or deodorant.

- **Particles of matter attract each other**: Particles of matter have forces of attraction which bind them together. This force of attraction between particles of the same substance is called **cohesion**. In the case of particles of different substances, it is known as **adhesion**. For example, if you beat a piece of iron nail, a cube of ice and a piece of chalk with a hammer, you will notice that chalk and ice easily break into smaller particles, but the iron nail does not . This is because the iron particles have a strong force of attraction, while that among of the particles of chalk and ice is weaker.

Competition Edge

Diffusion: The spreading out and mixing of a substance with another substance due to the motion of its particles is known as diffusion. It is a property of matter which is based on the motion of its particles.

Diffusion is the flow of molecules from the side with higher concentration to the side with lower concentration. For example, the smell of food being cooked spreads throughout the cooking area as diffusion of food gases occurs in the air.

Osmosis: Osmosis is the flow or movement of solvent particles from a dilute solution (less solute) to a concentrated solution (more solute) through a semi-permeable membrane. For example, raisins swell in water but shrink in sugar water.

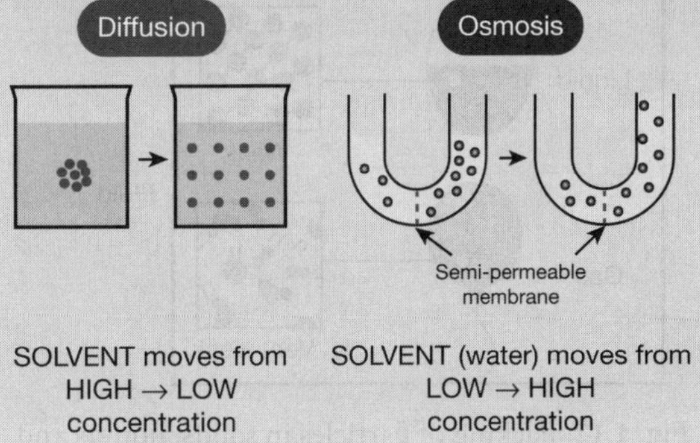

TEST YOUR KNOWLEDGE

1. Which of the following can be labelled as matter?
   Table, air, love, chocolate, smell, hate, almonds, thought, cold, soft drink, smell of perfume.

Solution: Table, air, chocolate, smell, almonds, soft drink and smell of perfume are classified as matter.

2. A diver can cut through water in a swimming pool. Which property of matter does this observation show?

Solution: The phenomenon of cutting through water by the diver shows that matter has space between its particles.

1.2.5 States of Matter

As we have discussed, on the basis of physical properties, matter can exist in solid, liquid and gas forms. First, let us understand the packing of particles and force of attraction present between the particles.

The packing of particles in solids, liquids and gases not only determines their shape, but also explains their properties. From Fig. 1.12, it is clear that in solids, the constituent particles are very closely packed (1 Å, 100 PM or 1010 m apart). This means that particles will have the strongest force of attraction and the least space between them; therefore, there will be almost no movement, with very low compressibility, high rigidity and density.

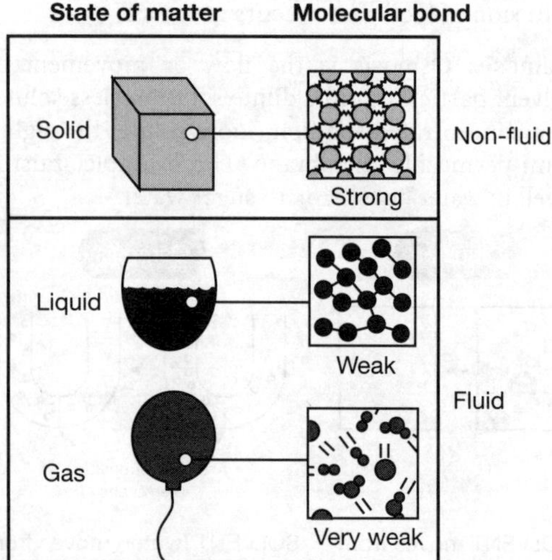

Fig. 1.12 Packing of particles in solids, liquids and gases

In liquids, the constituent particles are less closely packed (10–1000 Å) which means a lower force of attraction than in solids. This is why liquids show fluidity or movement and have less density and hardness than solids.

In gases, the constituent particles are quite far apart or the interparticle space is very large, which results in weak force of attraction but maximum fluidity and compressibility and the least density. Solids and liquids are called the **condensed phases** of matter due to there being less space between the constituent particles.

According to the kinetic theory of matter, particles are in continuous motion and have kinetic energy which is not the same for solids, liquids and gases. In fact, they show decreasing order of kinetic energy: gas > liquid > solid.

Rigidity refers to an inflexible or unbending nature, while fluid means flow of matter. For example, pieces of stone show rigidity while liquids like milk and tea and gases like chlorine and hydrogen sulfide show fluidity (Fig. 1.13).

(a) Solid stones are rigid

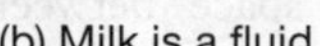

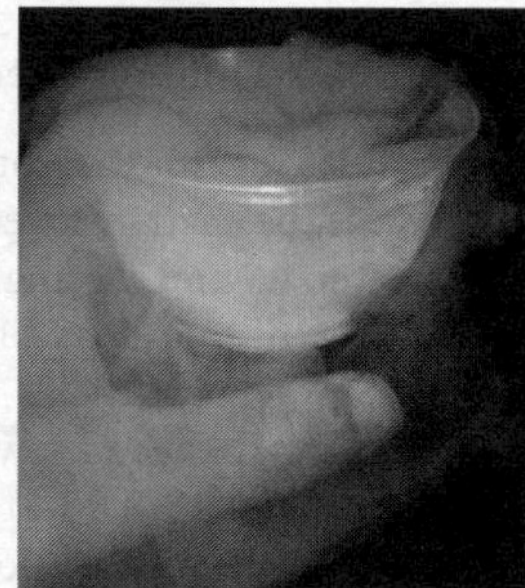

(b) Milk is a fluid   (c) Gas is a fluid

Fig. 1.13 Rigidity and fluidity in solids, liquids and gases

Now let us taker a closer look at the solid, liquid and gaseous states (Table 1.1).

Table 1.1 Comparison of characteristic properties of solids, liquids and gases

Property	Solid	Liquid	Gas
Shape	Definite	Attains the shape of the container, but does not necessarily occupy all of it	Attains the shape of the container by occupying the entire space available to it
Volume	Definite	Definite	Attains the volume of the container
Compressibility	Almost nil	Very low	Very large
Fluidity or rigidity	Rigid	Fluid	Fluid
Density	High	Low	Very low
Diffusion	Generally does not diffuse	Diffuses slowly	Diffuses rapidly
Free surfaces	Any number of free surfaces	Only one free surface	No free surface

Solid State

The solid state is one of the four fundamental states of matter. A solid is defined as a form of matter that possesses rigidity, is incompressible and has definite volume. The molecules in a solid are closely packed together and contain the least amount of kinetic energy. A solid is characterised by incompressibility, structural rigidity, high mechanical strength and resistance to a force applied to the surface. In a solid, the constituent species are closely packed or held together by strong forces; so, in between the constituents, there are very small gaps and they cannot move or flow. For example, ice, wood, coal, stone, iron, steel and brick (Fig. 1.14).

Fig. 1.14 Examples of solid state

Properties of solids: Solids are characterised by the following properties:

- **Shape and Volume:** A solid has a fixed shape and a fixed volume as the constituents (atoms, ions or molecules) are arranged in a definite pattern held together by strong intermolecular forces with the least possible interparticle space. This regular arrangement gives them a definite geometry and volume. For example, a pan in our hand or a table in a classroom not only have a fixed shape but also distinct boundaries.

  It is interesting to note that sugar or salt does not appear to have a fixed shape but it is considered as a solid as the shape of the individual crystals of sugar or salt remain fixed, even if they are placed in vessels having different shapes.

- **Rigidity:** Solids show rigidity, which means they have the tendency to maintain their shape when outside forces are applied. However, they may break when dropped or hammered. Some solids like rubber bands can change shape when force is applied, but they regain their original shape when the forced is removed.

- **Compressibility:** Solids are rigid and cannot be compressed easily as the constituents are closely packed; the intermolecular distance or space is very low and cannot be further reduced.

- **Hardness and Density:** Solids are quite hard and they have high density since they have a closely packed arrangement of particles. A solid cannot fill its container completely and does not show fluidity.

- **Diffusion:** A solid cannot diffuse into another solid as very strong intermolecular forces of attraction are present and the interatomic particle distance is very small. So solid particles can only vibrate around their fixed position and cannot move away from it.

  For example, if we write something on a blackboard and leave it for a considerable period of time, we will find that it becomes quite difficult to clean the blackboard afterwards. This is because some of the particles of chalk have diffused into the surface of the blackboard. If two metal blocks are bound together tightly and kept undisturbed for a few years, the particles of one metal will have diffused into the other metal.

- **Melting point**: Solids have a high melting point as a huge amount of energy is required to overcome the strong intermolecular forces present in their constituents.

- **Motion**: Solids do not flow, so they cannot acquire the shape of the vessel in which they are kept. They do not show any motion as the constituents are held at their respective positions by strong intermolecular forces. They can only vibrate.

TEST YOUR KNOWLEDGE

1. Give reasons for the following:
   (a) A wooden chair should be called a solid.
   (b) We can easily move our hand in the air, but to do the same through a solid block of wood, we would need to be a karate expert.

Solution:
(a) There is a strong force of attraction between the molecules of wood and the intermolecular space is the least. So, a wooden chair has a definite shape and volume and it should be called a solid.
(b) Air molecules are extremely far from each other due to the negligible force of attraction between them. So, our hand gets sufficient space to move in air and we also displace some air molecules by applying force. However, a solid block of wood has closely packed molecules, so there is no question of moving the hand through it in the absence of a suitable force in the proper direction.

Liquid State

A liquid is one of the four fundamental states of matter. It is defined as a form of matter that possesses fluidity, less compressibility and which has definite volume but no definite shape. A liquid is a nearly incompressible fluid that conforms to the shape of its container but retains a constant volume independent of pressure. In liquids, the molecules are more closely held than in gases due to strong intermolecular forces (but less than in solids). Liquids have more vacant space than solids. In liquids, molecules can move or flow.

For example, water, milk, fruit juice, ink, groundnut oil, kerosene and petrol (Fig. 1.15).

Liquid

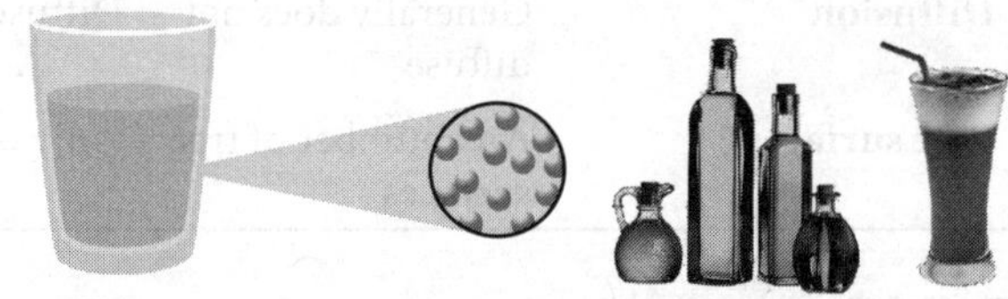

Fig. 1.15 Examples of liquid state

Properties of liquids: Liquids are characterised by the following properties:

- **Shape and Volume:** A liquid will have a fixed volume but cannot have a fixed shape. A liquid takes the shape of the vessel in which it is stored as the liquid molecules are not held strongly enough to have a fixed position or definite geometry (Fig. 1.16). However, the intermolecular force is strong enough for them to have a definite volume.

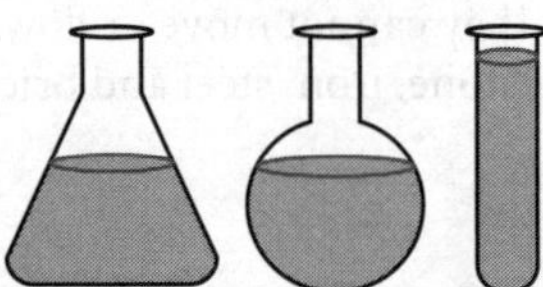

Fig. 1.16 Liquid adopts the shape of the vessel

- **Rigidity:** Liquids are not rigid but can flow as the interparticle distance is greater and the force of attraction is weaker than in solids. However, fluidity differs from one liquid to another. For example, water flows faster than honey or glycerol.

- **Compressibility:** Like solids, liquids cannot be compressed much because there is not much vacant space; however, they can be compressed more than solids but less than gases.

- **Hardness and Density:** Liquids have moderate-to-high density. They are usually less dense than solids but more dense than gases.
- **Diffusion:** Liquids are more diffusible than solids but less than gases as the vacant spaces between their molecules is more than that of solids but less than that of gases. They do not fill their container completely. Diffusion in liquids is slower than in gases as the particles in liquids move slower than the particles in gases. The rate of diffusion in liquids is much faster than that in solids because the particles in a liquid move much more freely and have greater spaces between them as compared to particles in solids. The rate of diffusion differs from one liquid to another. For example, when water and alcohol are mixed, both diffuse into each other, forming a homogeneous solution.
- **Melting point:** Liquids have lower melting points than solids but higher than that of gases as the intermolecular forces are weaker than in solids but stronger than in gases.
- **Motion:** Liquids generally flow or show motion easily as there is enough space between their molecules. However, their motion is less than in gases.
- **Evaporation:** Liquid molecules can evaporate and change into vapour at room temperature as the kinetic energy of liquid molecules can overcome the intermolecular force present between the molecules. The rate of evaporation is not the same for all liquids as the kinetic energies of liquid molecules differ.

TEST YOUR KNOWLEDGE

1. Give two reasons to justify the following:
   (a) Water at room temperature is a liquid.
   (b) An iron almirah is a solid at room temperature.

Solution:
(a) Water is a liquid at room temperature because it has a tendency to flow. It takes the shape of the container in which it is stored, but its volume remains the same.
(b) An iron almirah is a solid at room temperature because its shape and volume are definite. It is hard and rigid. Its density is high.

2. Explain the following:
   (a) Liquids generally have lower density compared to solids. But you must have observed that ice floats on water.
   (b) When a liquid is transferred from a small vessel to a large vessel at the same temperature, what will be the effect on vapour pressure?

Solution:
(a) Liquids have lower density than solids. Water is also a liquid, so it should also have less density than ice. However, this is not so because of the cage-like structure of ice. That is, the presence of vacant spaces between water (H_2O) molecules when they link with each other in ice. The number of these spaces is comparatively less in water. Being more porous than water, ice is lighter than water and floats over the surface of the water.
(b) There will be no effect on vapour pressure when a liquid is transferred from a small vessel to a large vessel at the same temperature as vapour pressure does not depend on the size of the vessel.

Competition Edge

Vapour pressure: The pressure exerted by vapour in equilibrium with the liquid state at a given temperature is known as the vapour pressure of the liquid. It depends on the nature and temperature of the liquid but not upon the size of the vessel.

$$\text{Vapour pressure} \propto \text{Temperature} \propto \frac{1}{\text{Intermolecular force of attraction}}$$

Gaseous State

The gaseous state is one of the four fundamental states of matter. A gas is defined as a form of matter that possesses fluidity, high compressibility and which has neither definite shape nor definite volume. Gases are compressible and diffusible as their molecules have large spaces between them and are not closely packed due to weak interactions. A pure gas may be made up of individual atoms, elemental molecules of one type of atom or compound molecules comprising a variety of atoms. A gas mixture, such as air, contains a variety of pure gases. Examples of gases are air, oxygen, hydrogen, nitrogen and helium (Fig. 1.17).

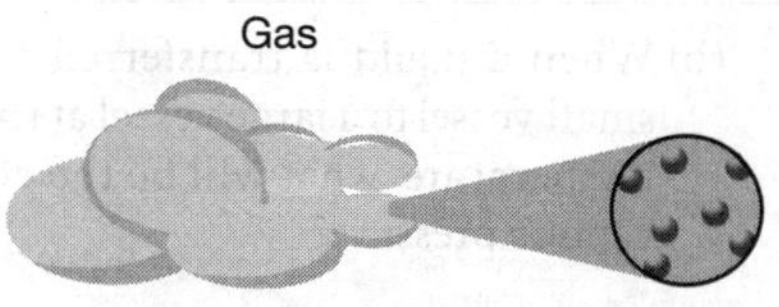

Fig. 1.17 Gas state

Properties of gases: Gases are characterised by the following properties:

- **Shape and Volume:** Gases have neither a fixed shape nor a fixed volume. They acquire the shape and volume of the vessel in which they are stored. This is because gaseous molecules have very weak interactions amongst each other and can thus move freely and occupy any amount of space. For example, when balloons are filled with air, the air simply takes the shape and volume of the balloon.
- **Rigidity:** Gases have the highest fluidity but the least rigidity as the constituent particles have the weakest interparticle forces of attraction and the maximum space between particles.
- **Compressibility:** Gases can be compressed easily by applying external pressure as the space between the gas molecules is large due to weak intermolecular forces. During compression, the gas molecules come closer to each other and occupy less space. For example, liquified petroleum gas (LPG) and compressed natural gas (CNG) are easily stored in small containers for domestic use and as a fuel in vehicles, respectively.
- **Density:** Gases have very low density and they are very light as the interatomic particle distance is very large. This means the particles of a gas are quite far apart, so the volume is high and density is low (less mass per unit volume value).
- **Diffusion:** Gases can diffuse easily as the gaseous molecules are weakly held and have enough kinetic energy to diffuse. The gases can fill their container completely. For example, if LPG leaks, its molecules can spread all around the kitchen and it can be detected by the smell.

  The rate of diffusion in gases is not the same and a lighter gas diffuses faster than a heavier gas; for example, hydrogen > bromine.
- **Diffusion of gases occurs against the law of gravitation:** In Fig. 1.9 shown earlier, the lighter gas hydrogen moves down while the heavier gas bromine moves up, which is against gravity. It can be seen that on removing the lid, both gases mix, giving a light brown colour.

- **Homogeneous nature:** Gases have similar composition in all parts, so they are homogeneous in nature. A gaseous mixture is always homogeneous as it has only the gas phase.
- **Liquification:** A gas can be liquified by cooling and by applying pressure.

Experiment to Show Compressibility in Solid, Liquid and Gas

In order to show that solids are almost incompressible, liquids are slightly compressible and gases are highly compressible, we perform the following experiment:

1. Take three 100-mL syringes and close their nozzles using rubber corks, as shown in Fig. 1.18.

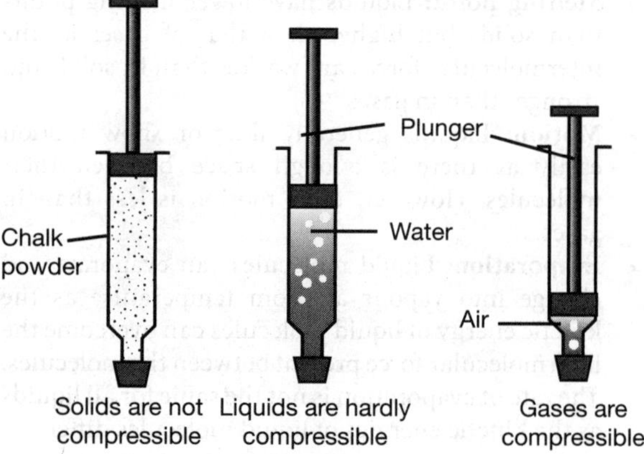

Fig. 1.18 Experiment to show compressibility in solid, liquid and gas

2. Remove the pistons from all the syringes and leave the first syringe as such.

3. Fill the second syringe with water and the third syringe with pieces of chalk.

4. Insert the pistons back into the syringes.

5. Try to compress the contents by pushing the piston in every syringe.

The piston of the syringe having gas (air) moves in when pressure is applied. It means that air gets compressed to a smaller volume, which means that gases are highly compressible.

The piston of the second syringe having liquid (water) moves slightly when pressure is applied, which means that liquids are slightly compressible.

The piston of the third syringe having a solid (chalk piece) does not move at all. This means that solids are completely incompressible.

Scales of Measuring Temperature

There are three scales using which we can measure the temperature of a system: Celsius scale (°C), Kelvin or absolute scale (K) and Fahrenheit (°F).

From Fig. 1.19, it is clear that freezing point and boiling point of water are taken as 0°C and 100°C, respectively, which means the thermometer with the Celsius scale is calibrated from 0°C to 100°C.

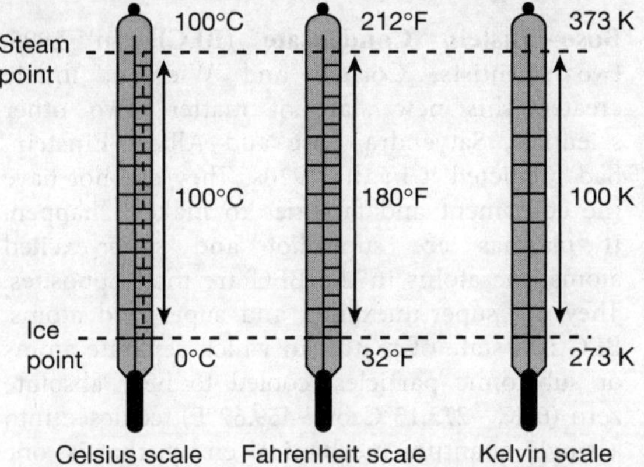

Fig. 1.19 Boiling point and freezing point

On the Fahrenheit scale, the freezing point and boiling point of water are taken as 32°F and 212°F, respectively, which means the thermometer with the Fahrenheit scale is calibrated from 32°F to 212°F.

The Celsius and Fahrenheit scales are related as follows:

$$\frac{F-32}{9} = \frac{C}{5}$$

For example, 14°F in °C is $\frac{14-32}{9} = \frac{C}{5}$

On solving, C = –10°C

Kelvin (K) is the SI unit of temperature in which the degree sign (°) is not used. For example, on the Kelvin scale, the freezing point and boiling points of water are 273 K and 373 K, respectively.

The Celsius and Kelvin scale can be related as:

Temperature in Kelvin (K) = t°C + 273 K

Temperature in °C = Temperature in Kelvin – 273 K

For example: –10°C in Kelvin is: –10° + 273 = 263 K

Temperature of 293 K in °C is: 293 – 273 = 20°C

Exertion of Pressure

Solids exert pressure only in the downward direction, liquids exert pressure downwards as well as to the sides, while gases exert pressure in all directions (Fig. 1.20).

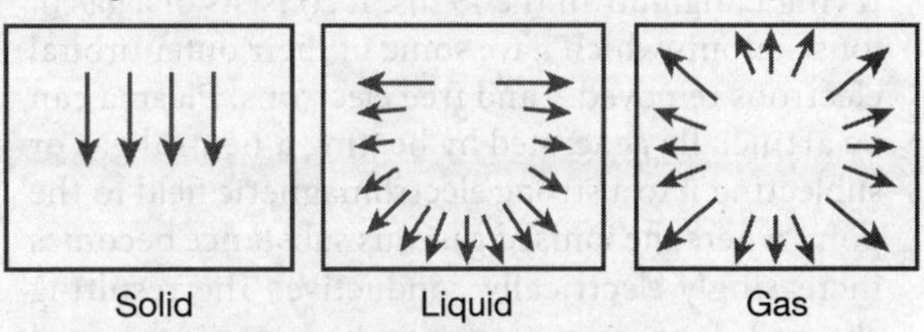

Fig. 1.20 Pressure exerted by solid, liquid and gas

Pressure can be given in terms of atmospheric pressure (atm), mm of mercury, bar, torr, etc.

1 atm = 760 mm of mercury = 760 torr = 1.013 bar.

TEST YOUR KNOWLEDGE

1. Give reasons for the following:
    (a) A gas completely fills the vessel in which it is kept.
    (b) A gas exerts pressure on the walls of the container.

Solution:

(a) The forces of attraction between the molecules of gases are negligible. So, molecules of gases occupy the maximum space available to them. High kinetic energy possessed by their molecules also helps.

(b) The motion of particles is random and occurs at very high speed in the gaseous state. Due to this random movement, the particles hit each other and also the walls of the container. The pressure exerted by the gas is due to this force exerted by the particles per unit area on the walls of the container.

2. (a) How can you separate gases from a gaseous mixture?
    (b) How can a gas be liquefied?

Solution:

(a) Gases can be separated from a gaseous mixture by diffusion as the rate of diffusion of each gas is different.

(b) A gas can be liquefied by cooling or lowering the temperature and by applying pressure. This is known as liquefaction of a gas.

Competition Edge

Plasma: Plasma is one of the four fundamental states of matter and was first described by chemist Irving Langmuir in the 1920s. It consists of a gas of ions – atoms which have some of their outer orbital electrons removed – and free electrons. Plasma can be artificially generated by heating a neutral gas or subjecting it to a strong electromagnetic field to the point where the ionised gaseous substance becomes increasingly electrically conductive. The resulting charged ions and electrons become influenced by long-range electromagnetic fields, making the plasma dynamics more sensitive to these fields than a neutral gas.

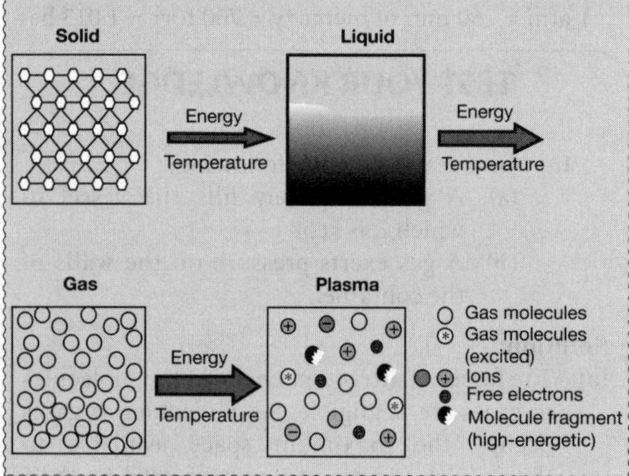

Examples of plasma state:

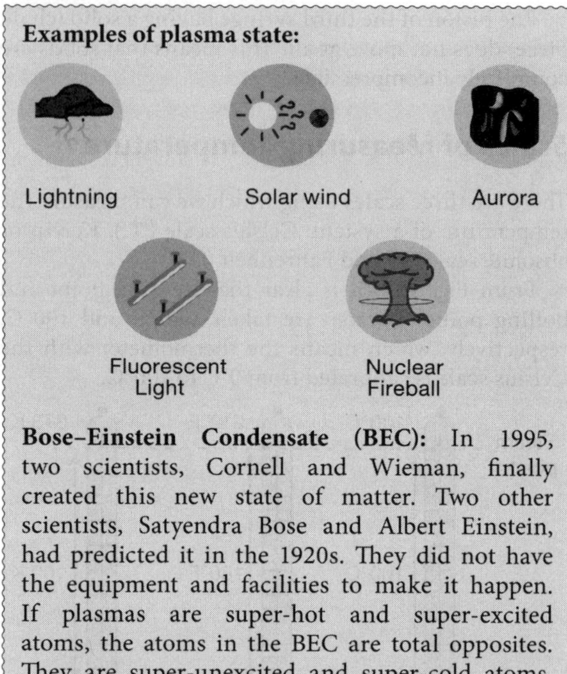

Bose–Einstein Condensate (BEC): In 1995, two scientists, Cornell and Wieman, finally created this new state of matter. Two other scientists, Satyendra Bose and Albert Einstein, had predicted it in the 1920s. They did not have the equipment and facilities to make it happen. If plasmas are super-hot and super-excited atoms, the atoms in the BEC are total opposites. They are super-unexcited and super-cold atoms. BEC is a state of matter in which separate atoms or subatomic particles, cooled to near absolute zero (0 K, −273.15°C or −459.67°F), coalesce into a single quantum mechanical entity; that is, one that can be described by a wave function on a near-macroscopic scale.

BEC		Solid	Liquid	Gas	Plasma

Low   Energy   High

Quick Review

- Everything in our universe is made up of material called matter.
- Anything which occupies space and which has mass is known as matter. The particles that make up matter are atoms or molecules. For example, food, water, air, and so on.
- Matter has mass, occupies space, shows inertia and is influenced by gravity.
- During all physical and chemical changes, the total mass of the matter before and after the change remains constant (Lavoisier's Law of Conservation of Mass).
- Diffusion and Brownian motion prove the existence of particles in matter and their motion.
- Diffusion is the flow of molecules from the side with higher concentration to the side with lower concentration.
- Osmosis is the flow or movement of solvent particles from the dilute solution (less solute) to the concentrated solution (more solute) through a semi-permeable membrane. For example, raisins swell in water.
- Intermolecular forces are the forces of attraction or repulsion that act between neighbouring particles (atoms, molecules or ions). For example, dispersion, dipole–dipole, hydrogen bonding, and so on.
- On the basis of their physical states, matter can be classified into four groups: solids, liquids, gases and plasma.
- Solids have the most closed packed arrangement of particles, are hard with a definite shape and volume and have the least compressibility and diffusibility.

- In a crystalline solid (quartz), the arrangement of constituents is of long-range order while in an amorphous solid (glass), there is only short range order.
- A liquid has fixed volume, no definite shape, is less compressible and diffusible than a gas and is less closely packed than solids but more than gases.
- A gas has no definite shape or volume but has more compressibility and diffusibility.
- Temperature in Kelvin (K) = $t°C + 273$ K.

Exercise 1.1

All questions marked * are practical based.

Section A: Multiple Choice Questions
(1 Mark)

1. 'Panch tatva' of life:
(a) Water, Air, God, Mother, Father
(b) God, Air, Water, Yield, Money
(c) God, Air, Water, Crop, Soil
(d) Air, Fire, Water, Earth, Sky

2. Every matter has its own ________ and ________.
(a) Shape, shadow
(b) Mass, weight
(c) Mass, volume
(d) Tough, brittleness

3. Which of the following is not a characteristic of matter?
(a) Matter is made up of extremely small particles
(b) There is no space between particles of matter
(c) The particles of matter move continuously
(d) The particles of matter attract each other

4. Which of the following cannot be considered a form of matter?
(a) Atom   (b) Water
(c) Humidity   (d) Electron

5. Which of the following statements is not true regarding the characteristics of matter?
(a) Particles of matter move randomly in all directions.
(b) The kinetic energy of the particles increases with a rise in temperature
(c) The kinetic energy of the particles remains the same at a particular temperature
(d) Particles of matter diffuse into each other on their own

6. Which is not a state of matter?
(a) Liquid   (b) Gas
(c) Solid   (d) Soil

7. Matter is classified on the basis of __________.
(a) Chemical properties
(b) Physical properties
(c) Biological properties
(d) Both (a) and (b)

8. Which is not a solid?
(a) Completely molten ice
(b) A book
(c) Completely cooked vegetables
(d) An ice bag

9. Any colour that can we detect with our eyes is:
(a) Due to air   (b) Due to light
(c) Due to visible light   (d) Both (b) and (c)

10. When potassium permanganate dissolves in water, it produces a:
(a) Blue colour   (b) Violet colour
(c) Orange colour   (d) Red colour

11. Smell is produced when we dissolve:
(a) Sugar in water
(b) Salt in water
(c) Dettol in water
(d) Potassium permanganate in water

12. Kinetic energy increases with:
(a) Increasing pressure
(b) Decreasing pressure
(c) Increasing temperature
(d) Both (a) and (b)

13. Intermixing of particles of two different types of matter on their own is called:
(a) Precipitate   (b) Mixing
(c) Diffusion   (d) Dilution

14. Particles move at the bottom of a glass due to:
(a) Potential energy   (b) Kinetic energy
(c) Dissolution   (d) Both (a) and (b)

15. Which of the following phenomena would increase on increasing the temperature?

(a) Diffusion, evaporation, compression of gases
(b) Evaporation, solubility, diffusion, compression of gases
(c) Evaporation, diffusion, expansion of gases
(d) Evaporation, compression of gases, solubility

16. On converting 25°C, 37°C and 68°C to Kelvin scale, the correct sequence of temperatures will be:
(a) 298 K, 310 K and 339 K
(b) 298 K, 310 K and 338 K
(c) 273 K, 278 K and 543 K
(d) 298 K, 310 K and 341 K

17. The boiling points of diethyl ether, acetone and n-butyl alcohol are 35°C, 56°C and 118°C, respectively. Which one of the following correctly represents their boiling points in the Kelvin scale?
(a) 329 K, 392 K and 308 K
(b) 308 K, 329 K and 391 K
(c) 308 K, 329 K and 392 K
(d) 306 K, 329 K and 391 K

18. In which of the following conditions would the distance between the molecules of hydrogen gas increase?
   (i) Increasing pressure on hydrogen contained in a closed container
   (ii) Some hydrogen gas leaking out of the container
   (iii) Increasing the volume of the container of hydrogen gas
   (iv) Adding more hydrogen gas to the container without increasing the volume of the container
(a) (i), (iii)   (b) (ii), (iii)
(c) (i), (iv)   (d) (ii), (iv)

19. Which of following options represents the correct set of statements?
   (i) The particles which make up matter are atoms or molecules.
   (ii) The evidence for the existence of particles in matter and their motion comes from the experiments on diffusion and Brownian motion.
   (iii) Bromine vapour is less heavy than air.
   (iv) When a crystal of potassium permanganate is placed in a beaker of water, the water slowly turns purple on its own.

(a) (i), (ii)   (b) (i), (ii), (iii)
(c) (i), (ii), (iv)   (d) (ii), (iv)

20. Match the following:

Column I	Column II
1. Solid	(i) Fluorescent tube, neon sign
2. Liquid	(ii) Air, LPG, CNG
3. Gas	(iii) Stone, ice, gold
4. Plasma	(iv) Alcohol, milk, water

(a) 1-(iv), 2-(iii), 3-(i), 4-(ii)
(b) 1-(iv), 2-(iii), 3-(ii), 4-(i)
(c) 1-(iii), 2-(iv), 3-(ii), 4-(i)
(d) 1-(i), 2-(ii), 3-(iii), 4-(iv)

Assertion–Reason Questions

Direction: In the following question two statements (Assertion) A and Reason (R) are given Mark.
(a) if A and R both are correct and R is the correct explanation of A;
(b) if A and R both are correct but R is not the correct explanation of A;
(c) A is true but R is false;
(d) A is false but R is true

Assertion	Reason
1. Air, fire and water are examples of matter.	1. They have mass and occupy space.
2. For a substance during the change of state, temperature remains constant.	2. The heat supplied increases both kinetic and potential energies of the particles.
3. Particles of matter are never at rest.	3. Particles of matter have kinetic energy.
4. In solids, the constituent particles are very closely packed.	4. The constituent particles of solids have strong forces of attraction.
5. Gases are highly compressible.	5. Solids have rigidity.
6. A punctured tyre becomes flat more easily in winter than in summer.	6. Rate of diffusion increases with increase in temperature.
7. Plasma consist of super-energetic and super-exited particles.	7. Fluorescent tubes and neon signs contain plasma.

Section B: Very Short Answer Questions (2 Marks)

1. Name two things required for a substance to be called as matter.

2. A substance has no mass; can we consider it as matter?

3. What are the five senses used to feel matter?

4. With what four aspects of matter is the science of chemistry concerned?

5. How can you say that air is matter?

6. Why are heat, sound and light not considered as matter?

7. Name the state of matter in which particles move around randomly because of very weak forces of attraction.

8. On a physical basis, matter is of two types. True or false?

9. On the basis of chemical properties, matter is divided into elements, compounds and mixtures. True or false?

10. Identify the name of the material which has no fixed shape but a fixed volume: sugar, wood, a piece of iron, oxygen and water.

11. Why do liquids mostly have lower density than solids?

12. Why does a gas fill a vessel completely?

13. A rubber band can change its shape on stretching. Will you classify it as a solid or not? Justify your answer.

14. Arrange the following substances in decreasing order of the force of attraction between their particles: oxygen, salt, milk.

15. Give two examples of practical applications based on high compressibility of gases.

16. Name the property of gases that helps aquatic plants and animals to survive in water.

17. A substance has finite volume but not a definite shape. What is the physical state of the substance?

18. Name the property of gases due to which it is possible to fill CNG in cylinders for use as fuel in cars.

19. A crystal of copper sulfate is dropped in a glass of water and is allowed to settle at the bottom. After some time, it is observed that a blue colour appears just above the solid crystal, and with the passage of time, all the water in the glass turns blue. Identify the characteristic of particles of matter associated with this observation.

Section C: Short Answer Questions (3 Marks)

1. List three characteristics of the particulate nature of matter.

2. (i) Define matter. Name the states of matter in which the forces between the constituent particles are (a) strongest, (b) weakest.

   (ii) When sugar and common salt are kept in different jars, they take the shape of the jars. Are they solid? Justify your answer.

3. Among solids, liquids and gases, which one has:
   (i) maximum force of attraction between the particles
   (ii) minimum space between particles

4. List any four characteristic properties of gases.

5. Ice, water and steam are the three states of a substance and not different substances. Justify it.

6. Write the physical state of matter that shows the property given below:
   (i) Most compressible form of matter.
   (ii) Rigid and incompressible.
   (iii) Has definite volume but no fixed shape.
   (iv) Has definite shape and volume.

7. A wooden chair is considered as a solid at room temperature. Give two reasons.

8. Arrange the three states of matter in increasing order of:
   (i) particle motion
   (ii) rate of diffusion

9. Mention any two properties of water to justify that water is a liquid at room temperature.

10. Why is the rate of diffusion of liquids more than that of solids?

11. Why are gases more compressible? Write two reasons for this.

12. (i) Write the full form of (1) LPG, (2) CNG.
    (ii) Give one use for each.

13. Give reasons for the following:
    (i) Solids are incompressible.
    (ii) Solids have negligible kinetic energy.

14. Give reasons for:
    (i) Why gases exert pressure on the walls of a container.
    (ii) Gases undergo diffusion very quickly.

Section D: Long Answer Questions
(5 Marks)

1. (i) Define the following terms: (1) Rigidity, (2) Compressibility, (3) Diffusion.
   (ii) Arrange the following in increasing order of (1) force of attraction, (2) intermolecular space: iron nail, kerosene and oxygen gas.

2. Compare in tabular form the properties of solids, liquids and gases with respect to: (i) shape, (ii) volume, (iii) compressibility, (iv) diffusion, (v) fluidity or rigidity.

3. (i) List any two properties that liquids have in common with gases.
   (ii) Give two reasons to justify why an iron almirah is a solid at room temperature.
   (iii) How can you justify that solids have rigidity?

4. (i) How can you say liquids are not rigid but have the property of flow?
   (ii) Give two factors that determine the rate of diffusion of a liquid in another liquid.
   (iii) Distinguish between solids and gases on the basis of following parameters:
      (1) Interparticle distance
      (2) Interparticle forces of attraction
      (3) Compressibility

5. Explain an activity that shows that particles of matter are very small.

6. (i) Why is oxygen called a gas? Give two reasons.
   (ii) Why is air more dense at sea level?
   (iii) Describe an activity using red ink, honey and water in beakers to show that the particles of matter move continuously.

What is the effect of increase in temperature on the movement of particles? Give reasons.

7. *(i) Shashwat dropped a crystal of potassium permanganate into two beakers X and Y having hot water and cold water, respectively. After keeping the beakers undisturbed for some time, what did he observe and why?
   (ii) State two characteristic properties each of: (i) solid, (ii) liquid and (iii) gas.
   (iii) What values are shown by Shashwat?

Section E: Case Study or Passage-Based Questions
(4 Marks)

1. Diffusion is the spread of particles into a medium or space. It is also defined as the intermixing of particles of different types of matter on their own. Gases have the highest diffusion rate in relation to liquids and solids. The phenomenon of diffusion is highly useful in detecting the presence of a gas; for example, leakage of LPG, spreading of perfume, etc. Diffusion is useful in liquids also and a less dense liquid diffuses more; for example, water diffuses more than honey.

(i) When 2 mL of Dettol is dissolved in 100 mL of water, the smell can be detected even on repeated dilution. Identify the physical nature of the matter.

(ii) What happens when you open a bottle of perfume?

(iii) If food is being cooked in the kitchen by Shanvi, name the process which spreads the smell.

(iv) When a drop of blue ink is put in water, the blue colour spreads and the whole solution becomes blue. Name the phenomenon due to which this happens.

(v) A gas jar 'P' having air is inverted over another jar 'Q' having a brown gas which is heavier than air. After some time, a brown colour is observed in the gas jar 'P'. Identify the phenomenon associated with this observation.

2. On the basis of physical properties, metal is further divided into solid, liquid and gas. They differ in physical state, hardness, fluidity, force of attraction between molecules, shape and volume. For example, in solids, the constituents are most closely packed while in gases they are loosely packed. Solids have definite shape and

volume while liquids have only definite volume. A gas has no definite shape or volume.

(i) A few substances are arranged in increasing order of 'forces of attraction' between their particles. Which one of the following represents the correct arrangement?
- (a) Salt, juice, air
- (b) Air, salt, oil
- (c) Oxygen, water, salt
- (d) Water, air, wind

(ii) Eraser, Book, Slate is solid matter which does not have?
- (a) Fixed volume
- (b) Definite shape
- (c) Distinct boundaries
- (d) Fixed velocity

(iii) The property of flow is unique to fluids. Which one of the following statements is correct?
- (a) Only gases behave like fluids
- (b) Only liquids are fluids
- (c) Gases and solids behave like fluids
- (d) Gases and liquids behave like fluids

(iv) Which among these will have no shape but a fixed volume?
- (a) An aluminum pot
- (b) An iron chisel
- (c) Mustard oil
- (d) A plastic bottle

(v) Which of following options represents the correct set of statements?
- (i) The molecules of a gas occupy all the space available
- (ii) The molecules in a liquid are arranged in a regular pattern
- (iii) The molecules in a gas exert negligibly small forces on each other, except during collisions
- (iv) The molecules in a solid vibrate about a fixed position
- (a) (i), (iii)
- (b) (ii), (iii)
- (c) (ii), (iii), (iv)
- (d) (i), (iii), (iv)

High Order Thinking Skills (HOTS) Questions

1. A small amount of gas is let into a large evacuated chamber.
- (i) How much of the chamber gets filled with the gas?
- (ii) What property of the gas helps it to do so?
- (iii) How will you differentiate between a gas and a vapour?

2. (i) Explain the following:
- (1) Sponge though compressible is a solid.
- (2) Rubber band though stretchable is a solid.
- (ii) Write the chemical name of dry ice. Justify its name. How is it stored?

3. State the reason for the following:
- (i) The smell of a lighted incense stick reaches you several metres away, but to get the smell from an unlighted incense stick, you need to go close to the stick.
- (ii) Naphthalene balls disappear with time without leaving any solid residue.
- (iii) Rajma (kidney bean) or whole gram (black or white chana) are kept in water for a few hours before cooking. Explain why.

4. (i) A diver can cut through water in a swimming pool. Which property of matter does this observation suggest? How?
- (ii) Liquids generally have lower density than solids, but ice floats on water. Give reasons why.

5. *Two millilitres of Dettol are added to a beaker containing 500 mL of water and stirred. State four observations that you make.

6. *Design an experiment to show that air contains water vapour.

7. (i) Osmosis is a special kind of diffusion. Comment.
- (ii) Ronit took two beakers X and Y containing hot water and cold water, respectively. In each beaker, he dropped a crystal of copper sulfate. He kept the beakers undisturbed. After some time, what did he observe and why?

Answers

Section A: Multiple Choice Questions

1. (d)	**2.** (c)	**3.** (b)	**4.** (c)
5. (c)	**6.** (d)	**7.** (d)	**8.** (a)
9. (c)	**10.** (b)	**11.** (c)	**12.** (c)
13. (c)	**14.** (b)	**15.** (c)	**16.** (d)
17. (b)	**18.** (b)	**19.** (c)	**20.** (c)

Assertion–Reason Questions

1. (a)	**2.** (c)	**3.** (a)	**4.** (a)
5. (b)	**6.** (d)	**7.** (b)	

Section B: Very Short Answer Questions

1. A substance must have mass and volume.

2. No.

3. Sight, Touch, Smell, Hearing and Taste.

4. Composition, structure, properties and changes.

5. Air is matter as its presence can be felt when the wind blows and it has mass and volume.

6. They do not have mass and volume.

7. Gaseous state.

8. False.

9. True.

10. Water (as liquids have no fixed shape but have definite volume).

11. As liquids have less force of attraction between molecules, that is, less mass and more volume as compared to solids, the density of a liquid is less.

12. As particles of gas have negligible force of attraction between them and possess high kinetic energy, they move very fast to completely fill the vessel in which they are kept.

13. A rubber band changes shape under force and regains the shape when the force is removed; so it is classified as a solid.

14. Salt > milk > oxygen.

15. CNG (compressed natural gas) and LPG (liquified petroleum gas).

16. Solubility in water; for example, oxygen in water.

17. Liquid.

18. Compressibility.

19. Particles of matter move continuously.

Section C: Short Answer Questions

1. (i) Particles of matter have space between them
   (ii) Particles of matter move continuously
   (iii) Particles of matter attract each other

2. (i) Anything which possesses mass and which occupies space is known as matter.
   (1) Solid – strongest, (2) Gas – weakest.
   (ii) As the shape of each individual sugar or salt crystal remains fixed, they are solids.

3. (i) Solids show maximum force of attraction between particles as they are closely packed.
   (ii) Solids have minimum space between particles as the particles are closely packed.

4. (i) Gases neither have definite shape nor definite volume.
   (ii) Gases are compressible.
   (iii) It exerts pressure on the walls of the container due to collision of molecules.
   (iv) Gases flow easily.

5. When ice is melted, water is produced and when water is heated, steam is produced. Conversely, when steam is cooled, water is produced and when water is cooled, ice is formed. Moreover, the chemical formula (H_2O) is the same for all the three states, that is, no new compound is formed during conversions.

6. (i) Gas
   (ii) Solid
   (iii) Liquid
   (iv) Solid

7. A wooden chair is solid at room temperature because:

(i) it has definite shape and volume, (ii) it cannot be compressed.

8. (i) Particle motion: solid < liquid < gas
   (ii) Rate of diffusion: solid < liquid < gas

9. Water is a liquid at room temperature as it takes the shape of the container in which it is stored and it shows fluidity also.

10. Rate of diffusion of liquids is higher than in solids because particles of liquid move freely and particles of liquid have greater spaces between them than solids.

11. Gases are more compressible due to weak forces of attraction and more intermolecular space.

12. (i) (1) LPG: Liquified petroleum gas
       (2) CNG : Compressed natural gas
   (ii) Use of LPG: Fuel at home.
        Use of CNG: Fuel for vehicles.

13. (i) Solids are incompressible because the particles are closely packed and there is almost no space for their movement.
   (ii) As the particles of solids do not have any intermolecular space and henced show almost no movement, they have negligible kinetic energy.

14. (i) The particles of a gas are free to move randomly in all possible directions and during the motion they collide with one another and also with the walls of the container. The pressure of the gas is due to their collisions with the walls of the container.
   (ii) Gases undergo diffusion very quickly as particles of a gas are loosely packed and move randomly due to the space between them; so they intermix with other particles present there.

Section D: Long Answer Questions

1. (i) (1) Rigidity: It is the property of matter to maintain shape against any applied external force.
       (2) Compressibility: It is the property of matter by virtue of which molecules of matter are brought closer to each other.

(3) Diffusion: The intermixing of particles of matter is known as diffusion.

(ii) (1) Force of attraction: Oxygen gas < Kerosene < Iron nail
    (2) Intermolecular space: Iron nail < Kerosene < Oxygen gas

2.

Property	Solids	Liquids	Gases
Shape	Definite shape	No definite shape	No definite shape
Volume	Definite volume	Definite volume	No definite volume
Compressibility	Non-compressible	Slightly compressible	Highly compressible
Diffusion	Can diffuse in liquids	Can diffuse in liquids	Can diffuse in other gases
Fluidity or rigidity	Rigid	Shows fluidity	Shows fluidity

3. (i) (1) Both gases and liquids do not have a fixed shape.
       (2) Both gases and liquids flow easily.
   (ii) (1) The shape of the almirah does not change when pressed. This means it is hard and rigid (character of solid).
        (2) It has a definite shape and high density (character of solid).
   (iii) Rigidity means that solids have the tendency to maintain shape when some force from outside is applied.

4. (i) Due to large interparticle distances and weak forces of attraction, liquids can flow and change shape. So they are not rigid but have fluidity.
   (ii) The two factors that determine the rate of diffusion of a liquid in another liquid temperature and pressure.
   (iii) Differences between solids and gases on the basis of their properties are as follows:

Solids	Gases
Interparticle space is small, so the distance is less.	Interparticle space is maximum, so the distance is more.
Interparticle force of attraction is maximum.	Interparticle force of attraction is minimum.
Solids are rigid and non-compressible.	Gases are not rigid and they are compressible.

5. (i) Take 2–3 crystals of copper sulfate and dissolve them in 100 mL of water.
   (ii) Take approximately 10 mL of this solution and put it into 90 mL of clear water.
   (iii) Take 10 mL of this solution and put it into another 90 mL of clear water.
   (iv) Keep diluting 5–8 times.

   Inference: A crystal of copper sulfate contains millions of tiny particles which keep on dividing into smaller and smaller numbers with each dilution. Here, the colour of copper sulfate changes with dilution.

6. (i) (1) Oxygen has neither fixed shape nor fixed volume.
       (2) Oxygen exerts pressure due to collision of the molecules on the walls of the containing vessel and among themselves.
   (ii) The pressure of air is maximum at sea level, which means air is more compressed, so it is more dense at sea level.
   (iii) Take two beakers filled with water, put a drop of red ink in one beaker and honey in the second beaker and leave them undisturbed. After some time, it can be observed that the colour of the ink spreads evenly throughout the water and the honey. This happens because the molecules keep on moving. When temperature is increased, the movement of particles become faster. This is due to increase in their kinetic energy.

7. (i) Potassium permanganate crystals diffuse faster in hot water because as the temperature increases, diffusion increases.
   (ii) Properties of solids: They have fixed shape and volume. They are rigid.

Properties of liquid: They have fixed volume but not fixed shape. They are not rigid.

Properties of gas: They neither have fixed shape nor fixed volume. They are highly compressible.

(iii) Shashwat showed his experimental and observational skills.

Section E: Case Study or Passage-Based Questions

1. (i) Particles of matter are very small which can be dissolved in water and due to diffusion its smell can be observed readily in air.
   (ii) Particles of perfume diffuse into the air and as a result can be detected even at a distance.
   (iii) Gases show the property of diffusing very fast due to the high speed of particles and large space between them.
   (iv) It takes place due to 'diffusion'.
   (v) This observation is based on 'diffusion'.

2. (i) (c) (ii) (a) (iii) (d) (iv) (c) (v) (d)

High Order Thinking Skills (HOTS) Questions

1. (i) The whole chamber gets filled with gas.
   (ii) The particles move about randomly at high speed.
   (iii) A substance is considered as a gas if its boiling point is below room temperature; for example CO_2, O_2, etc. If the normal physical state of a solid or liquid is converted by applying energy or on its own into gaseous state, it is known as vapour state. For example, on heating, mercury changes from liquid to gaseous state (mercury vapour).

2. (i) (1) As a sponge has minute holes in which air is trapped, on pressing it, air is expelled and we are able to compress it.
       (2) A rubber band changes its shape under force and regains shape when force is removed.
   (ii) The chemical name of dry ice is solid carbon dioxide (CO_2). It looks just like

ice but is absolutely different. Solid CO_2 converts directly to gaseous state on decreasing pressure to one atmosphere without passing through the liquid state. It is stored at high pressure.

3. (i) Particles of matter move continuously. The rate of movement of particles increases with temperature; therefore, the smell of a lighted incense stick reaches us several metres away due to diffusion.

   (ii) Because it sublimes, due to which it directly converts into vapour and disappears without leaving any solid.

   (iii) When rajma or whole gram is kept in water for a few hours, osmosis occurs. It is easier to cook these swollen kidney beans or whole grams as less heat is needed for cooking.

4. (i) Particles of water are held together by weak forces of attraction. So the diver is able to cut through water in the swimming pool.

   (ii) Ice molecules have gaps in between, so the volume is more. Water molecules are packed more closely together, so the volume is less. In ice, the mass per unit volume $M/V = D$ density of ice is less than the density of water; hence being lighter, ice floats on water.

5. (i) Level of water remains the same.

6. (i) Take ice cold water in a metallic tumbler and keep it in the open air for some time.

   (ii) After some time, water droplets can be seen on the outer surface of the tumbler.

   (iii) The water vapour present in air on coming in contact with the cold metallic surface of the tumbler loses energy and is converted to liquid state.

   (iv) This shows that air has water vapour.

7. (i) Diffusion is the process in which molecules of a substance move from the place of higher concentration to the place of lower concentration. But during osmosis, solvent or the water molecules move from concentrated solution (more) to dilute solution (less) through a semi-permeable membrane. Hence, osmosis can be considered as a special kind of diffusion.

   (ii) The solutions in both beakers turned blue after some time. But the colour change was observed earlier in beaker X containing hot water as compared to beaker Y containing cold water. This took place due to the faster rate of diffusion at a higher temperature.

(ii) A uniform mixture is formed / A true solution is obtained.

(iii) The solution becomes white in colour.

(iv) Smell can be detected even on repeated dilution.

1.2.6 Change of State of Matter

A substance may exist in any of the three states of matter (solid, liquid or gas) depending upon the temperature and pressure conditions. By changing the temperature and pressure, the state of a substance can be changed. The space between the constituent particles, forces of attraction and kinetic energy of the particles changes, so change of state takes place.

For example, a solid on heating usually changes into a liquid, which on further heating changes into a gas as the space between the particles and the kinetic energy increases while the force of attraction decreases. Similarly, a gas on cooling condenses into a liquid which on further cooling changes into a solid due to decreasing space between the particles, lower kinetic energy and higher force of attraction (Fig. 1.21).

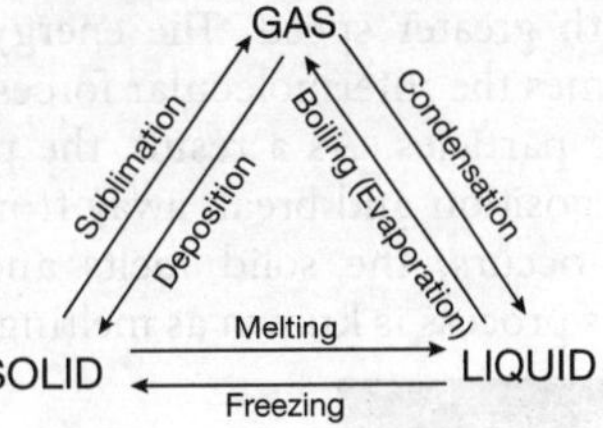

Fig. 1.21 Change in state of matter

The most familiar and common example is water. It exists in all three states: as ice (solid), as water (liquid) and as water vapour (gas). Ice and may be melted to form water which on further heating changes into steam. These changes can also be reversed on cooling (Fig. 1.22).

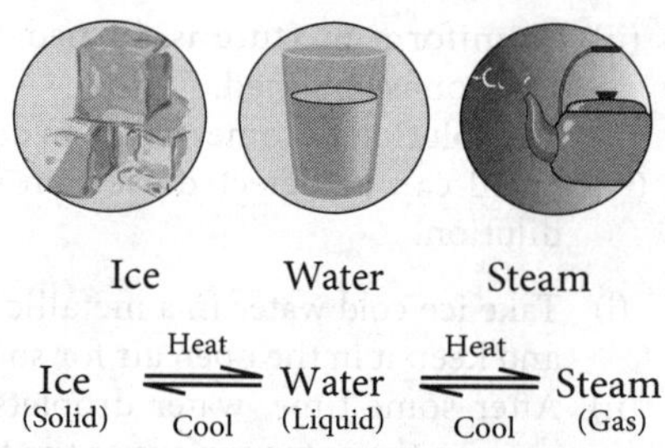

Fig. 1.22 Change in state of water

Effect of Temperature Change

By increasing the temperature (heating), a solid can be converted into a liquid, and a liquid can be converted into a gas or vapour. By decreasing the temperature (cooling), a gas can be converted into a liquid, and a liquid can be converted into a solid.

Solid to liquid (melting): The process by which the solid state of a substance is changed into its liquid state on heating is known as melting or fusion.

The temperature at which a solid substance melts and changes into a liquid at atmospheric pressure is known as the **melting point** of the substance. The melting point of a solid is a measure of the force of attraction between its particles. The higher the melting point of a solid, the greater will be the force of attraction between its particles. For example, the melting point of ice = 0°C or 273 K, melting point of wax = 63°C (Fig. 1.23) or 336 K and melting point of iron = 1535°C or 1808 K.

When a solid is heated, its particles absorb heat energy and they become more energetic; the kinetic energy of the particles increases. Due to increase in kinetic energy, the particles start vibrating more strongly with greater speed. The energy supplied by heat overcomes the intermolecular forces of attraction between the particles. As a result, the particles leave their mean position and break away from each other. When this occurs, the solid melts and a liquid is formed. This process is known as melting.

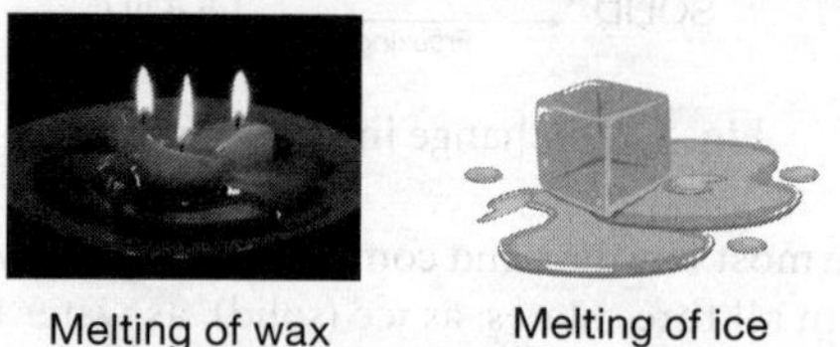

Fig. 1.23 Melting of wax and ice

Factors affecting the melting point: Although every substance has a specific melting point under atmospheric conditions, it can be affected by the following factors:

- **Effect of pressure:** The effect of pressure on the melting point varies according to the nature of the solid. The melting points of solids which expand during melting, increase with increase in pressure as this opposes expansion. For example, Cu, Au, Ag, paraffin wax. The melting points of solids which contract during melting, decrease with increase in pressure as this favours contraction. For example, ice, brass, cast iron. Ice is a solid which contracts during melting. Therefore, when pressure is applied on two ice cubes, the ice at the interface melts and when the pressure is released, it solidifies again. The ice cubes re-join at this spot. This phenomenon is known as **regelation**.
- **Effect of impurities:** If impurities are associated with a solid, its melting point decreases. This causes it to melt at a lower temperature. For example, in the case of rose metal [alloy of Sn (mp 239.1°C), pb (mp 327 °C), Bi (mp 271°C)], melting occurs at 94.5°C.

Experiment to show that temperature remains constant during melting

1. Take nearly 100 g of ice in a beaker and suspend a laboratory thermometer so that its bulb is in contact with the ice, as shown in Fig. 1.24.

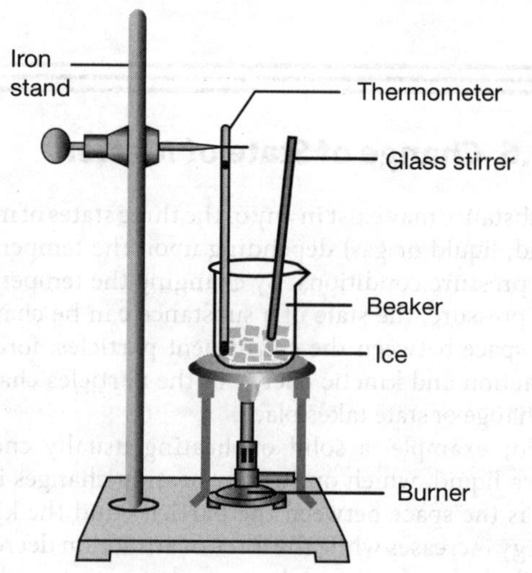

Fig. 1.24 Melting of ice

2. Note the temperature of the ice, which is 0°C or 273 K.

3. Heat the beaker on a low flame and note the temperature when melting of ice starts.

4. Keep recording the temperature of the melting ice every minute.

As heating continues, more ice melts to form water; however, the temperature is still 0°C; as long as there is ice in the beaker, the temperature will remain as 0°C. Now it is clear that there is no rise in temperature during the melting of ice. However, when all of the ice has melted into water, the temperature will start increasing.

Latent heat: The word '*latent*' means 'hidden'. The heat energy that needs to be supplied to change the state of a substance is known as its latent heat. It does not raise the temperature but it needs to be supplied to change the state of a substance since every substance has some forces of attraction between its particles that hold them together. These forces of attraction need to be broken by supplying heat. The latent heat does not increase the kinetic energy of the particles of the substance. There are two types of latent heat:

- **Latent heat of fusion:** The heat needed to convert a solid substance into its liquid state is known as latent heat of fusion. In other words, the latent heat of fusion of a solid is the quantity of heat in joules required to convert 1 kilogram of the solid to liquid, without any change in temperature. For example, the latent heat of fusion of ice is 3.34×10^5 J/kg or 80 cal/g. Now it is clear that at 0°C, both water and ice exist together; however, the particles of water have more energy with respect to the particles of ice. That is why at 0°C, only ice changes to water as the heat energy is equal to the latent heat of fusion of ice, or we can say that at 0°C, ice is more effective in cooling a substance than water.

- **Latent heat of vaporisation:** The heat needed to convert a liquid substance into its vapour state is known as latent heat of vaporisation. In other words, the latent heat of vaporisation of a liquid is the quantity of heat in joules required to convert 1 kilogram of the liquid to its vapour or gaseous state, without any change in temperature. For example, the latent heat of vaporisation of water is 22.59×10^5 J/kg or 540 cal/g.

Liquid to gas (boiling or vaporisation): The process by which the liquid state of a substance is changed into its gaseous state on heating is known as boiling or vaporisation (Fig. 1.25).

Fig. 1.25 Vaporisation of a liquid

The temperature at which a liquid boils and changes rapidly into a gas at atmospheric pressure is called the **boiling point** of the liquid. The boiling point of a liquid is a measure of the force of attraction between its particles. The higher the boiling point, the greater will be the force of attraction. For example, the boiling point of water = 100°C or 373 K, the boiling point of alcohol = 78°C or 351 K and the boiling point of mercury = 357°C or 630 K.

In a liquid, most of the particles are close together. When we supply heat energy to the liquid, the particles of the liquid start vibrating fast. Some of the particles become so energetic that they can overcome the attractive forces of the particles around them. Hence, they become free to move and escape from the liquid. When this happens, the liquid evaporates, that is, starts changing into gas. At boiling point, the particles of a liquid have sufficient kinetic energy to overcome the forces of attraction holding them together and separate into individual particles. The liquid then boils to form a gas.

Factors affecting the boiling point: Although every liquid substance has a specific boiling point under atmospheric conditions, it can be affected by the following factors:

- **Effect of pressure:** On increasing the pressure, the boiling point of the liquid increases. For example, at higher altitudes, water boils at a lower temperature than 373 K. In a pressure cooker, water is heated in a closed vessel in a confined space. So when steam is formed, the pressure increases. As a result, the boiling point of water also increases and the food gets cooked faster and earlier. It is also clear from this that cooking of food takes longer on the mountains and less in pressure cookers.

- **Effect of impurities:** If impurities are associated with a liquid, its boiling point increases and it will boil at a higher temperature. For example, water containing common salt (NaCl) boils at a higher temperature than 373 K. A 60% aqueous solution of urea increases the boiling point of water by 0.52°C or K.

Experiment to show that temperature remains constant during boiling

1. Take nearly 100 g of water in a beaker and suspend a laboratory thermometer so that its bulb is in contact with the water, as shown in Fig. 1.26.

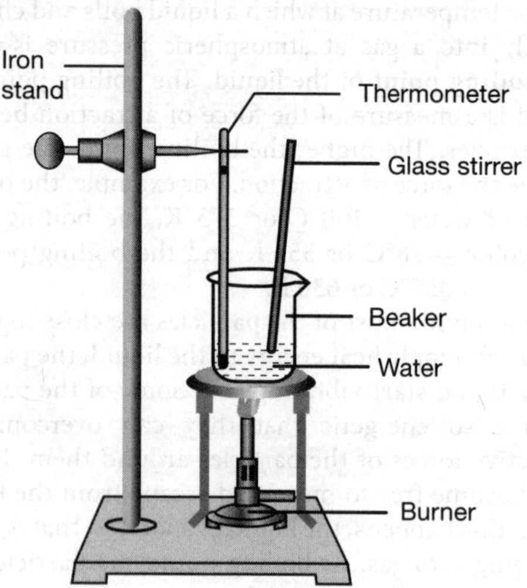

Fig. 1.26 Boiling of water

2. Note the temperature of water which is 100°C or 373 K.

3. Start heating the beaker on a flame and note the temperature when boiling of water starts.

4. Continue recording the temperature of the boiling water every minute.

As heating is continued, more water vaporises to form steam; however, the temperature is still 100°C. As long as there is water in the beaker, the temperature will remain as 100°C. Now it is clear that there is no rise in temperature during the boiling of water. However, when all the water has been boiled into steam, the temperature will start increasing.

Now it is clear that at 100°C, both water and steam exist together. However, the particles of steam have more energy with respect to the particles of water.

This is why at 100°C, only water changes to steam as heat energy is equal to the latent heat of vaporisation of water (22.59×10^5 J/kg or 540 cal/g). Alternatively, we can say that at 100°C, steam is more effective than boiling water for heating purposes.

Liquid to vapour (evaporation): The process of converting a liquid into vapour at any temperature below its boiling point is known as evaporation (Fig. 1.27). It is a surface phenomenon as it is confined only to surface molecules. It is a slower process than boiling (Table 1.2) and always causes cooling.

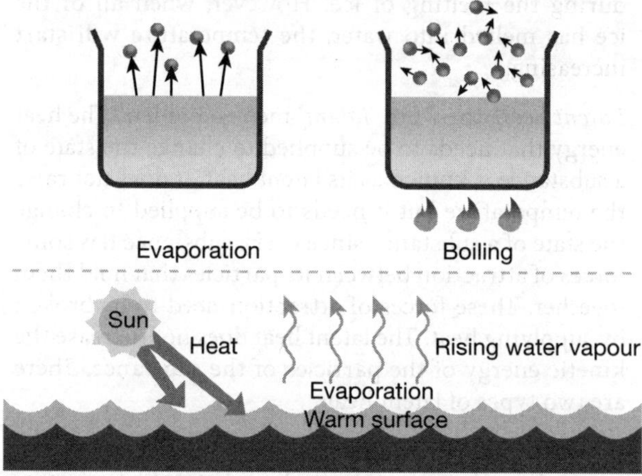

Fig. 1.27 Evaporation

Table 1.2 Differences between boiling and evaporation

Boiling	Evaporation
Boiling can take place only at a particular temperature for a liquid.	Evaporation can occur at any temperature.
Boiling is a bulk phenomenon. That is, bubble formation occurs even below the surface.	Evaporation is a surface phenomenon in which bubble formation occurs only on the surface of the liquid.
Heat is always required for boiling.	Not always necessary.
No cooling is caused during boiling.	Cooling is always observed during evaporation.

Factors affecting evaporation

- **Temperature:** The rate of evaporation increases with the temperature. With an increase in temperature, more particles acquire enough kinetic energy to enter the vapour state.

Rate of evaporation ∝ Temperature

  For example, drying of clothes takes place more rapidly in summer than in winter. Water evaporates more quickly when it is heated; as the water boils, it turns into steam.

- **Surface area:** The rate of evaporation increases with the surface area of the liquid.

Rate of evaporation ∝ Surface area

  For example, if the same liquid is kept in a test tube and in a china dish, the liquid kept in the latter will evaporate more rapidly as more of its surface area is exposed to air.

- **Humidity:** Humidity is the amount of water vapour present in air. The air around us cannot hold more than a definite quantity of water vapour at a given temperature. If the amount of water in air is already large or the humidity is more, the rate of evaporation decreases. Hence, the rate of evaporation increases with decrease in humidity in the atmosphere. For example, drying of clothes on a humid day takes longer.

Rate of evaporation ∝ 1/Humidity

- **Pressure:** The rate of evaporation also increases with a decrease in the pressure of the liquid as the molecules prefer to move from the side with the higher pressure to the side with the lower pressure.

- **Wind speed:** The rate of evaporation also increases with an increase in the speed of the wind. With an increase in the speed of wind, the particles of water vapour move away with the wind, resulting in a decrease in the amount of vapour in the atmosphere. For example, clothes dry faster on a windy day.

Cooling produced due to evaporation: Liquid molecules can absorb energy from the surroundings, overcome the force of attraction and change into vapour form. Here, the surroundings lose energy and become cold. This principle is quite useful in our daily activities. Some examples are as follows:

- **Cotton clothes produce a cooling effect in summer:** During summer, wearing cotton clothes is better as it keeps us cool. This is because during summer we sweat a lot and cotton being a good absorber of water easily absorbs sweat and exposes it to air.

- **A cool sensation is felt when alcohol is poured on the palm:** On pouring some acetone (nail polish remover or thinner) on our palm, we feel cool as the energy needed for evaporation is taken from the palm.

- **Cooling of water is possible in earthen pots (pitchers) during summer:** There are pores in the earthen pot through which the liquid inside the pot evaporates. This evaporation makes the water inside the pot cool.

- **Water droplets can be seen on the outer surface of a glass containing ice cold water:** This due to the fact that water vapour present in the air, on coming in contact with the cold surface of the glass (tumbler), loses energy and gets condensed, which we see as water droplets.

- **Perspiration keeps our body cool in summer:** In summer, due to higher temperatures, our body gives out sweat and when the sweat evaporates it absorbs heat from our body and helps us feel cool.

- **Sprinkle water on the roof or open ground in summer:** By doing this, water evaporates by absorbing heat from the roof or the ground and the surrounding air. Thus, due to loss of heat, the ground or the roof becomes cool and we also feel cool and comfortable.

- **Surgeons spray ether on the skin before starting minor surgery:** Surgeons spray ether on the skin of the patient before starting minor surgery as ether evaporates quickly by absorbing heat or energy from the skin. Thus, the temperature of the skin is reduced or almost makes the skin numb. Due to this numbness, the patient does not feel much pain during minor cuts needed for surgery.

Liquid to solid (freezing): The process of converting a liquid into a solid by cooling is known as freezing. Freezing means solidification. Freezing is the reverse of melting. So, the freezing point of a liquid is the same as the melting point of its solid form. For example, the melting point of ice = 0°C or 273 K and the freezing point of water = 0°C or 273 K.

When a liquid substance is subjected to cooling, heat is extracted from the liquid. Therefore, the kinetic energy of the molecules decreases which results in decrease in temperature. On reaching a certain temperature, further extraction of heat from the liquid results in decrease in potential energy instead of decreasing the kinetic energy. As a decrease in potential energy leads to an increase in intermolecular forces of attraction due to decrease in intermolecular space, the molecular arrangement of the liquid changes to that of a solid.

Factors affecting the freezing point: Like the melting point, the freezing point is also affected by temperature and the presence of impurities. For example, the presence of impurities decreases the freezing of the liquid to some extent. This concept is used in making freezing mixtures. We can add substances like urea, sodium chloride, calcium chloride, sodium chloride, etc. For example, by mixing 3 parts of ice with 1 part of common salt, we get a freezing mixture which can lower the temperature to –21°C. A 30% solution of calcium chloride in water freezes at 218 K or –55°C, that is, the addition of $CaCl_2$ lowers the freezing of water by 55°C. That is why in cold countries $CaCl_2$ or NaCl is spread on roads to clear snow or ice. Such freezing mixtures are used for the preservation of foodstuffs such as meat, fish, and so on, as well as vaccines!

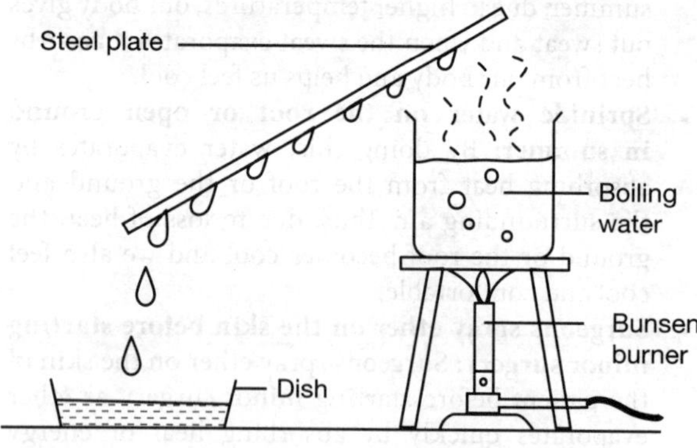

Fig. 1.28 Condensation

Gas to liquid (condensation): The process of converting a gaseous substance or vapour into its liquid state is known as condensation (Fig. 1.28). It can be carried out by cooling the gas below a particular temperature. When heat is extracted from the gas, the kinetic energy of the molecules decreases so the temperature decreases. When a sufficiently low temperature is reached, further extraction of heat from the gas does not reduce the kinetic energy. As a result, an increase in the intermolecular forces of attraction brings the molecules closer; thus, at this point, the gas passes into the liquid state.

It is interesting to note that condensation is the reverse of vaporisation. For example, the boiling point of water is 100°C at 1 atm pressure and for steam, the temperature of condensation is also 100°C at 1 atm pressure. The difference is in terms of heat as during vaporisation, 540 cal/g of heat is absorbed while during condensation the same amount of heat is released. Using this concept of heat, we can say that burns caused by steam are much more severe than those caused by

boiling water at the same temperature as steam releases more energy than water.

Solid to gas (sublimation): By now you would have understood that most of the substances around us change their state, from solid to liquid and from liquid to gas by absorbing heat (latent heat). However, we may have seen many substances which on heating directly change from solid to gas and vice versa on cooling, without passing through the intervening liquid state. Hence, the process of conversion of a solid directly into its gaseous or vapour form on heating is known as sublimation, and of the vapour form into solid on cooling is known as deposition or desublimation. The solid substance which undergoes sublimation is said to sublime while the solid deposit of a substance which has sublimed is known as the sublimate. Sublimation can be represented as:

$$\text{Solid} \underset{\text{Cooling}}{\overset{\text{Heating}}{\rightleftharpoons}} \text{Vapour (or gas)}$$

It is important to note that only those substances whose vapour pressure become equal to the atmospheric pressure (1 atm) below their respective melting point can show sublimation. For example, iodine, dry ice, ammonium chloride, camphor, benzoic acid, naphthalene and anthracene.

Experiment to demonstrate sublimation

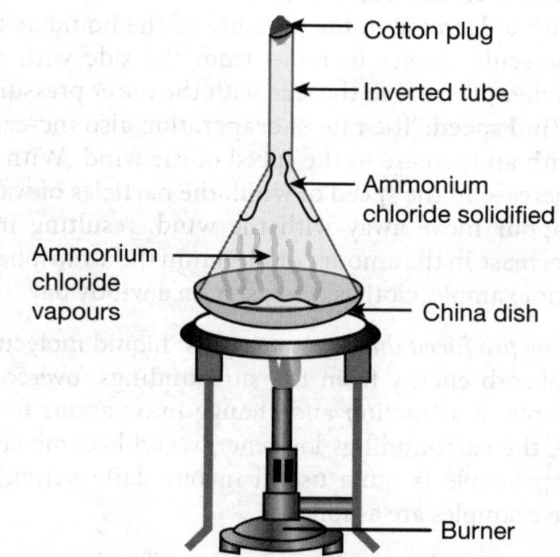

Fig. 1.29 Experiment to demonstrate sublimation

To show sublimation experimentally, let us take the example of ammonium chloride, as shown in Fig. 1.29.

1. Take some impure ammonium chloride in a clean and dry china dish and cover the dish with a perforated filter paper. Also place an inverted funnel on the filter paper. It is important to close the stem of the funnel with a cotton plug.

2. Heat the china dish with solid ammonium chloride.

It directly changes into ammonium chloride vapours. When the hot ammonium chloride vapour is cooled, it directly changes into solid ammonium chloride. Now pure solid ammonium chloride is deposited on the inner walls of the funnel, which can be easily scraped out using a knife.

Applications of sublimation

- To remove non-volatile impurities from the substances which can be sublime, like naphthalene.
- In frost-free refrigerators, ice on the walls of the freezer can be sublimated by circulating warm air through the compartment during the defrost cycle.
- Freeze-dried fruits or foods obtained by sublimation can be stored for a longer duration, keeping all nutritional values intact.
- In very cold cities where snow does not melt, sublimation is used for direct vaporisation.

Effect of Pressure Change

The three states of matter differ in the intermolecular forces and intermolecular distances between the constituent particles (atoms, ions, molecules). Gases are said to be compressible since the space between the gaseous particles decreases on applying pressure. A substance may exist in any of the three different states of matter depending upon the conditions of temperature and pressure.

- If the melting point of a substance is above the room temperature under atmospheric pressure, it is said to be a solid.
- If the boiling point of a substance is above the room temperature under atmospheric pressure, it is classified as a liquid.
- If the boiling point of the substance is below the room temperature under atmospheric pressure, it is called a gas.

Now to convert a gas into a liquid or to convert a liquid into a solid, the interparticle distance must be decreased or the magnitude of interparticle forces must be increased and this can be accomplished as follows:

- **By applying pressure:** Let us take a certain volume of a gas in a cylinder fitted with a piston, as shown in Fig. 1.30. On compressing the gas by pushing in the piston, if the pressure applied is sufficient, the gas will compress to a small volume. This is because the particles come closer together or the force of attraction increases or the interparticle distance decreases; hence, the gas gets liquified.

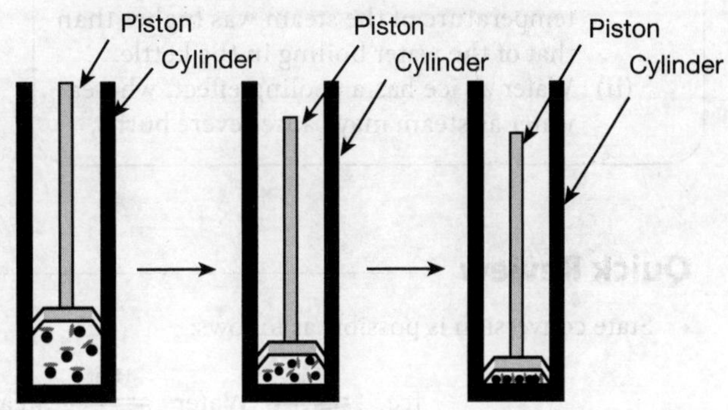

Fig. 1.30 Effect of pressure change

- **By lowering the temperature:** On lowering the temperature, the kinetic energy of the particles decreases. So the particles are packed closer together as the force of attraction increases or the interparticle distance decreases. Hence, the gas gets liquified or close to being liquified.

  Gases like ammonia, carbon dioxide and sulfur dioxide can be liquified easily by applying pressure or by cooling. However, carbon dioxide on cooling under high pressure is sublimated into solid carbon dioxide; dry ice can undergo sublimation and that is why it does not wet the surface on which it is kept (it is also used as a refrigerant under the name Drikold).

- **By applying pressure and lowering the temperature (liquification):** When pressure is applied and temperature is reduced, gases can be converted into liquids or gases will be liquefied. This process is known as liquification. Every gas needs a certain minimum temperature for passing into the liquid state. It means that above this minimum temperature, even on applying pressure, the gaseous state cannot be transformed into the liquid state. This is known as the **critical temperature**. Every gas has a specific critical temperature. For example, the critical temperatures of CO_2 is 31.1°C and H_2 is −165°C.

  Liquification of a gas $\propto$ Critical temperature

TEST YOUR KNOWLEDGE

1. Explain these observations:
   (i) Khushi was making tea in a kettle. Suddenly, she felt intense heat from a puff of steam gushing out of the spout of the kettle. She wondered whether the temperature of the steam was higher than that of the water boiling in the kettle.
   (ii) Water as ice has a cooling effect, whereas water as steam may cause severe burns.

Solution:
   (i) The temperature of both boiling water and steam is 100°C, but steam has more energy because of its latent heat of vaporisation. Hence, steam is hotter than boiling water.
   (ii) Water in the form of ice has low energy since water freezes at a lower temperature. When ice comes in contact with the body, it draws heat from the body and produces a cooling effect. In the case of steam, the water molecules have high energy. The high energy of steam is transformed into heat and may cause severe burns.

Quick Review

- State conversion is possible as follows:

$$\text{Ice (Solid)} \underset{\text{Cool}}{\overset{\text{Heat}}{\rightleftharpoons}} \text{Water (Liquid)} \underset{\text{Cool}}{\overset{\text{Heat}}{\rightleftharpoons}} \text{Steam (Gas)}$$

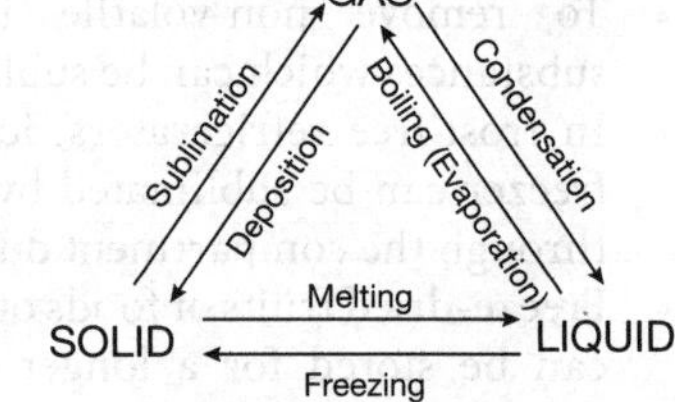

- The temperature at which a solid substance melts and changes into a liquid at atmospheric pressure is known as the melting point of the substance and this process is known as melting.
- The temperature at which a liquid boils and changes rapidly into a gas at atmospheric pressure is called the boiling point of the liquid and this process is called vaporisation.
- The process of converting a liquid into vapour at any temperature below its boiling point is known as evaporation.
- The temperature at which a liquid is converted into a solid at atmospheric pressure is known as the freezing point of the liquid.
- The process of conversion of a gaseous substance into its liquid state is known as condensation.
- The process of conversion of a solid substance directly into its gaseous or vapour form on heating is known as sublimation. The process of conversion of a vapour directly into its solid form on cooling is known as desublimation or deposition.
- The heat energy which needs to be supplied to change the state of a substance is known as its latent heat.
- The heat needed to convert a solid substance into its liquid state is known as the latent heat of fusion. For example, the latent heat of fusion of ice is 3.34×10^5 J/kg or 80 cal/g.
- The heat needed to convert a liquid substance into its vapour state is known as the latent heat of vaporisation. For example, the latent heat of vaporisation of water is 22.5×10^5 J/kg or 540 cal/g.
- When we apply pressure and reduce the temperature, gases can be converted into liquids, that is, gases get liquefied. The process of conversion of a gas into a liquid by increasing pressure or decreasing temperature is known as liquification.

Exercise 1.2

All questions marked * are practical based.

Section A: Multiple Choice Questions
                                                    (1 Mark)

1. The physical state of water at 25°C, 0°C and 100°C, respectively, are:

   (a) Liquid, solid, gas
   (b) Solid, liquid, gas
   (c) Liquid, gas, solid
   (d) Gas, solid, liquid

2. To find the boiling point of water, three students used water, at 0°C, at room temperature and

lukewarm water, respectively. On comparing their observations, it would be found that the boiling point of water was:
(a) Same in all cases
(b) Different in all three cases
(c) Maximum in the case of lukewarm water
(d) Least in water at 0°C

3. To determine the boiling point of water, the bulb of the thermometer should:
(a) Be completely dipped in water
(b) Be just above the surface of the water
(c) Be near the mouth of the flask
(d) Just touch the bottom of the flask

4. Under which conditions can we boil water at room temperature?
(a) At low pressure
(b) At high pressure
(c) At atmospheric pressure
(d) In any of these conditions

5. The least count of a laboratory thermometer is generally:
(a) 0.1°C   (b) 0.5°C
(c) 2°C   (d) 1°C

6. At melting point temperature:
(a) Only ice is present
(b) Only water is present
(c) Both ice and water are present
(d) Cannot be said

7. Which of the following is the correct method of finding the melting point of ice?

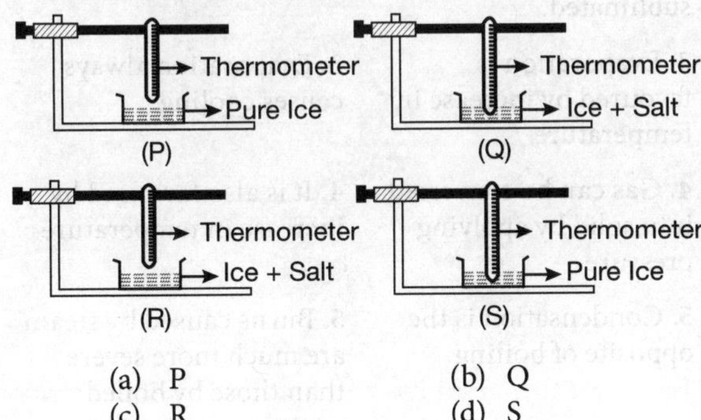

(a) P   (b) Q
(c) R   (d) S

8. When a solid starts melting, further heat energy which is supplied:
(a) Is lost to the surroundings as such
(b) Is absorbed as latent heat of fusion by the solid
(c) Increases the kinetic energy of the particles of the liquid
(d) Increases the temperature of the liquid

9. In an experiment to find the melting point of ice, the correct reading of the thermometer is noted when.
(a) Ice starts melting
(b) Temperature becomes constant
(c) Temperature starts increasing
(d) Whole of the ice melts

10. While determining the melting point of ice, Ronit used a glass stirrer. The purpose of this is to:
(a) Increase the kinetic energy
(b) Keep the temperature uniform
(c) Help the fusion process
(d) Decrease the kinetic energy

11. A beaker contains 50 g of ice-and-water mixture. The temperature of this mixture is:
(a) Less than 0°C   (b) 0°C
(c) More than 0°C   (d) 4°C

12. Of the following substances, which does not undergo sublimation?
(a) Sucrose   (b) Camphor
(c) Iodine   (d) Naphthalene

13. Which of the following substances cannot be separated by sublimation?
(a) Potassium chloride
(b) Ammonium chloride
(c) Camphor
(d) Iodine

14. Which one of the following figures describes the process of solid vapour?

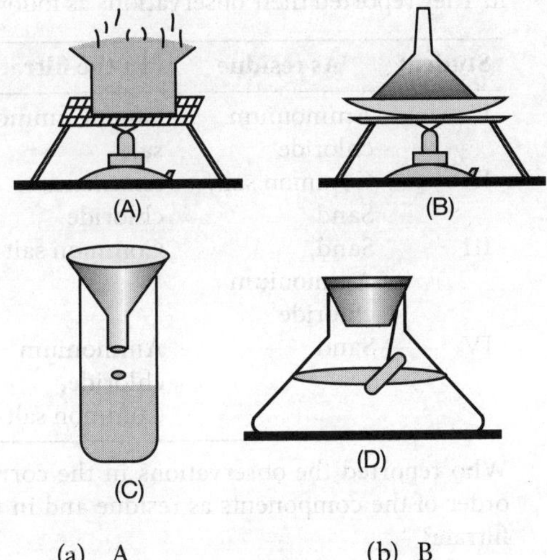

(a) A   (b) B
(c) C   (d) D

15. Which of the following is known as dry ice?

(a) O_2 (b) N_2
(c) Solid CO_2 (d) Liquid H_2

16. The reverse of vaporisation is:
(a) Sublimation   (b) Solidification
(c) Condensation   (d) Fusion

17. The heat needed for a liquid to convert to gas is known as:
(a) Latent heat of fusion
(b) Latent heat of vaporisation
(c) Heat of sublimation
(d) Any of these

18. Which of the following examples is based on evaporation?
(a) Pouring of acetone on palm
(b) Wearing of cotton cloth in summer
(c) Sprinkling of water on the roof
(d) All of these

19. Which option represents the set having correct statements:
(i) Sublimation is an endothermic (heat needed) process
(ii) Ammonium chloride and camphor undergo sublimation
(iii) Perspiration keeps our body cold
(iv) Cooling is caused during boiling
(a) (i), (ii)   (b) (i), (ii), (iii)
(c) (i), (iii), (iv)   (d) (i), (ii), (iii), (iv)

20. Four students took a mixture of sand, common salt and ammonium chloride in beakers, added water, stirred the mixture well and then filtered it. They reported their observations as follows:

Student	As residue	In the filtrate
I	Ammonium chloride	Sand, Common salt
II	Common salt, Sand	Ammonium chloride
III	Sand, Ammonium chloride	Common salt
IV	Sand	Ammonium chloride, Common salt

Who reported the observations in the correct order of the components as residue and in the filtrate?
(a) I   (b) II
(c) III   (d) IV

21. Match the following:

Column I	Column II
1. Solid to liquid	(i) Sublimation
2. Liquid to gas	(ii) Condensation
3. Gas to liquid	(iii) Vaporisation
4. Solid to gas	(iv) Fusion

(a) 1-(iv), 2-(iii), 3-(i), 4-(ii)
(b) 1-(iv), 2-(iii), 3-(ii), 4-(i)
(c) 1-(iii), 2-(iv), 3-(ii), 4-(i)
(d) 1-(i), 2-(ii), 3-(iii), 4-(iv)

Assertion–Reason Questions

Direction: In the following question two statements (Assertion) A and Reason (R) are given Mark.
(a) if A and R both are correct and R is the correct explanation of A;
(b) if A and R both are correct but R is not the correct explanation of A;
(c) A is true but R is false;
(d) A is false but R is true

Assertion	Reason
1. Surgeons often spray ether on the patient's skin before minor surgery.	**1.** Applying ether on the skin makes the skin numb by releasing energy from it.
2. Potassium permanganate can be sublimated.	**2.** In sublimation, solid changes into gas directly.
3. Evaporation is favoured by increase in temperature.	**3.** Evaporation always causes cooling.
4. Gas can be converted into solid by applying pressure.	**4.** It is also favoured by increase in temperature.
5. Condensation is the opposite of boiling.	**5.** Burns caused by steam are much more severe than those by boiled water.
6. Evaporation is a surface phenomenon.	**6.** Evaporation causes no cooling.
7. Latent heat of fusion of ice is 80 Kcal/kg.	**7.** Latent heat of vaporisation of water is also 80 Kcal/kg.

Section B: Very Short Answer Questions   (2 Marks)

1. Select the substances which do not have the property of sublimation from the following: Potassium permanganate, Copper sulfate, Naphthalene, Camphor, Sugar

2. Mention two factors that need to be varied to liquify atmospheric gases.

3. What is meant by latent heat of fusion?

4. Molecules of water have more energy as compared to molecules of ice at the same temperature. Justify this statement.

5. State one difference between gas and vapour.

6. The boiling point and freezing point of water are 100°C and 0°C, respectively. Convert these temperatures to K.

7. Why does the palm of your hand feel cold when you pour some acetone on it?

8. The boiling point of alcohol is 78°C. What is the corresponding temperature on the Kelvin scale?

9. Name the phenomenon of conversion of a liquid into vapour at a temperature below its boiling point.

10. The kinetic energy of particles of water in three vessels A, B and C are E_A, E_B and E_C, respectively, and $E_A > E_B > E_C$. Arrange the temperatures TA, TB and TC of water in the three vessels in increasing order.

11. What happens to the heat energy which is supplied to a solid once it starts melting?

12. What happens to the boiling point of a liquid when atmospheric pressure decreases?

13. What happens to the melting point of ice when pressure is increased?

14. Naphthalene balls disappear with time without leaving any residue. Why?

15. Explain whether the following statement is true or false:
"Sublimation occurs only when the solid is heated".

16. Define the term sublimation. Write the names of any two substances which sublimate.

Section C: Short Answer Questions   (3 Marks)

1. What is the effect of change of pressure on the physical state of matter? Explain with an example of a gas.

2. Explain the inter-conversion of three states of matter in terms of force of attraction and kinetic energy of the molecules.

3. The following triangle exhibits inter-conversion of the three states of matter. Complete the triangle by labelling the arrows marked A, B, C and D.

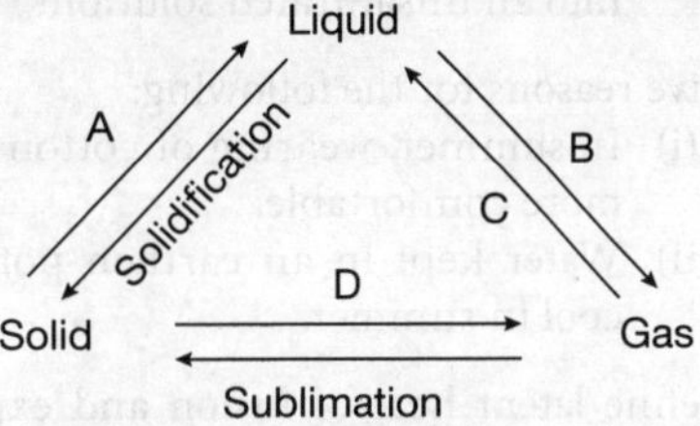

4. (i) Enumerate the changes that take place in matter during change of state.
   (ii) When a solid melts, its temperature remains the same. Give reasons.

5. Why does the temperature of a substance remain constant during its melting point or boiling point?

6. Carbon dioxide was taken in an enclosed cylinder and compressed by applying pressure:
   (i) Which state of matter will we obtain after completion of the process?
   (ii) Name and define this process.
   (iii) What is the common name of the product obtained in the above process?

7. Steam produces more severe burns than boiling water. Why?

8. Answer in one word:
   (i) The process by which a solid directly changes into gas without liquifying.
   (ii) Energy required to change 1 kg of a liquid to gas at atmospheric pressure at its boiling point.
   (iii) The property of gases which makes it possible to inflate a large number of balloons from a small cylinder of hydrogen gas.

9. How does the following affect the rate of liquid of vaporisation or evaporation?

(i) Surface area (ii) Temperature
(iii) Humidity

10. Explain the following:
    (i) How can we liquify gases?
    (ii) Why do clothes take more time to dry on a rainy day?

11. (i) After winters, people pack off their woollens and keep naphthalene balls in them. With the passage of time, these balls become smaller in size. Why does this happen? What type of change is involved during this process?
    (ii) How can you convert a saturated solution into an unsaturated solution?

12. Give reasons for the following:
    (i) In summer, wearing of cotton clothes is more comfortable.
    (ii) Water kept in an earthen pot becomes cool in summer.

13. Define latent heat of fusion and explain why heat energy is required to melt a solid.

14. (i) Explain with examples from your daily life where cooling is caused by evaporation.
    (ii) People in villages use earthen pots to get cool water. Explain why water remains cool in earthen pots.

15. (i) Define latent heat of vaporisation.
    (ii) Give reasons for the following:
        (1) You feel cold when you pour some nail polish remover on your palm.
        (2) In summer, sitting under a fan makes us comfortable.

16. Explain why temperature remains the same during melting of ice.

17. *(i) Wax is heated in a china dish. How will the following change during heating:
        (1) Kinetic energy of the particle
        (2) Inter-particle distance
    (ii) The melting points of three substances A, B, C are 52°C, 175°C and 80°C. Arrange them in decreasing order of the inter-particle force of attraction. Give reasons for your answer.

18. *With the help of a well-labelled diagram (with two labels) explain how solid ammonium chloride converts directly to gaseous state on heating?

19. *(i) In an experiment to determine the melting point of ice in a laboratory, what form of ice should be used? When should the reading of the thermometer be noted?
    (ii) Mention any two important precautions to be taken here during the experiment.

20. *(i) In an experiment to determine the boiling point of water, state the reasons for the following precautions:
        (1) The bulb of the thermometer should not touch the sides of the beaker.
        (2) While boiling water, pumice stones should be added.
    (ii) Why do we fix a two holed-cork in the round-bottomed flask while determining the boiling point of water?

Section D: Long Answer Questions
(5 Marks)

1. (i) Mention the physical state of water at (1) 100°C, (2) 0°C.
   (ii) Convert the following temperature into Celsius scale:
       (1) 298 K   (2) 310 K   (3) 285 K

2. Explain with examples from your daily life where cooling is caused due to evaporation.

3. Answer the following questions:
   (i) Between boiling and evaporation, which is a surface phenomenon?
   (ii) In the absence of a refrigerator, butter is kept wrapped in a wet cloth in summer. Why?
   (iii) Why does ice-cream feel colder than water at the same temperature?

4. (i) Define evaporation.
   (ii) Write the ways in which evaporation differs from boiling.
   (iii) Explain two factors affecting evaporation.

5. Account for the following:
   (i) When sugar crystals dissolve in water, the level of water does not rise appreciably.
   (ii) Doctors advise us to put strips of wet cloth on the forehead of a person having high fever.
   (iii) Naphthalene balls disappear with time without leaving any solid residue.
   (iv) Dogs generally hang out their tongue in summer.

6. While heating ice in a beaker with a thermometer suspended in it, a student recorded the following observations:

Time (in minutes)	0	1	2	3	4	5	6	7	8	10	15	20	25	30	35
Temperature (in °C)	-3	-1	0	0	5	8	12	15	19	22	30	50	73	100	100

Based on the above observations, answer the following questions:

(i) State the change(s) observed between 2 minutes and 3 minutes, and name the process involved.

(ii) Between 30 minutes and 35 minutes, the temperature remains constant. State the reason for this. Name the heat involved in the process and define it.

7. *(i) Give reasons for the following:

(1) Why do we see droplets of water on the outer surface of a glass containing ice cold water?

(2) On a hot sunny day, people sprinkle water on the roof or open ground.

(ii) Describe an activity with a labelled diagram to illustrate the effect of increase of temperature on ice.

8. *During an experiment, Shashwat observes changes in temperature at different intervals as follows:

(a) For the first time interval from 't_1' to 't_2' seconds, the temperature first gradually dropped from 30°C to 0°C.

(b) For the next time interval from 't_2' to 't_3' seconds, the temperature remained constant until all the ice melted.

(c) In the next time interval from 't_3' to 't_4' seconds, the temperature increased gradually from 0°C to 100°C.

(i) Draw a graph to show the abovementioned changes.

(ii) Name the property by virtue of which the temperature remained constant in observation (b).

(iii) What should be the subsequent readings of temperature on further heating after 't_4' seconds of Shashwat's observation?

Section E: Case Study or Passage-Based Questions   (4 Marks)

1. The conversion of a liquid into gas is possible by heating or decreasing the pressure. However, sometimes this phenomenon occurs on its own. For example, we can see that when water is left uncovered in a dish, it slowly changes into vapour without reaching its boiling point. This phenomenon is known as evaporation and it has lot of applications in our daily life as it has a cooling effect.

(a) Are evaporation and boiling the same? If not, why?

(b) Between nylon and cotton clothes, which will be more comfortable during summer and why?

(c) State all the factors that affect the rate of evaporation of water and also state how these factors affect it.

(d) Give reasons why we are able to sip hot tea or milk faster from a saucer than from a cup?

(e) Why is ice rubbed on burnt skin?

2. Boiling point is the temperature at which a liquid starts boiling at atmospheric pressure. Every pure liquid has a fixed value of boiling point. Boiling point can be measured by using a beaker, thermometer, burner, glass, stirrer, etc. During boiling point measurement, a few precautions are necessary.

(a) Two students Shanvi and Manvi are asked to arrange the apparatus to determine the boiling point of water. They arranged the apparatus as shown below:

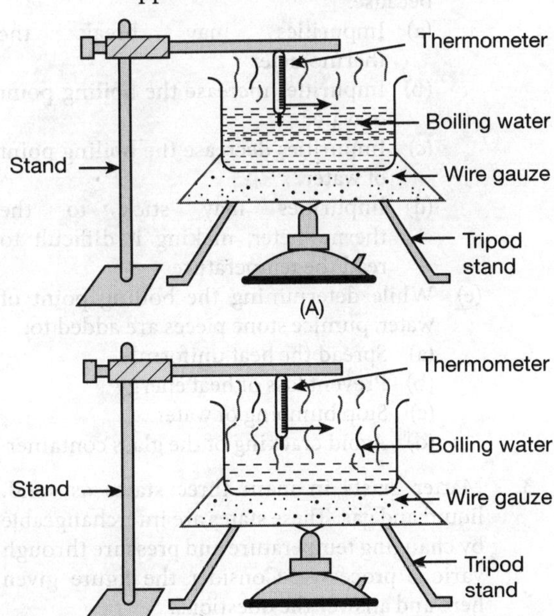

The diagram in which the apparatus is correctly arranged is:
(a) (A) only
(b) (B) only
(c) Both (A) and (B)
(d) Neither (A) nor (B)

(b) When liquid starts boiling, further heat is supplied, which is:
(a) Absorbed as latent heat of vaporisation by the liquid
(b) Absorbed to convert the liquid into solid
(c) Absorbed to increase the temperature
(d) Lost to the surroundings

(c) In determining the boiling point of water, it is advisable to put the bulb of the thermometer in the steam rather than in water, so as to:
(a) Reduce the error due to atmospheric pressure.
(b) Make sure that the boiling point obtained is accurate even when the water sample contains non-volatile dissolved impurities
(c) Reduce the error due to expansion of glass because of heat
(d) Obtain the boiling point in a shorter time

(d) While determining the boiling point of water, we should always use distilled water because:
(a) Impurities may break the thermometer
(b) Impurities increase the boiling point of water
(c) Impurities decrease the boiling point of water
(d) Impurities may stick to the thermometer, making it difficult to read the temperature

(e) While determining the boiling point of water, pumice stone pieces are added to:
(a) Spread the heat uniformly
(b) Prevent loss of heat energy
(c) Stop bumping of water
(d) Avoid cracking of the glass container

3. Matter exists in main three states as solid, liquid and gas. These states are interchangeable by changing temperature and pressure through various processes. Consider the figure given here and answer the questions.

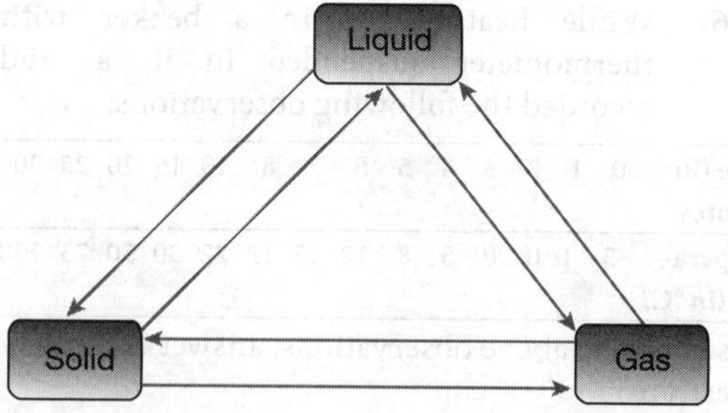

(a) The conversion of solid state to liquid state is known as:
(a) Fusion
(b) Distillation
(c) Vapourisation
(d) Sublimation

(b) The conversion of liquid state to gas state is known as:
(a) Fusion
(b) Distillation
(c) Vapourisation
(d) Sublimation

(c) The conversion of solid state to gas state directly is known as:
(a) Fusion
(b) Distillation
(c) Vapourisation
(d) Sublimation

(d) The liquification of a gas is favoured by:
(a) Increase of temperature
(b) Decrease of temperature
(c) Increase of pressure
(d) Both (b) and (c)

(e) Which option represents the set having correct statements:
(i) For converting solid into gas, high temperature is needed
(ii) For converting solid into liquid, high pressure is needed
(iii) For converting liquid into gas, low pressure is needed
(iv) For converting liquid into gas, high temperature is needed
(a) (i), (ii)
(b) (i), (ii), (iii)
(c) (i), (iii), (iv)
(d) (i), (ii), (iii), (iv)

High Order Thinking Skills (HOTS) Questions

1. About 10 mL of water was taken in a test tube and a china dish. These samples were kept under different conditions as follows:
(i) Both the samples are kept under a fan.
(ii) Both the samples are kept inside a cupboard.
State in which case evaporation will be faster? Give reasons to support your answer. How will

the rate of evaporation change if the above activity is carried out on a rainy day? Justify your answer.

2. The temperature–time graph given below shows the heating curve for pure wax. After studying the graph, answer the following questions:

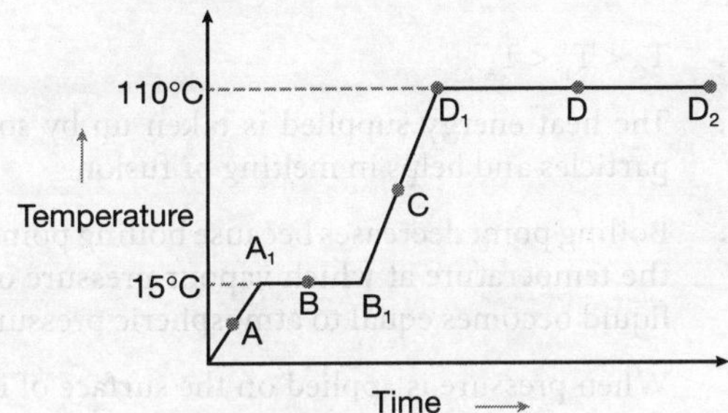

(i) What is the physical state of the substance at the points A, B, C and D?

(ii) What is the melting point of the substance?

(iii) What is its boiling point?

(iv) Which portions of the graph indicate that change of state is taking place?

(v) Name the terms used for heat absorbed during change of states involved in this process.

3. *(i) Why are pieces of pumice stone placed in the container before heating water while determining the boiling point of water in the laboratory? Explain briefly.

(ii) To find the boiling point of water, three students Khushi, Shanvi and Misti used distilled water at 0°C, at room temperature and lukewarm, respectively. Compare the boiling point of water observed by the three students and give reasons for your answer.

4. (i) Which of the following will give you more severe burns and why?
(1) Steam at 373 K (2) Water at 373 K

(ii) Identify and explain the factor responsible for changed rate of evaporation in the following situations:
(1) While putting clothes for drying, we spread them out.
(2) Water coolers are not effective on a rainy day.

5. *One day, when Misti was playing with her mother's cosmetics, she felt cold when a liquid bottle broke and some liquid fell on her hand.

She washed her hand and immediately asked her mother why she felt cold. Her mother answered her.

(i) Can you guess what that liquid was?

(ii) Why did Misti felt cool when the liquid fell on her hand?

(iii) What values are shown by Misti's mother?

6. *Shanvi went to her grandmother's village, where she found most of the ladies preparing dal in a saucepan. She advised them to use a pressure cooker to cook food faster.

(i) How is the pressure cooker helpful in cooking food faster?

(ii) What values are displayed by Shanvi?

(iii) How is the pressure cooker useful?

7. *Shashwat is fond of mountaineering. In his summer holidays, he planned to climb a mountain with his friends. While climbing, one of his friends got hurt and required hot water for treatment. He quickly took out his burner to boil the water. He found that on the mountain, water took much less time to boil than on the ground.

(i) Why does water boil at a lower temperature at higher altitudes?

(ii) What values are displayed by Shashwat?

(iii) Give one more example related to this phenomenon.

8. *One day, Nikunj fell ill and had high fever. The doctor advised his mother to put wet strips of cloth on his forehead. His mother took the doctor's advice properly and put wet strips of cloth on Nikunj's forehead. The process was very helpful in bringing down his fever.

(i) Why are wet strips of cloth placed on the forehead of a person suffering from high fever?

(ii) What values are displayed by Nikunj's mother?

(iii) Name the phenomenon involved in this process.

9. *(i) Amit wants to wear his favourite shirt to a party, but the problem is that it is still wet after a wash. Mention three steps with reasons that Amit can take to dry it faster?

(ii) It is a hot summer day; Mukul and Ronit are wearing cotton and nylon clothes, respectively.

(1) Who do you think would be more comfortable and why?

(2) In the rainy season, we feel sticky

and uncomfortable even under the fan. Why?

Answers

Section A: Multiple Multiple Choice Questions

1. (a)	2. (a)	3. (a)	4. (c)
5. (b)	6. (c)	7. (d)	8. (b)
9. (b)	10. (b)	11. (b)	12. (a)
13. (a)	14. (b)	15. (c)	16. (c)
17. (b)	18. (d)	19. (b)	20. (d)
21. (b).			

Assertion-Reason Questions

1. (a)	2. (d)	3. (b)	4. (c)
5. (b)	6. (c)	7. (c)	

Section B: Very Short Answer Questions

1. Potassium permanganate, Copper sulfate and Sugar

2. Temperature and pressure

3. The amount of heat energy required to change 1 kg of a solid into liquid at atmospheric pressure at its melting point is known as the latent heat of fusion.

4. Water has more energy than ice at the same temperature because particles in water have absorbed more energy during the change of state.

5. **Gas:** It is a stable state as compared to vapour. Examples: O_2, H_2.
   Vapour: It is an unstable state. On normal cooling, vapour changes into liquid state.

6. $100°C + 273 = 373\ K$ $0°C + 273 = 273\ K$

7. It will evaporate, taking latent heat of vaporisation from the palm. The hand loses heat and gets cooled.

8. $78 + 273 = 351\ K$

9. Evaporation.

10. $T_C < T_B < T_A$

11. The heat energy supplied is taken up by solid particles and helps in melting or fusion.

12. Boiling point decreases because boiling point is the temperature at which vapour pressure of a liquid becomes equal to atmospheric pressure.

13. When pressure is applied on the surface of ice, the change into liquid state is assisted. Thus, melting point decreases.

14. Naphthalene being volatile converts from solid to gas directly by a process called sublimation. Therefore, no solid residue is left after some time as it takes the heat from the surroundings and sublimes.

15. False, it can occur at room temperature also. For example, naphthalene balls sublime at room temperature.

16. Sublimation is the change of solid directly into the gaseous state without passing through the liquid state. For example, ammonium chloride and naphthalene.

Section C: Short Answer Questions

1. The physical state of matter can be changed by changing the pressure and temperature. For example, by lowering temperature and increasing pressure, gases can be changed into liquids. This happens with gases as there is a lot of space between the particles, and upon applying high pressure, the particles come close to each other which upon cooling get liquified.

2. During the inter-conversion of a solid to a liquid and liquid to a gas, on increasing the temperature, the kinetic energy of the molecules increases and the force of attraction decreases and vice versa.

3. A – Fusion   B – Vaporisation
   C – Condensation   D – Sublimation

4. (i) When temperature increases, the kinetic energy of the molecules increases and the force of attraction between the molecules decreases and the state of matter changes.

   (ii) This is because the heat supplied to the matter is utilised in changing the state by overcoming the force of attraction.

5. During melting, latent heat of fusion and during boiling, latent heat of vaporisation overcome the inter-particle force of attraction; therefore, temperature remains constant.

6. (i) Solid.
   (ii) Sublimation: The process of conversion of solid directly into vapour, and vice versa.
   (iii) Dry ice

7. When water at 373 K is converted into steam at 373 K, it absorbs energy equal to the latent heat of vaporisation. Thus, steam at 373 K has more heat energy than water at 373 K and hence steam produces more severe burns than boiling water.

8. (i) Sublimation
   (ii) Latent heat of vaporisation
   (iii) Compressibility of hydrogen gas

9. (i) Surface area: As surface area increases, more particles are converted into vapour, and the rate of evaporation increases.
   (ii) Temperature: With increase in temperature, the kinetic energy of the particles of liquid also increases and the rate of evaporation increases.
   (iii) Humidity: As humidity in air increases, the rate of evaporation decreases.

10. (i) We can liquify gases by applying pressure and reducing temperature.
    (ii) As on a rainy day, the amount of water vapours present in air as humidity is already high, the rate of evaporation decreases.

11. (i) Sublimation, naphthalene balls sublimate and becomes smaller in size. It is a physical change.
    (ii) By adding large quantities of the solvent to the solution.

12. (i) As cotton clothes are good absorbers of sweat, during evaporation of sweat, heat is lost by the body which makes us feel cool.

   (ii) On a hot day, evaporation of water from the pot through its pores becomes faster. During evaporation, heat is taken from the water.

13. Latent heat of fusion is the amount of heat energy needed to change 1 kg of solid into liquid at atmospheric pressure at its melting point. Heat energy is needed to melt a solid because there exist forces of attraction between the molecules and heat energy is essential to overcome the forces of attraction.

14. (i) When a liquid evaporates, it draws the latent heat of vaporisation from anything which it touches. This causes cooling. For example, sweating and water in an earthen pot.
    (ii) There are pores in an earthen pot through which the liquid inside the pot evaporates. This evaporation makes the water inside the pot cool. In this way, water kept in an earthen pot becomes cool during summer.

15. (i) The amount of heat energy required to change 1 kg of a liquid to gas at atmospheric pressure at its boiling point is called latent heat of vaporisation.
    (ii) (1) Particles gain heat energy from the palm and evaporate, causing the palm to feel cool.
         (2) When we sit under a fan during summer, rate of evaporation of sweat increases due to increase in wind speed. Sweat takes heat away from the body, leaving us cool.

16. The temperature remains constant during melting of ice even though heat is supplied regularly as the temperature is supplied for changing the state of matter. The heat of temperature is consumed by the particles of ice and they vibrate faster and break the force of attraction and become liquid.

17. (i) (1) Kinetic energy will increase.
         (2) Inter-particle distance will increase.
    (ii) B, C, A.

    Higher melting point indicates that inter-particle force is stronger. So more heat is required by particles to break away from bonds.

18. Take some camphor or ammonium chloride. Crush it and put it in a china dish. Put an inverted funnel over the china dish. Put a

cotton plug on the stem of the funnel. Now heat slowly. It is observed that NH_4Cl sublimes and gets deposited on the inner core of funnel.

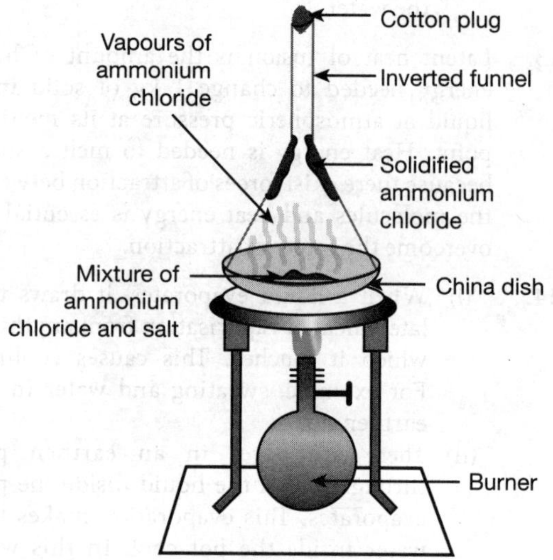

19. (i) Here, crushed ice preferably made from distilled water. The reading of the thermometer should be noted when all of the ice melts and the temperature becomes constant.
     (ii) The two precautions are
        • The bulb of the thermometer should be completely inside the crushed ice.
        • The solution should be stirred regularly to keep the temperature uniform.

20. (i) (1) Sides of the beaker are at slightly higher temperature.
         (2) To avoid bumping.
     (ii) We fix a two holed-cork in the round-bottomed flask in order to fix the thermometer through one of the holes in the cork and the glass tube through the other.

Section D: Long Answer Questions

1. (i) (1) Liquid and gas   (2) Liquid and solid.
    (ii) Temperature in Kelvin = Temperature in °C + 273
    Temperature in °C + 273 = Temperature in Kelvin − 273
    (i) 298 − 273 = 25°C   (ii) 310 − 273 = 37°C
    (iii) 285 − 273 = 12°C

2. Here are few examples of our daily life where cooling is caused due to evaporation.

(i) In summer, trees absorb more water and minerals from the soil as the rate of transpiration increases.
(ii) Cotton being a good absorber of water helps in absorbing the sweat and exposing it to the atmosphere for easy evaporation.
(iii) Rain water evaporates in the heat, leaving the road dry.
(iv) Water evaporates from the roof taking heat from the surroundings on a hot summer day, leaving roof cool.
(v) Acetone evaporates by taking heat from the palm, leaving it cool.

3. (i) Evaporation is a surface phenomenon in which particles from the surface gain enough energy to overcome the forces of attraction present in the liquid and change into vapour state.
    (ii) Due to the wet cloth, the temperature is comparatively lower than room temperature. As a result, butter does not melt when wrapped in wet cloths.
    (iii) Ice cream at 273 K will take latent heat from the medium to convert itself into liquid at 273 K and then into liquid at higher temperature, but in case of water, such conditions are not possible. This is why ice cream feels colder than water.

4. (i) Refer to Section 1.2.6, Change of State of Matter.
    (ii) Differences between boiling and evaporation:

Boiling	Evaporation
Boiling occurs at a particular temperature, that is, boiling point of that liquid.	Evaporation takes place when the liquid is placed in an open container at any temperature below its boiling point.
Heating takes place during boiling.	Cooling takes place during evaporation.

    (iii) (1) When temperature is increased, the rate of evaporation increases because with the increase in temperature, more particles get enough kinetic energy to go into the vapour state.
          (2) When humidity is decreased, the rate of evaporation increases and vice versa.

5. (i) Particles of sugar crystals occupy the space between the particles of water.

(ii) The excess heat from the body is taken by high latent heat of vaporisation of water. As a result, body temperature decreases.

(iii) Sublimation, naphthalene is converted into vapour.

(iv) Evaporation of saliva causes cooling.

6. (a) Ice converts to water and the process is called fusion.

(b) Heat is used for conversion of state.

Reason: Due to latent heat of vaporisation. The latent heat of vaporisation is the amount of heat required to convert a unit mass of liquid into its vapour state, without a change in its temperature at its boiling point.

7. (i) (1) Water vapour of air condenses because it loses energy when it comes in contact with a cool surface.

(2) Evaporation leads to decrease in temperature because water takes heat for evaporation from the floor.

(ii)

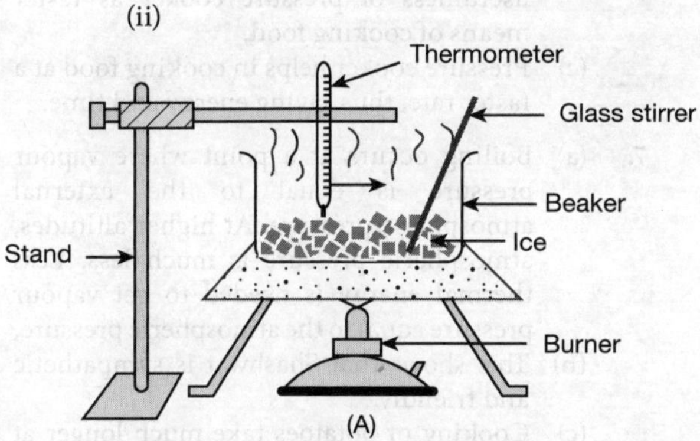

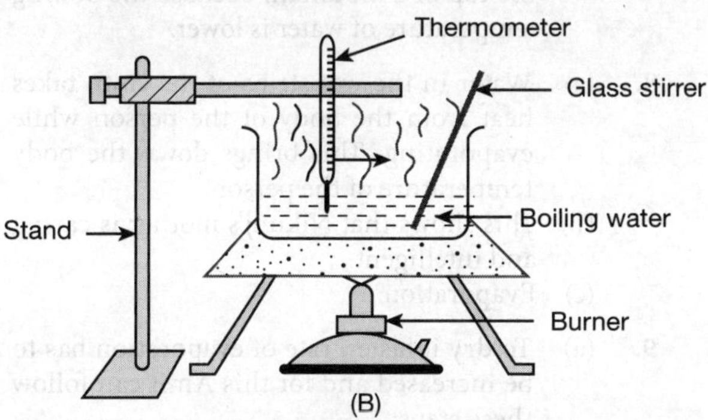

(1) Take about 150 g of ice in a beaker and suspend a laboratory thermometer so that its bulb is in contact with the ice.

(2) Start heating the beaker on a low flame.

(3) Note the temperature when the ice starts melting.

(4) Note the temperature when all the ice has become water.

(5) Record your observations for this conversion of solid to liquid state.

(6) Now, put a glass rod in the beaker and heat while stirring till the water starts boiling.

(7) Keep a careful eye on the thermometer reading till most of the water has vaporised.

Observation: As the ice reaches its melting point, it starts getting converted into water.

8.
(i)

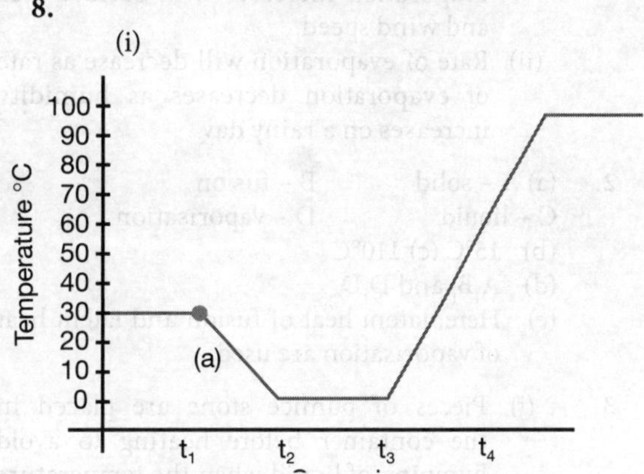

(ii) Latent heat of fusion.

(iii) The temperature will remain constant at 100°C.

Section E: Case Study or Passage-Based Questions

1.

(a) Not exactly the same as boiling of liquid takes place at its boiling point only, whereas evaporation can occur at any temperature.

(b) Cotton is a better absorber of water than nylon. So, in summer, cotton clothes absorb sweat, which on evaporation cause a cooling sensation in the body.

(c) Rate of evaporation increases with increase in surface area, increase in temperature, decrease in humidity and increase in speed of wind.

(d) Evaporation is directly proportional to surface area. The surface area of a saucer is greater, so evaporation is more.

(e) Ice is rubbed on burnt skin so that excess of heat on the skin can be taken as latent heat of fusion. As a result, the temperature decreases and relief from pain is possible.

2. 1. (a)   2. (a)   3. (b)   4. (b)   5. (c)

3. 1. (a)   2. (c)   3. (d)   4. (d)   5. (c)

High Order Thinking Skills (HOTS) Questions

1. (i) Evaporation will be faster when the china dish is kept under a fan as the rate of evaporation increases with surface area and wind speed.

(ii) Rate of evaporation will decrease as rate of evaporation decreases as humidity increases on a rainy day.

2. (a) A – solid   B – fusion
C – liquid   D – vaporisation
(b) 15°C (c) 110°C
(d) A_1B_1 and D_1D_2
(e) Here, latent heat of fusion and latent heat of vaporisation are used.

3. (i) Pieces of pumice stone are placed in the container before heating to avoid bumping of liquid when the temperature increases. On boiling, water releases energy as bubbles. If the bubbles do not form, it can develop lot of heat and possibly explode. Addition of stones given a lot of surface area for bubbles to form and release the energy gradually.

(ii) The value of the boiling point does not depend on the temperature of the liquid when pressure is kept constant. Thus, all the three students will observe the same boiling point.

4. (i) Steam at 373 K will give more severe burns. This is because steam has more heat content because of latent heat of vaporisation. When it touches our body, it gives this extra amount of heat, causing more severe burns.

(ii) (a) Spreading the clothes for drying increases the surface area which helps it to dry faster as the rate of evaporation increases with increase in surface area.

(b) As humidity is inversely proportional to evaporation, on a rainy day, humidity increases, which decreases the rate of evaporation.

5. (i) The liquid is a nail polish remover which contains ether or acetone.

(ii) Ether evaporates by taking heat energy from the hand (body). That is why she felt cold.

(iii) She exhibits knowledge, carefulness and an educating nature towards her daughter.

6. (a) The pressure in the enclosed volume above the liquid reaches much greater values than atmospheric pressure; thus, the temperature of boiling water within the cooker is greater than the normal boiling temperature. That is why food cooks much faster.

(b) Shanvi is trying to perform social work by educating village women about the usefulness of pressure cooker as faster means of cooking food.

(c) Pressure cooker helps in cooking food at a faster rate, thus saving energy and time.

7. (a) Boiling occurs at a point where vapour pressure is equal to the external atmospheric pressure. At higher altitudes, atmospheric pressure is much less. Less thermal energy is needed to get vapour pressure equal to the atmospheric pressure.

(b) This shows that Shashwat is sympathetic and friendly.

(c) Cooking of potatoes take much longer at the top of a mountain because the boiling temperature of water is lower.

8. (a) Water in the wet strips of the cloth takes heat from the body of the person while evaporating. This brings down the body temperature of the person.

(b) This shows that Nikunj's mother is caring and intelligent.

(c) Evaporation.

9. (a) To dry it faster, rate of evaporation has to be increased and for this Amit can follow these steps:

(a) He can spread it to increase the surface area.

(b) He can iron it to increase the temperature.

(c) He can spread it under a fan to increase wind speed.

(b) (i) Mukul will be more comfortable as in cotton clothes, sweat will evaporate faster by taking the latent heat of vapourisation from the body and thus making it comfortable by taking away the body's heat faster.

(ii) In the rainy season, humidity is high. So the rate of evaporation is slower.

Activities with Discussion and Conclusion

Activity 1.1 (To demonstrate that particles of matter have spaces between them)

1. Take a 100-mL beaker.

2. Fill half the beaker with water and mark the level of the water.

3. Dissolve some salt/sugar with the help of a glass rod.

4. Observe any change in water level.

Now Answer
- What do you think has happened to the salt?
- Where does it disappear to?

- Does the level of water change?

Discussion: Matter is composed of small or tiny particles and between these particles, there are empty spaces known as voids. When common salt or sugar is added to water, the particles of salt or sugar get into these empty spaces. As a result, the salt or sugar dissolves in water. If only a small amount of salt or sugar is dissolved, the water level does not rise, but if a large amount of salt or sugar (5 g in 50 mL of water) is dissolved, the water level rises by nearly 2–3 mL.

Conclusion: The particles of matter have spaces or voids between them.

Activity 1.2 (To demonstrate that particles of matter are very small)

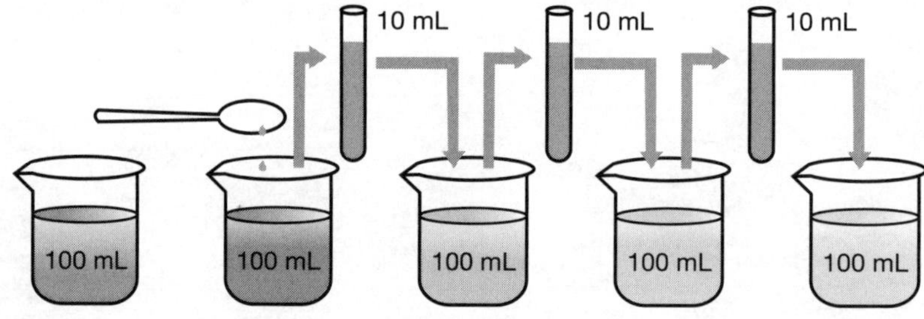

1. Take 2–3 crystals of potassium permanganate and dissolve them in 100 mL of water.

2. Take out approximately 10 mL of this solution and put it into 90 mL of clear water.

3. Take out 10 mL of this solution and put it into another 90 mL of clear water.

4. Keep diluting the solution 5 to 8 times.

Now Answer
- Is the water still coloured?

Discussion: A crystal of potassium permanganate has millions of tiny particles which keep on dividing into smaller and smaller numbers with each dilution, thereby making the colour lighter and lighter.

Conclusion: Matter is composed of extremely small or tiny particles which cannot be seen even with a powerful microscope. This means what we actually see is an aggregate of tiny particles (molecules).

Activity 1.3 (To demonstrate that particles of matter move continuously)

1. Put an unlit incense stick in a corner of your classroom. How close do you have to be so as to

detect its smell?

2. Record your observations.

3. Now light the incense stick. What happens? Do you detect the smell sitting at a distance?

4. Record your observations.

Discussion: The particles of matter move continuously, which means they have kinetic energy. As the temperature rises, the kinetic energy increases and the particles move faster. When the incense stick is not lit, the temperature is low and the kinetic energy of the particles of incense is also low. In other words, particles of incense do not mix with the particles of air rapidly. That is why you have to go near the incense stick to detect its smell. However, when the incense stick is lit, the temperature rises and so the kinetic energy of the incense particles also increases. In other words, the particles of the incense now move rapidly and thus intermix with the particles of the air rapidly. That is why you can detect the smell of incense even when you are sitting at a distance.

Conclusion: The particles of matter are never at rest but move continuously. As the temperature increases, their kinetic energy increases and they move even faster (speed $\propto$ temperature).

Activity 1.4 (To demonstrate that the rate of intermixing depends upon the forces of attraction between the particles of matter)

1. Take two glasses/beakers filled with water.

2. Place a drop of blue or red ink slowly and carefully along the sides of the first beaker and honey in the same way in the second beaker.

3. Leave them undisturbed in your house or in a corner in the classroom.

4. Record your observations.

Now Answer
- What do you observe immediately after adding the ink drop?
- What do you observe immediately after adding a drop of honey?
- How many hours or days does it take for the colour of ink to spread evenly throughout the water?

Discussion: As the forces of attraction between the particles of ink are weak, they move rapidly. As a result, particles of ink rapidly get into the spaces between the particles of water and are evenly distributed. However, the forces of attraction between the particles of honey are strong, so they move slowly. As a result, it takes a longer time for the particles of honey to get into the spaces between the particles of water and be evenly distributed throughout the water.

Conclusion: Particles of matter move continuously, but their average speed at any particular temperature depends upon the forces of attraction between them. The stronger the forces of attraction, the lower the average speed, while the weaker the forces of attraction, the greater the average speed.

Activity 1.5 (To demonstrate that the rate of intermixing increases with temperature)

1. Drop a crystal of copper sulfate or potassium permanganate into a glass of hot water and another containing cold water. Do not stir the solution. Allow the crystals to settle at the bottom.

Now Answer
- What do you observe just above the solid crystal in the glass?
- What happens as time passes?
- What does this suggest about the particles of solid and liquid?

- Does the rate of mixing change with temperature? Why and how?

Discussion: Due to strong forces of attraction, the particles of a solid virtually cannot move and hence remain fixed in their respective positions. However, the particles of a liquid like water move continuously and thus possess some kinetic energy, even in the cold. Due to their kinetic energy, the particles of water overcome the forces of attraction between the particles of solid crystal. As a result, the particles of water move in between the spaces of the particles of the solid crystal and hence the crystal starts dissolving in cold water. Consequently, the particles of water lying immediately above the crystal start acquiring a blue colour due to dissolution of $CuSO_4$ or purple colour due to dissolution of $KMnO_4$. As time passes, more and more particles of water can now pass into particles of $CuSO_4$ or $KMnO_4$ and the colour starts deepening and spreading into more water. Ultimately, the whole solution is coloured.

As the temperature rises, the kinetic energy of both the solid and the liquid particles increases. Due to greater kinetic energy at higher temperature, the forces of attraction between the solid particles weaken. Further, due to greater kinetic energy, the particles of the liquid move faster and hence can more easily overcome the weakened forces of attraction between the particles of the solid. As a result, the rate of intermixing increases and the solid dissolves more quickly in hot water.

Conclusion: As with increase in temperature, the kinetic energy of the particles increases so the rate of intermixing increases, and hence, the solid dissolves more quickly in hot water than in cold water.

Activity 1.6 (To determine the strength of attractive forces between particles of different kinds of matter)

Source: Sara0404, 2019, https://commons.wikimedia.org/wiki/File:Odia_Tribal_Folk_Dance,_Odisha,_India.jpg. Licensed under: CC-BY-SA-4.0

1. Play this game in the field. Make four groups and form human chains.

2. The first group should hold each other from the back and lock arms, like the dancers in the figure.

3. The second group should hold hands to form a human chain.

4. The third group should form a chain by touching each other with only their fingertips.

5. Now, the fourth group of students should run around and try to break the three human chains one by one into as many small groups as possible.

Now Answer

1. Which group was the easiest to break? Why?

2. If we consider each student as a particle of matter, then in which group do the particles hold each other with maximum force?

Discussion: The first group of students that formed the human chain was most strongly held and their human chain was the most difficult to be broken. The second group of students that formed the human chain by holding their hands was the next strongly held while the students of the third group that formed the human chain by just touching their fingers was the most weakly held, and hence, their human chain was the easiest to break.

Conclusion: The particles of matter which are most strongly held are the most difficult to be separated while the particles of matter which are most weakly held are the most easy to be separated.

Activity 1.7 (To demonstrate the strength of the attractive forces between particles of different kinds of matter)

1. Take an iron nail, a piece of chalk and a rubber band.

2. Try breaking them by hammering, cutting or stretching.

Now Answer

- In which of the above three substances do you think the particles are held together with greater force?

Discussion: It is most difficult to break the iron nail, so the particles of the iron nail are held together by the strongest force, followed by the piece of chalk, while the particles of the rubber band are held by the weakest forces of attraction.

Conclusion: The particles of matter attract each other and the strength of this force of attraction differs from one kind of matter to another.

Activity 1.8 (To demonstrate the strength of the attractive forces between particles of different kinds of matter)

1. Open a water tap and try breaking the stream of water with your fingers.

Now Answer

- Were you able to cut the stream of water?
- What could be the reason behind the stream of water staying together?

Discussion: When we try to cut the stream of water with our fingers, the stream cannot be cut as the particles of water attract one another strongly and hence tend to remain together.

Conclusion: Particles of matter attract one another.

Activity 1.9 (To study the properties of solids)

1. Collect the following articles: a pen, a book, a needle and a piece of thread.

2. Sketch the shape of the above articles in your notebook by moving a pencil around them.

Now Answer

- Do all these articles have definite shape, distinct boundaries and fixed volume?
- What happens if they are hammered, pulled or dropped?
- Are these articles capable of diffusing into each other?
- Try compressing them by applying force. Are you able to compress them?

Discussion: All the articles listed above are solids as they are rigid, they are not compressible and have definite shapes, distinct boundaries and fixed volumes.

Conclusion: Particles of a solid are rigid and not compressible. They have definite shapes, distinct boundaries and fixed volumes.

Activity 1.10 (To demonstrate that liquids have definite volume but no definite shape)

1. Collect the following:
   (a) Water, cooking oil, milk, juice, a cold drink.
   (b) Containers of different shapes.

2. Draw a 50-mL mark on these containers using a measuring cylinder from the laboratory.

Now Answer
- What will happen if these liquids are spilt on the floor?
- Measure 50 mL of any one liquid and transfer it into different containers one by one. Does the volume remain the same?
- Does the shape of the liquid remain the same?
- When you pour the liquid from one container into another, does it flow easily?

Discussion: All these liquids can flow when spilt on the floor. The volume of the liquid remains the same, that is, 50 mL, when it is transferred from one container to the other. The shape of a liquid is not fixed as it takes up the shape of the container in which it is stored. Further, when the liquid is poured from one container to another, it flows.

Conclusion: Liquids have definite volume but do not have a definite shape. When they are poured from one container to another, they flow easily and can take up the shape of the container in which they are stored.

Activity 1.11 (To study the compressibility of solids, liquids and gases)

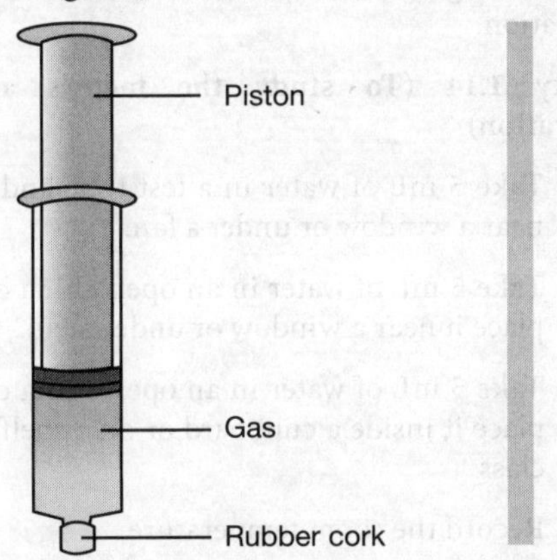

1. Take three 100-mL syringes and close their nozzles with rubber corks.

2. Remove the pistons from all the syringes.

3. Leaving one syringe untouched, fill water in the second and pieces of chalk in the third.

4. Insert the pistons back into the syringes. You may apply some Vaseline on the pistons before inserting them into the syringes for smooth movement.

Now Answer
- Try to compress the content by pushing the piston in each syringe. What do you observe? In which case was the piston easily pushed in?

- What do you infer from your observations?

Discussion: The syringe which was left untouched, that is, the first syringe which contained air, was the one in which the piston was easily pushed in. The piston of the second syringe which contained water was pushed in only a little while the piston of the third syringe which contained chalk pieces could not be pushed in at all. Hence, air is easily compressible, water is almost incompressible while chalk pieces are fully incompressible.

Conclusion: The spaces between the particles of gases are the largest, intermediate in liquids and shortest in the case of solids. It means gases are highly compressible, liquids are almost incompressible while solids are completely incompressible.

Activity 1.12 (To demonstrate that temperature remains constant during change of state)

1. Take about 150 g of ice in a beaker and suspend a laboratory thermometer so that its bulb is in contact with the ice.

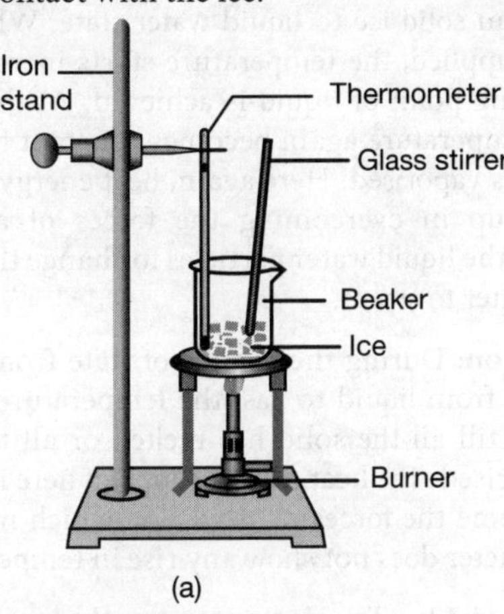

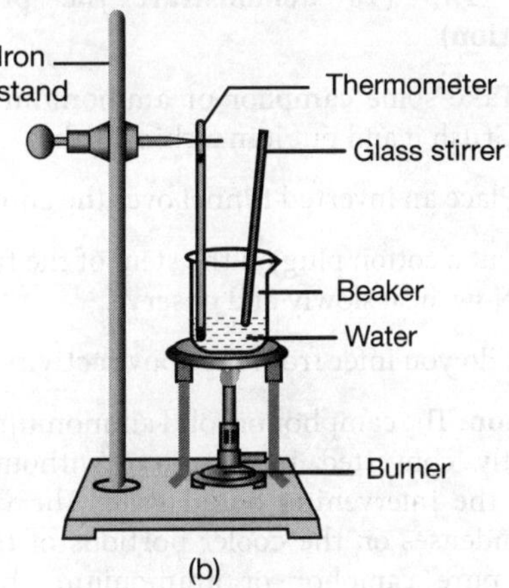

2. Start heating the beaker on a low flame.

3. Note the temperature when the ice starts melting.

4. Note the temperature when all the ice has been converted into water.

5. Record your observations for this conversion of solid to liquid state.

6. Put a glass rod in the beaker and heat while stirring till the water starts boiling.

7. Keep a careful eye on the thermometer reading till most of the water has vaporised.

8. Record your observations for the conversion of water in the liquid state to gaseous state.

Discussion: The temperature at which ice starts melting is known as the melting point of ice. This temperature remains constant till all the ice has melted. The heat energy supplied here is used up in overcoming the forces of attraction between the ice particles to change them from solid ice to liquid water state. When more heat is supplied, the temperature starts increasing till the boiling point of liquid is achieved. At the boiling point, temperature again becomes constant till all the liquid has vaporised. Here again, heat energy supplied is used up in overcoming the forces of attraction between the liquid water particles to change them from liquid water to steam.

Conclusion: During the change of state from solid to liquid or from liquid to gas, the temperature remains constant till all the solid has melted or all the liquid has vaporised. The heat energy supplied here is used up to overcome the forces of attraction which means the thermometer does not show any rise in temperature.

Activity 1.13 (To demonstrate the process of sublimation)

1. Take some camphor or ammonium chloride. Crush it and put it in a china dish.

2. Place an inverted funnel over the china dish.

3. Put a cotton plug on the stem of the funnel. Now, heat slowly and observe.

• What do you infer from the above activity?

Discussion: The camphor or solid ammonium chloride is directly converted into vapour without passing through the intervening liquid state. The vapour, in turn, condenses on the cooler portions of the funnel to give pure camphor or ammonium chloride. It

means camphor and ammonium chloride undergo sublimation.

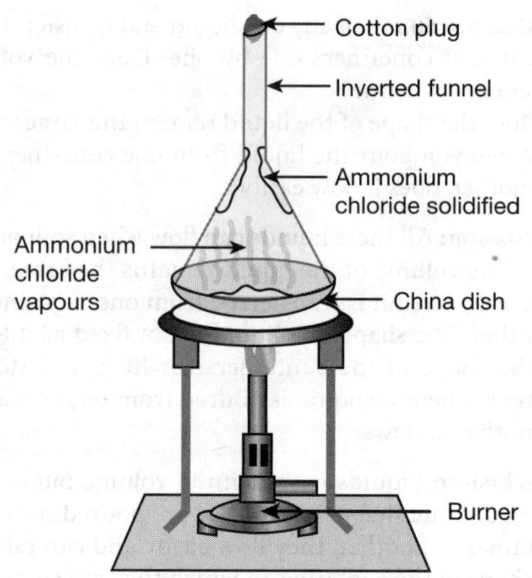

Conclusion: A change of state directly from solid to gas without changing into liquid (vice versa) is known as sublimation.

Activity 1.14 (To study the factors affecting evaporation)

1. Take 5 mL of water in a test tube and place it near a window or under a fan.

2. Take 5 mL of water in an open china dish and place it near a window or under a fan.

3. Take 5 mL of water in an open china dish and place it inside a cupboard or on a shelf in your class.

4. Record the room temperature.

5. Record the time or days taken for the evaporation process in the above cases.

6. Repeat the above steps on a rainy day and record your observations.

• What do you infer about the effect of temperature, surface area and wind velocity (speed) on evaporation?

Discussion: As the surface area of water exposed to the atmosphere is minimum in the case of a test tube, it takes a long time (2/3 days) for 5 mL of water to evaporate. Although the surface area of 5 mL of water taken in two open china dishes is the same, water in the dish placed near the window or under the fan evaporates more quickly than the water in the dish placed inside a cupboard or on a shelf in your classroom.

All these three steps will take longer time on a rainy day as the humidity will be high.

Conclusion: The rate of evaporation depends upon the surface area exposed to the atmosphere, wind velocity and the humidity of air.

Exemplar Problems

Short Answer Questions

1. A sample of water under study was found to boil at 102°C at normal temperature and pressure. Is the water pure? Will this water freeze at 0°C? Comment.

2. A student heats a beaker containing ice and water. He measures the temperature of the contents of the beaker as a function of time. Which of the following would correctly represent the result? Justify your choice.

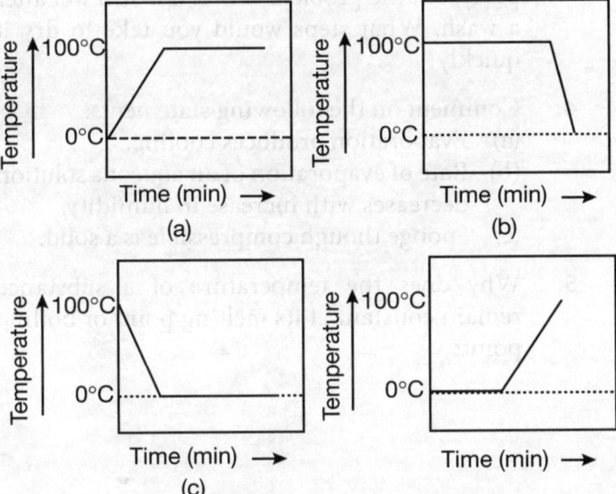

3. Fill in the blanks:
   (a) Evaporation of a liquid at room temperature leads to a _______ effect.
   (b) At room temperature, the forces of attraction between the particles of solid substances are _______ than those which exist in the gaseous state.
   (c) The arrangement of particles is less ordered in the _______ state. However, there is no order in the _______ state.
   (d) _______ is the change of gaseous state directly to solid state without going through the _______ state.
   (e) The phenomenon of change of a liquid into the gaseous state at any temperature below its boiling point is called _______.

4. Match the physical quantities given in column A to their SI units given in column B.

Column (A)	Column (B)
(a) Pressure	(i) Cubic metre
(b) Temperature	(ii) Kilogram
(c) Density	(iii) Pascal
(d) Mass	(iv) Kelvin
(e) Volume	(v) Kilogram per cubic metre

5. The non-SI and SI units of some physical quantities are given in column A and column B, respectively. Match the units belonging to the same physical quantity.

Column (A)	Column (B)
(a) Degree Celsius	(i) Kilogram
(b) Centimetre	(ii) Pascal
(c) Gram per centimetre cube	(iii) Metre
(d) Bar	(iv) Kelvin
(e) Milligram	(v) Kilogram per metre cube

6. 'Osmosis is a special kind of diffusion'. Comment.

7. Classify the following into osmosis/diffusion:
   (a) Swelling up of a raisin on keeping in water
   (b) Spreading of virus on sneezing
   (c) Earthworm dying on coming in contact with common salt
   (d) Shrinking of grapes kept in thick sugar syrup
   (e) Preserving pickles in salt
   (f) Spreading of smell of cake being baked throughout the house
   (g) Aquatic animals using oxygen dissolved in water during respiration

8. Water as ice has a cooling effect, whereas water as steam may cause severe burns. Explain these observations.

9. Alka was making tea in a kettle. Suddenly she felt intense heat from the puff of steam gushing out of the spout of the kettle. She wondered whether the temperature of the steam was higher than that of the water boiling in the kettle. Comment.

10. A glass tumbler containing hot water is kept in the freezer compartment of a refrigerator (temperature < 0°C). If you could measure the temperature of the contents of the tumbler, which of the following graphs would correctly represent the change in its temperature as a function of time?

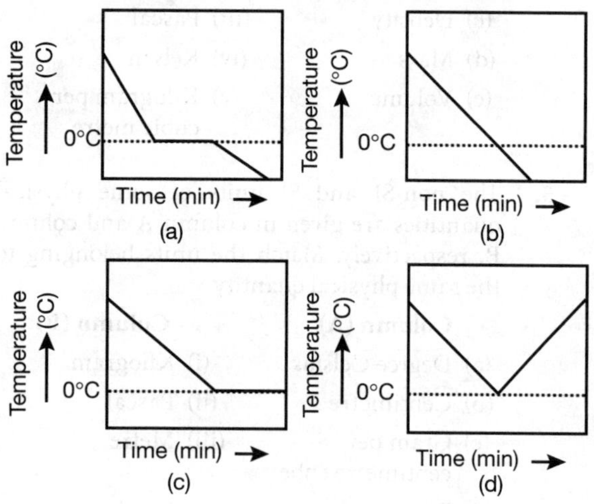

11. Look at the figures and suggest in which of the vessels A, B, C or D the rate of evaporation will be highest? Explain.

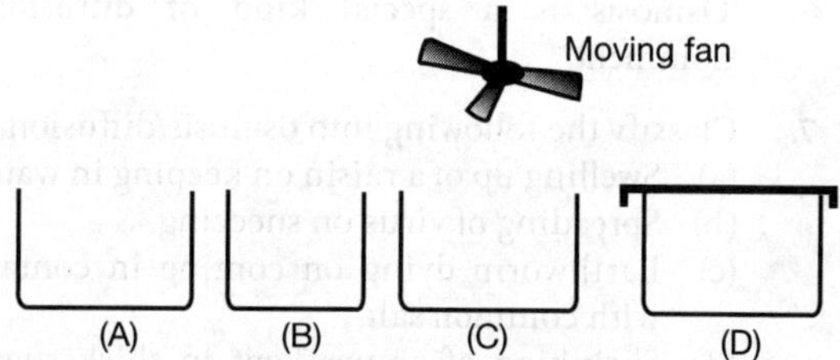

12. (a) Conversion of solid to vapour is called sublimation. Name the term used to denote the conversion of vapour to solid.
    (b) Conversion of solid state to liquid state is called fusion; what is meant by latent heat of fusion?

Long Answer Questions

1. You are provided with a mixture of naphthalene and ammonium chloride by your teacher. Suggest an activity to separate them with a well-labelled diagram.

2. It is a hot summer day. Priyanshi and Ali are wearing cotton and nylon clothes, respectively. Who do you think would be more comfortable and why?

3. You want to wear your favourite shirt to a party, but the problem is that it is still wet after a wash. What steps would you take to dry it quickly?

4. Comment on the following statements:
   (a) Evaporation produces cooling.
   (b) Rate of evaporation of an aqueous solution decreases with increase in humidity.
   (c) Sponge though compressible is a solid.

5. Why does the temperature of a substance remain constant at its melting point or boiling point?

Answers

Short Answer Questions

1. The sample of water boils at a higher temperature which shows that water is not pure. Due to impurities present in it, water boils at a higher temperature. This water will freeze below 0°C.

2. In (d) as ice and water are in equilibrium, the temperature would be zero. When we heat the mixture, energy supplied is utilised in melting the ice and the temperature does not change till the ice melts because of latent heat of fusion. On further heating, the temperature of the water would increase. Hence (d) is the correct option.

3. (a) Cooling (b) Stronger
(c) Liquid, gaseous (d) Sublimation, liquid
(e) Evaporation

4. (a) (iii) (b) (iv) (c) (v) (d) (ii) (e) (i)

5. (a) (iv) (b) (iii) (c) (v) (d) (ii) (e) (i)

6. In diffusion, the particles move from higher concentration to lower concentration without separation by a semi-permeable membrane. In osmosis, the particles move from lower concentration to higher concentration, that is, solvent to solution, when the two solutions are separated by a semi-permeable membrane. It means osmosis is a special kind of diffusion involving movement of particles.

7. (a) Osmosis (b) Diffusion
(c) Osmosis (d) Osmosis
(e) Osmosis (f) Diffusion
(g) Diffusion

8. Water in the form of ice has low energy since water freezes at a lower temperature. When ice comes in contact with the body, it draws heat from the body and gives a cooling effect. In the case of steam, the water molecules have high energy. The high energy of steam is transformed as heat and may cause severe burns.

9. The temperature of both boiling water and steam is 100°C, but steam has more energy because of latent heat of vaporisation. Hence, steam is hotter than boiling water.

10.

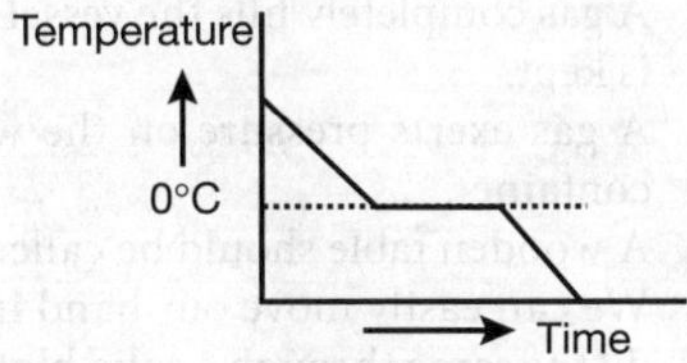

In (a), the hot water in the glass tumbler kept in the freezer will first become cold and the temperature will drop till 0°C. At 0°C, water loses heat equal to latent heat of fusion till all the water freezes to form ice at 0°C. During this change of state from liquid to solid, the temperature remains constant. On further cooling, the temperature of ice slowly falls with time.

11. The rate of evaporation depends on the surface area of the container. The greater the surface area, the higher the evaporation. It also depends on the speed of the wind. If the speed of the wind is more, more particles will evaporate from the surface. Hence, (c) in which both these factors, surface area and moving fan, are present will have maximum rate of evaporation.

12. (a) Sublimation
(b) Latent heat of fusion is the amount of heat required to change 1 kg of solid into liquid at atmospheric pressure at its melting point.

Long Answer Questions

1.

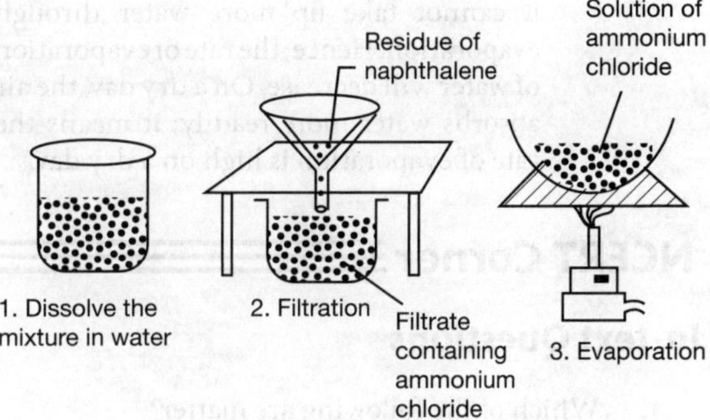

A mixture of naphthalene and ammonium chloride can be separated as follows:
Step 1: Put the mixture in a beaker and add water to it. Stir with a glass rod. Ammonium chloride being soluble in water gets dissolved, leaving behind the insoluble naphthalene.
Step 2: Filter the solution. Naphthalene remains on the filter paper while ammonium chloride is obtained as filtrate.
Step 3: Evaporate the filtrate to get back ammonium chloride.

2. Priyanshi is wearing cotton clothes which are more comfortable in summer because cotton absorbs the sweat which causes cooling on evaporation. Ali is wearing nylon clothes which do not absorb sweat. Hence, Ali will be uncomfortable.

3. The process of drying the shirt can be speeded up in the following ways:
   (a) Spread the shirt to increase the surface area which will increase the rate of evaporation.
   (b) Put it in the sun to increase the temperature to increase the rate of evaporation.
   (c) Keep it under the fan to increase the wind speed which increases the rate of evaporation.

4. (a) Evaporation is a surface phenomenon. The particles from the surface of the liquid take energy from the surroundings and change into vapour, which results in decrease in the energy of the surroundings. So a cooling effect is produced during evaporation.
   (b) The amount of water present in the air is known as humidity. If the water vapour in air is already present in large amounts, it cannot take up more water through evaporation. Hence, the rate of evaporation of water will decrease. On a dry day, the air absorbs water more readily; it means the rate of evaporation is high on a dry day.

(c) A sponge is a solid but it has minute pores in which air is trapped. These pores make the sponge a soft material. When sponge is pressed, the air present in the pores comes out and the sponge is compressed.

5. When a substance melts, it absorbs heat for the conversion of solid state into liquid state. As we continue heating, the heat supplied is used up in converting the solid state into liquid state by overcoming the forces of attraction between the particles and there is no change in temperature till the whole solid is converted into liquid. This heat absorbed by the solid which does not resist in increase in temperature is known as latent heat of fusion. When a liquid is heated, it starts being converted into vapour. Further heat given to the liquid is used in changing the state and there is no increase in the temperature till the liquid starts boiling. This heat is known as latent heat of vaporisation. It means the temperature of a substance remains constant at its melting point or boiling point until all of the substance melts or boils.

NCERT Corner

In-text Questions

1. Which of the following are matter?
   Chair, air, love, smell, hate, almonds, thought, cold, cold-drink, smell of perfume.

2. Give reasons for the following observation:
   The smell of hot sizzling food reaches you several metres away, but to detect the smell from cold food you have to go close to it.

3. A diver is able to cut through water in a swimming pool. Which property of matter does this observation show?

4. What are the characteristics of particles of matter?

5. The mass per unit volume of a substance is called density. (Density = Mass/Volume). Arrange the following in order of increasing density: Air, exhaust from chimneys, honey, water, chalk, cotton and iron.

6. (a) Tabulate the differences in the characteristics of states of matter.
   (b) Comment upon the following:
   Rigidity, compressibility, fluidity, filling a gas container, shape, kinetic energy and density

7. Give reasons:
   (a) A gas completely fills the vessel in which it is kept.
   (b) A gas exerts pressure on the walls of the container.
   (c) A wooden table should be called a solid.
   (d) We can easily move our hand in air but to do the same through a solid block of wood, we need to be a karate expert.

8. Liquids generally have lower density as compared to solids. But you must have observed that ice floats on water. Find out why.

9. Convert the following temperature to Celsius scale
   (a) 300 K   (b) 573 K

10. What is the physical state of water at
    (a) 250°C   (b) 100°C?

11. For any substance, why does the temperature remain constant during change of state?

12. Suggest a method to liquefy atmospheric gases.

13. Why does a desert cooler cool better on a hot dry day?

14. How does the water kept in an earthen pot become cool during summer?

15. Why does our palm feel cold when we put some acetone or petrol or perfume on it?

16. Why are we able to sip hot tea or milk faster from a saucer than from a cup?

17. What type of clothes should we wear in summer?

Exercises

1. Convert the following temperatures to the Celsius scale:
   (a) 293 K   (b) 470 K

2. Convert the following temperatures to the Kelvin scale:
   (a) 25°C   (b) 373°C

3. Give reasons for the following observations:
   (a) Naphthalene balls disappear with time without leaving any residue.

   (b) We can detect the smell of perfume sitting several metres away.

4. Arrange the following substances in increasing order of forces of attraction between the particles: Water, sugar, oxygen

5. What is the physical state of water at (a) 25°C, (b) 0°C, (c) 100°C?

6. Give two reasons to justify:
   (a) Water at room temperature is a liquid.
   (b) An iron almirah is a solid at room temperature.

7. Why is ice at 273 K more effective in cooling than water at the same temperature?

8. What produces more severe burns, boiling water or steam?

9. Name A, B, C, D, E and F in the following diagram showing change in its state.

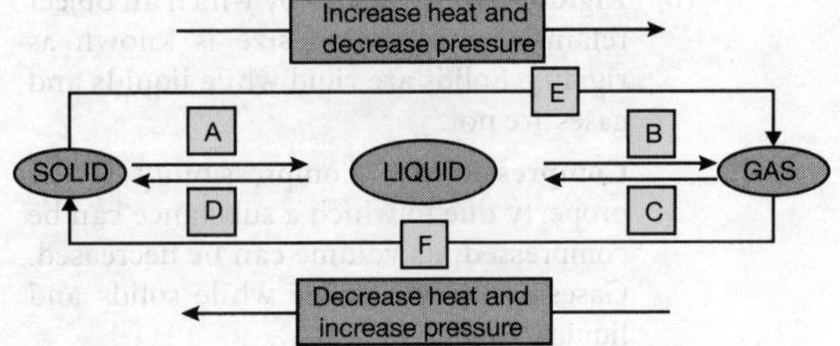

Answers

In-text Questions

1. Anything which has weight and occupies space is matter.
   Chair, air, smell, almonds, cold-drink and smell of perfume

2. Evaporation is directly proportional to temperature; this means hot food evaporates easily. Diffusion of hot food vapour in air is very fast and can reach a distant place within a very short time.

3. The phenomenon of cutting through water by the diver shows that matter has space between its particles.

4. Characteristics of particles of matter are given below:
   (i) Particles of matter have space between them.
   (ii) Particles of matter are continuously moving.
   (iii) Particles of matter have a force of attraction between them.
   (iv) Particles of matter are very small in size.

5. We can solve this question by keeping this concept in mind. The correct order of density for gas, liquid and solid is: Gas < Liquid < Solid

   Thus, $\xrightarrow{\text{Increasing order of Density}}$

   Air, exhaust from chimneys (Gas)
   Water, honey (Liquid)
   Cotton, chalk, iron (Solid)

6. (a)

Solid	Liquid	Gas
(i) Solids have definite volume.	**(i)** Liquids also have definite volume.	**(i)** Gases do not have definite volume. Their volume varies with the container in which they are stored or kept.
(ii) Solids do not tend to flow.	**(ii)** Liquids tend to flow.	**(ii)** Gases also tend to flow
(iii) Solids are rigid.	**(iii)** Liquids are not rigid.	**(iii)** Gases are not rigid.
(iv) Generally solids have a definite shape, with very few exceptions like sponge, rubber band, etc.	**(iv)** Liquids do not have a definite shape. They take the shape of the container.	**(iv)** Gases do not have a definite shape.
(v) Solids are generally incompressible, with very few exceptions like sponge, rubber band, etc.	**(v)** Liquids are almost incompressible.	**(v)** Gases are compressible.

(b) **Rigidity:** The property by which an object retains its shape and size is known as rigidity. Solids are rigid while liquids and gases are not.

Compressibility: Compressibility is the property due to which a substance can be compressed; its volume can be decreased. Gases are compressible while solids and liquids are not.

Fluidity: The flowing tendency of a substance is known as fluidity. Gases and liquids are fluids while solids are not.

Filling a gas container: A large volume of gas can be filled in a gas container by compressing it under very high pressure.

Shape: The property of having a definite geometry is known as the shape of a particular substance. Solids have a definite shape whereas gases and liquids do not.

Kinetic energy: The energy possessed by a moving object or by the moving molecules is known as kinetic energy. On increasing the temperature, the kinetic energy of a substance (or its molecules) also increases. Molecules of gases possess the highest kinetic energy.

Density: The mass per unit volume of a substance is known as density.

$$\text{Density} = \frac{\text{Mass}}{\text{Volume}}$$

7. (a) As the force of attraction between the molecules of gases is negligible, the molecules of gases occupy the maximum space available to them. High kinetic energy possessed by their molecules also helps in this.

(b) The motion of particles is random and has very high speeds in the gaseous state. Due to this random movement, the particles hit each other and also the walls of the container. The pressure exerted by the gas is due to this force exerted by these particles per unit area on the walls of the container.

(c) There is a strong force of attraction between the molecules of wood and the intermolecular space is the least. So, a wooden table has a definite shape and volume and it should be called a solid.

(d) Air molecules are very far from each other due to negligible force of attraction between them. So, our hand gets sufficient space to move in air and we also displace some air molecules by applying force. But a solid block of wood has closely packed molecules. So there is no question of the movement of a hand through it, in the absence of suitable force in the proper direction.

8. Liquids have lower density than solids. Water is also a liquid, so it should also have less density than ice. But this case is not so and the reason for the same is the cage-like structure of ice or the presence of vacant spaces between water (H_2O) molecules when they linked in ice. The number of these spaces is comparatively less in water. Being more porous than water, ice is lighter than water and floats over the surface of water.

9. Using given formula, we can convert the Kelvin temperature to Celsius.

$$T\ K - 273 = t°C$$

   (a) 300 K – 273 = 27°C
   (b) 573 K – 273 = 300°C

10. As the boiling point of water is 100°C
   (a) At 250°C, the state of water will be steam or water vapour.
   (b) At 100°C, there will be a transition of liquid state into gaseous state. So, at this temperature, the state is/may be liquid as well as gaseous.

11. During change of state of a substance, the temperature remains constant. This can be understood with the help of an example. When a solid is heated to its melting point, the temperature first rises and becomes constant when reaches its melting point. Now, on further heating, the heat energy provided to the substance helps to break the attraction force between the solid molecules. This heat is known as latent heat. That is why the temperature does not rise.

12. By applying pressure and reducing the temperature, atmospheric gases can be liquefied.

13. A desert cooler functions on the basis of evaporation. The rate of evaporation increases with increase in temperature and decrease in humidity. As evaporation increases when the day is hot and dry, the desert cooler functions better.

14. A large number of tiny pores are present on the surface of the earthen pot (matka). The water stored in the earthen pot evaporates faster through these pores due to the increased exposed surface area. As the process of evaporation causes cooling, the stored water inside the earthen pot becomes cool.

15. Acetone, petrol, perfume, etc., being volatile, evaporate very fast when exposed as larger surfaces. During the process, they absorb the required latent heat of vaporisation from the palm. So the process causes cooling and the palm feels cool.

16. A liquid has a larger surface area in a saucer than in a cup. Hence, it evaporates faster and cools faster in a saucer than in a cup. For this reason, we are able to sip hot tea or milk faster from a saucer than from a cup.

17. We should wear light-coloured cotton clothes in summer because:
   (i) Cotton is a good absorber of water (sweat). It provides more surface area for the sweat to evaporate.
   (ii) Light colours absorb less heat.

Exercises

1. In order to covert temperature from Kelvin to Celsius scale, we have to subtract 273 from the given value because K – 273 = °C.
   (a) 293 K – 273 = 20°C, (b) 470K – 273 = 197°C

2. To convert temperature from Celsius to Kelvin scale, add 273 to the given values because

$$°C + 273 = K$$

   (a) 25°C + 273 = 298 K, (b) 373°C + 273 = 646 K.

3. (a) Naphthalene is a substance which directly changes from solid to gas on heating by the process of sublimation. So, the naphthalene balls disappear with time as they sublime due to the heat of the surroundings.
   (b) The smell (aroma) of perfume reaches several metres away due to the fast diffusion of the gaseous perfume particles through the air.

4. The forces of attraction are strongest in solids and weakest or negligible in gases. Sugar is a solid, water is in liquid form and oxygen is a gas. So the order of forces of attraction is oxygen < water < sugar.

5. At 0°C, water (liquid) begins to get converted to its solid form (ice) and at 100°C, water (liquid)

starts to change into water vapour. Between 0° and 100°C, it remains in liquid state.

Hence, (a) Liquid state, (b) Solid or/and liquid state (Transition state) and (c) Liquid or/and gaseous state (Transition state).

6. (a) Water is a liquid at room temperature because it has a tendency to flow. It takes the shape of the container in which it is stored, but its volume remains the same.

    (b) An iron almirah is a solid at room temperature because its shape and volume are definite. It is hard and rigid and its density is high.

7. Ice at 273 K has less energy than water. Water has additional latent heat of fusion. So at 273 K, ice is more effective at cooling than water.

8. Steam causes more severe burns than boiling water. It releases extra heat or latent heat which it has already taken during vaporisation.

9. A = Melting or fusion, where solid changes into liquid.

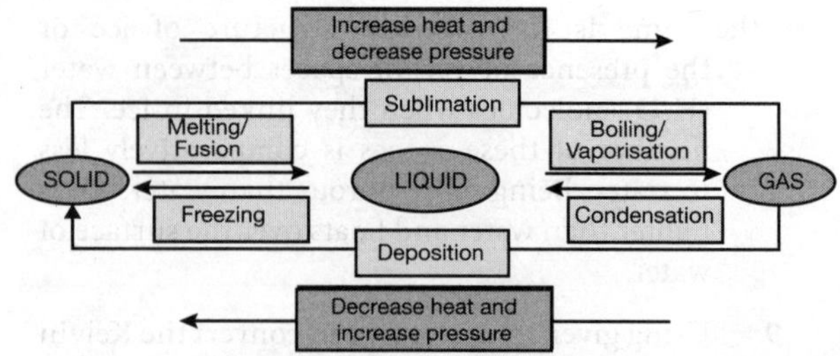

B = Evaporation or vaporisation, where liquid changes into gas.

C = Condensation or liquification where gas changes into liquid.

D = Freezing or solidification, where liquid changes into solid.

E = Sublimation, where solid directly changes into gas without passing through the liquid state.

F = Sublimation, where gas changes into solid without passing through liquid state.

CHAPTER 2

Is the Matter Around Us Pure?

Learning Objectives

After studying this unit, you will be able to:

- Describe matter and pure substances
- List the elements and their types
- Explain compounds and mixtures
- Describe solutions and their types
- Explain physical and chemical changes
- List the various methods of separation of mixtures

2.1 INTRODUCTION

Is the matter around us pure or not? It is an interesting question. In order to understand it, let us take an example. Suppose we have some salt and some soil in two different plates or containers and we want to know which one is pure and which one is impure. What should we do? We can use a magnifying glass to examine the salt and soil. The colour, shape and size of all the particles of salt are the same or almost the same. However, in soil, we can find clay particles and some other objects like dead insects and dead grass particles. This means that salt has only one type of particle; so it is pure. However, soil is impure.

For us, 'pure' signifies no adulteration. However, from a scientist's perspective, salt and soil are actually mixtures of different substances and hence, they are not pure! For example, milk is actually a mixture of water, fat and proteins. When a scientist claims that something is pure, it means that all the constituent particles of that substance are the same in their chemical nature. A **pure substance** consists of only one type of species (particles). When we look around us, we see that most matter exists as mixtures of two or more pure components. For example, sea water, minerals and soil are all mixtures. Thus, the matter around us is of two types: pure and mixture. A **mixture** has two or more types of particles.

In Chapter 1, we learnt about the three states of matter (solid, liquid and gas). Before going into the chemical nature of matter, it is important for us to understand the scientific meaning of the term chemical substance. In scientific terms, a **substance** is a type of matter that cannot be separated into other kinds of matter by any physical process. Then, that substance is a pure form of matter and not a mixture of many different kinds of matter. Most things that we use in our day-to-day life are mixtures. Pure substances are rare. For example, dissolved sugar can be separated from its solution by using a physical process such as evaporation or distillation. Thus, sugar is a pure substance and cannot be separated by any physical process into its constituent parts. Similarly, common salt (sodium chloride), silver, gold, iron, mercury, magnesium oxide and hydrochloric acid are examples of pure substances (Fig. 2.1).

2.2 PURE SUBSTANCES

A pure substance has only a one type of particle and they are always homogeneous. A pure substance cannot be separated into other kinds of matter by any physical process. It is characterised by a fixed composition as well as a fixed boiling point and melting point. Examples are hydrogen, nitrogen, oxygen, iron, copper, silver and gold.

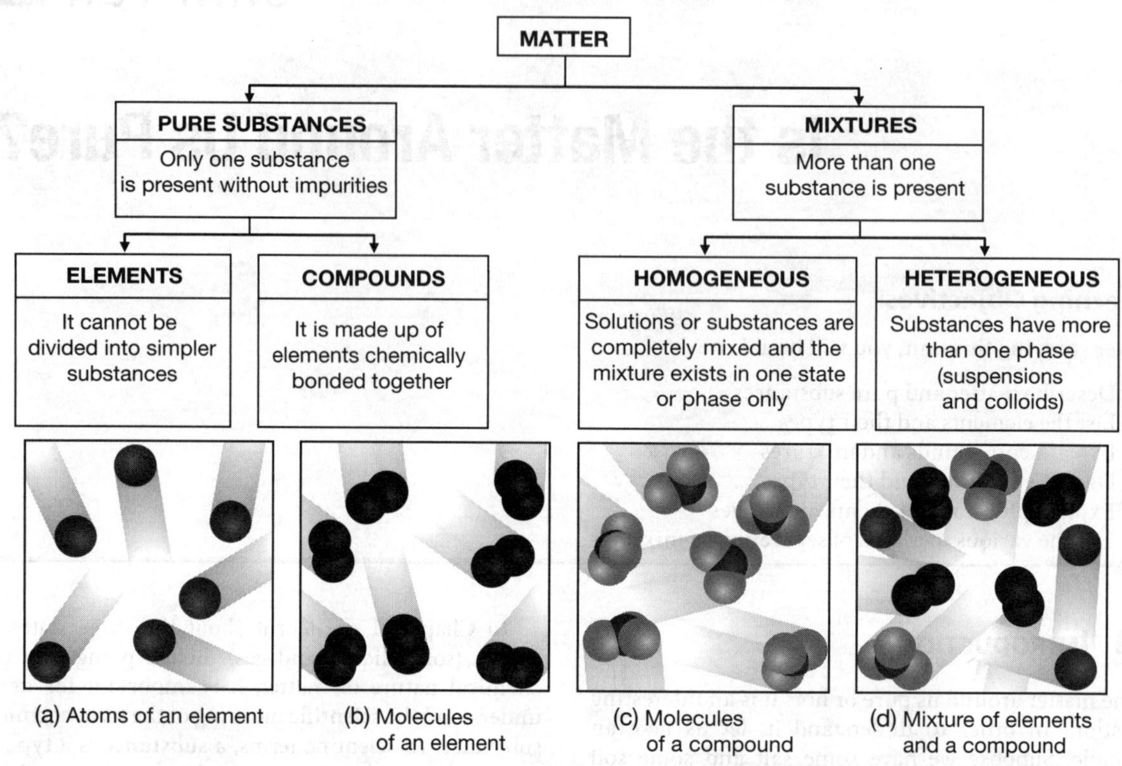

Fig. 2.1 Pure substances and mixtures

Pure substances can be further divided into two types: elements and compounds.

2.2.1 Elements

The term element was first used by Robert Boyle, and Lavoisier was the first scientist to define it. An element is a substance that cannot be split into two or more simpler substances by the usual chemical methods of applying heat, light and electric energy. This is because an element is made up of only one type of atom. For example, oxygen is an element as it cannot be split into two or more simpler substances by applying heat, light and electricity. If we consider gold and silver, we would find that they are made up of only gold or silver atoms, respectively (Figs 2.2 and 2.3).

Fig. 2.2 Gold is an element made up of only gold atoms.

Fig. 2.3 Silver is an element made up of only silver atoms.

Currently, there are nearly 118 elements, out of which 90 are naturally occurring while the remaining 28 are synthetic or artificially made. Every substance has one or more of these elements. For example, sugar contains carbon, hydrogen and oxygen.

Elements can be solid, liquid or gaseous at room temperature. For example, carbon, sodium and aluminium are solids; mercury and bromine are liquids; while hydrogen, oxygen and nitrogen are gases at room temperature. It is important to note that most elements (101) are solid, 11 are gases and 6 are liquid. Some of the common elements are given in Table 2.1.

Table 2.1 Some elements and their symbols

Element name	Symbol	Element name	Symbol	Element name in Latin	Symbol
Aluminium	Al	Hydrogen	H	Copper (Cuprum)	Cu
Arsenic	As	Iodine	I	Lead (Plumbum)	Pb
Barium	Ba	Fluorine	F	Potassium (Kalium)	K
Zinc	Zn	Chlorine	Cl	Antimony (Stibium)	Sb
Cadmium	Cd	Nitrogen	N	Silver (Argentum)	Ag
Calcium	Ca	Oxygen	O	Tin (Stannum)	Sn
Manganese	Mn	Phosphorus	P	Sodium (Natrium)	Na
Chromium	Cr	Sulfur	S	Gold (Aurum)	Au
Cobalt	Co	Bromine	Br	Mercury (Hydrogyrum)	Hg
Magnesium	Mg	Neon	Ne	Iron (Ferrum)	Fe

Types of Elements

Elements can be divided into three types: metals, non-metals and metalloids.

- **Metals**: A metal is an element that is malleable and ductile and which can conduct electricity. All metals are solids, except mercury, which is a liquid. Examples are aluminium, iron, copper, silver, gold and zinc. The properties of metals are as follows:
  - **Physical state:** Metals are solids at room temperature. For example, iron, copper, aluminium, silver and gold are solids at room temperature. Only mercury is in a liquid state at room temperature. Gallium and cesium become liquid at slightly above room temperature (around 303 K).
  - **Appearance:** Metals are shiny or lustrous and can be polished. This property is known as metallic lustre. It makes metals quite useful in jewellery and decorative pieces. The shiny surface of metals also makes them good reflectors of light. For example, silver.
  - **Sonorous nature:** Metals are sonorous; they make a ringing sound when struck. Due to their sonorous nature, they are used in making wires or strings for stringed musical instruments such as violins, guitars and sitars, and in plate-type musical instruments such as cymbals and ringing bells.

- **Ductile and malleable nature:** Metals are ductile, so they can be drawn or stretched into thin wires. However, not all metals are equally ductile (gold and silver are the most ductile). Copper and aluminium are highly ductile and can be drawn into thin wires used in electrical wiring. Metals are malleable, so they can be beaten into thin sheets without breaking by using hammers. Not all metals are equally malleable. For example, gold and silver are very malleable. Aluminium is quite malleable and can be converted into thin sheets (aluminium foil) used for packing food items. Copper is highly malleable and its sheets are used in making utensils.
- **Conductors:** Metals are good conductors of heat and electricity. Conduction of heat is known as thermal conductivity. Silver is the best conductor of heat and has the highest thermal conductivity. Metals like iron and mercury have low electrical conductivity due to greater resistance to the flow of current. Lead and mercury have poor heat conductivity. The conducting nature of metals is used as follows:
  - Copper and aluminium are very good conductors of heat and electricity. So they are used in making cooking utensils, water boilers, etc.
  - Electric wires are made of copper and aluminium as they are very good conductors of electricity.

- **Hardness and tensile strength:** Metals are generally hard; however, all metals are not equally hard. The hardness varies but usually they cannot be cut with a knife. The exceptions are lithium, sodium and potassium, which are quite soft and can be cut with a knife. Metals are usually strong and have high tensile strength, so they can hold large weights without breaking. For example, iron (as steel) is very strong and has high tensile strength; so it is used in the construction of railway lines, girders, machines, vehicles, bridges, buildings and chains.
- **High values of boiling point and melting point:** Metals generally have high melting points and high boiling points, so most metals melt and vaporise at high temperatures. For example, iron has a melting point of 1535°C or 1808 K; solid iron melts and turns into liquid iron on heating to 1808 K. Copper also has a high melting point of 1083°C or 1260 K. Tungsten has the highest melting point of 3422°C or 3695 K. Alkali metals (sodium, potassium and gallium) have low melting points.
- **Density:** Metals have high density, so they are heavy substances. For example, the density of iron metal is 7.8 g/cm³. Iridium (Ir) and osmium (Os) have the highest density while lithium (Li) has the lowest density.
- **Colour:** Metals are usually silver or grey in colour, with the exception of copper (reddish-brown) and gold (yellow).

- **Uses:** Metals and their alloys (homogenous mixture of metals) are widely use in our daily life. For example, in making cooking utensils, electric appliances, vehicles, etc (Fig. 2.4).
- **Non-metals:** A non-metal is an element that is neither malleable nor ductile and which cannot conduct electricity (with the exception of graphite (C)). All non-metals are gases or solids, except bromine which is a liquid non-metal at room temperature. Examples (Fig. 2.5) are hydrogen (H), nitrogen (N), carbon (C), oxygen (O), phosphorus (P), sulfur (S), chlorine (Cl), bromine (Br), iodine (I), helium (He), neon (Ne), argon (Ar), krypton (Kr) and xenon (Xe).

The physical properties of non-metals are the opposite of those of metals:

- **Physical state:** At room temperature, they may be solid (carbon, sulfur, phosphorus), liquid (bromine) or gas (hydrogen, nitrogen, oxygen, inert gases).
- **Appearance:** They are not lustrous or shiny but are dull in appearance as they do not have a shining surface. For example, sulfur, phosphorus.
- **Non-sonorous nature:** They are not sonorous; they do not make any ringing sound on being struck.
- **Non-ductile and non-malleable nature:** They are not ductile and cannot be drawn into wires; they are non-malleable and cannot be converted into sheets by hammering. However, solid non-metals are brittle and split into pieces on being hammered.

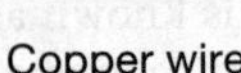

Copper wire   Aluminum foil   Gold ornament   Aluminum cooker   Sonorous bell   Metal (W) Filament   Metal wires

Fig. 2.4 Uses of metals

Carbon   Sulfur   Iodine   Bromine   Chlorine

Fig. 2.5 Examples of non-metals

- **Non-conductors:** They are bad conductors of heat and electricity. However, graphite shows electrical conductance due to the presence of free electrons in its hexagonal sheet-like structure, which is why it is used in making electrodes.
- **Soft and low tensile strength:** They are generally soft, not strong, and have low tensile strength. Being soft, they can be cut with a knife, with the exception of diamond (allotrope of carbon) which is one of the hardest natural substances.
- **Low values of boiling point and melting point:** They have comparatively low melting points and boiling points. This means they can be easily melted and vapourised at low temperatures. For example, sulfur can be melted at 119°C. Due to their low boiling point, most non-metals exist as gases at room temperature. However, graphite has a very high melting point (of nearly 3700°C).
- **Density:** Non-metals have low density; they are light substances. For example, sulfur has a density of 2 g/cm³. However, iodine has a higher density.
- **Colour:** They may be colourless or may have different colours. For example, hydrogen and oxygen are colourless, phosphorus is white, sulfur is yellow, chlorine is yellow-green and bromine is red-brown.
- **Uses:** Non-metals have a lot of applications in our life. For example, all of life on this earth is made up of carbon compounds like carbohydrates, proteins, fats and vitamins.
  - Hydrogen is used for the hydrogenation of oils into ghee and formation of ammonia which is used as a refrigerant and in making fertilisers.
  - Oxygen is essential for breathing to sustain life and also for all combustion (burning) processes.
  - Graphite is used in making electrodes and pencils while diamonds are used as a glass cutter and in making jewellery.
- **Metalloids:** The elements which show some properties of metals and some other properties of non-metals are known as metalloids. Their properties are intermediate between the properties of metals and non-metals. Metalloids are also known as semi-metals. Examples are arsenic (As), selenium (Se), antimony (Sb), boron (B), silicon (Si) and germanium (Ge). Metalloids are neither good conductors of electricity nor insulators; they are generally used as semiconductors.

The differences between metals and non-metals are listed in Table 2.2.

Table 2.2 Differences between metals and non-metals

Metals	Non-metals
(i) They are generally solids at room temperature.	(i) They may be solids, liquids or gases at room temperature.
(ii) They are lustrous or shiny and can be polished.	(ii) They are non-lustrous or dull and cannot be polished.
(iii) They are sonorous and make a ringing sound when struck.	(iii) They are not sonorous.

(Continued)

Table 2.2 (Continued)

Metals	Non-metals
(iv) They are malleable and ductile and they can be hammered into thin sheets and drawn into thin wires.	(iv) They are brittle and are neither malleable nor ductile.
(v) They are strong and tough and also have high tensile strength.	(v) They are not strong and have low tensile strength.
(vi) They are good conductors of heat and electricity.	(vi) They are bad conductors of heat and electricity.

TEST YOUR KNOWLEDGE

1. Classify the given elements as non-metal, metal or metalloid:
   Br, Ge, S, Sn, Pb, Al, B, Ga, Ar, Sb

Solution: Non-metals: Br, S, Ar
Metals: Sn, Pb, Al, Ga
Metalloids: Ge, B, Sb

2. What is the effect of temperature on metallic conductance?

Solution: On increasing the temperature, metallic conductance decreases as mobile electrons are pushed away due to the vibration of the slippery kernels.

2.2.2 Compounds

A compound is a substance made up of two or more elements chemically combined in a fixed proportion by mass. A compound is formed as a result of chemical reactions between the constituent elements. It is important to note that the properties of compounds are different from those of the elements from which they are formed. For example:

- Methane (CH_4) is a compound made up of two elements, carbon and hydrogen, chemically combined in a fixed proportion of 3 : 1 by mass (12 u carbon and 4 u hydrogen).
- Water (H_2O) is a compound made up of two elements, hydrogen and oxygen, chemically combined in a fixed proportion of 1 : 8 by mass (2u hydrogen and 16 u oxygen).
- Carbon dioxide (CO_2) is a compound made up of two elements, carbon and oxygen, in a fixed proportion of 3 : 8 by mass (12 u carbon and 32 u oxygen).
- Calcium carbon or limestone or marble ($CaCO_3$) is a compound made up of three elements, calcium, carbon and oxygen, in a fixed proportion of 10 : 3 : 12 by mass (40 u calcium, 12 u carbon and 48 u oxygen).

Types of Compounds

On the basis of their properties, compounds can be further divided into three classes: acids, bases and salts. For example, hydrochloric acid is an acid, potassium hydroxide is a base whereas potassium chloride is a salt; sulfuric acid is an acid, sodium hydroxide is a base whereas sodium sulfate is a salt.

Features of Compounds

- A compound is always homogeneous in nature.
- In a compound, the constituents are present in definite proportions by mass.
- The properties of a compound are different from those of its constituents.
- A compound has a fixed melting point and boiling point.
- The constituents of a compound cannot be separated by simple physical processes.
- The formation of a compound is generally accompanied by evolution of energy in the form of heat or light. For example, burning of a candle gives CO_2 and water, and liberates heat and light energy.

Criteria of Purity of a Compound: A pure compound is characterised by a fixed melting point if it is in solid form and a fixed boiling point if it is in liquid form. For example, pure naphthalene has a melting point of 80°C or 353 K and in pure form it must melt at this temperature, but if it melts at a temperature difference of 5–10°C, it is impure. Similarly, pure water boils at 100°C or 373 K, while sea water boils at a temperature higher than this, which means that sea water is impure.

2.3 MIXTURES

Mixtures comprise more than one kind of pure form of matter; that is, a mixture is a substance which consists of two or more elements or compounds not chemically combined together. All solutions are mixtures and the various substances present in a mixture are known as the 'constituents or components of the mixture'. For example, lemonade is a mixture of water, lemon juice, sugar and salt.

A substance cannot be separated into other kinds of matter by a physical process. Sodium chloride is a substance and cannot be separated by any physical process into its chemical constituents. Similarly, sugar is a substance because it contains only one kind of pure matter and its composition is the same throughout.

A mixture has two or more different types of particles possessing different chemical natures. A mixture may be homogeneous or heterogeneous. All mixtures are impure substances and a mixture does not have a fixed composition or a fixed melting point and boiling point. A sew examples of mixtures are sugar or salt in water, air, alcohol and water, soda water, soft drinks, vinegar, milk, butter, cheese, shaving cream and smog.

2.3.1 Types of Mixtures

Depending on the nature of the components that form the mixture, we can have different types of mixtures. Generally, mixtures are of two types: homogeneous and heterogeneous.

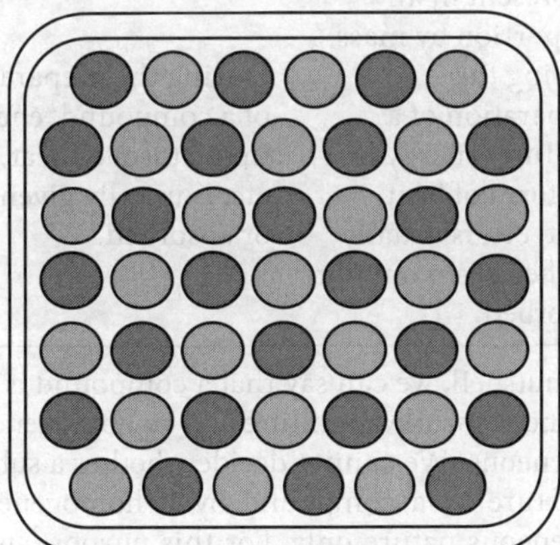

Fig. 2.6 Homogeneous mixtures

- **Homogeneous**: Mixtures in which the substances are completely mixed together and are indistinguishable from one another and form only one phase are known as homogeneous mixtures. All homogeneous mixtures are known as solutions (Fig. 2.6). They may be further divided into three types:

- **Solid–liquid mixture:** A mixture of sugar in water is a homogeneous mixture as all the parts of the sugar solution have the same sugar–water composition and appear to be equally sweet and there is no visible boundary of separation between sugar and water particles in a sugar solution (that is, one-phase system).

- **Solid–solid mixture:** Alloys (mixture of two or more metals) such as steel (80% Fe + 12% Cr + 8% Ni), brass (59.2% Cu + 36.4% Zn + 3.26% Pb) and bronze.

- **Liquid–liquid mixture:** Rectified spirit (alcohol + gasoline), mixture of benzene and toluene (Fig. 2.7).

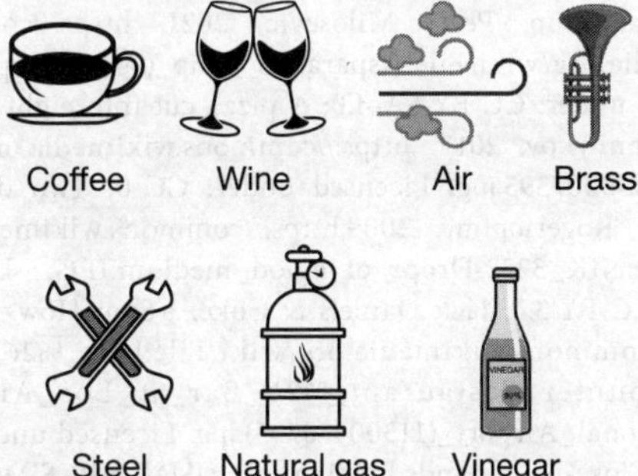

Fig. 2.7 Some homogeneous substances

- **Heterogeneous**: Miixtures in which all the substances remain separate and one substance is spread throughout the other substance as small particles, droplets or bubbles are known as heterogeneous mixtures (they have distinct phases). For example, a mixture of sugar and sand is a solid–solid heterogeneous mixture as different parts of this mixture will have different sugar–sand compositions and there is a visible boundary of separation between sugar and sand particles (Fig. 2.8). Suspensions of solids in liquids are also examples of solid–liquid heterogeneous mixtures. A mixture having two or more immiscible liquids is also a case of liquid–liquid heterogeneous mixture. Most of the mixtures are heterogeneous except alloys. For example: milk, paint, glass, blood, soap solution, sugar, gunpowder (sulfur, charcoal and potassium nitrate) and sand mixture.

Fig. 2.8 Some heterogeneous substances

Source: Kelloggs's Corn Flakes, with milk, Th78blue, 2022, https://commons.wikimedia.org/wiki/File:Kellogg%27s_Corn_Flakes,_with_milk.jpg. Licensed under: CC-BY-SA-4.0; asparagus soup, Petar Milosevic, 2021, https://commons.wikimedia.org/wiki/File:Asparagus_soup_(spargelsuppe).jpg. Licensed under: CC-BY-SA-4.0; A pizza cut into eight slices..., igorovsyannykov, 2013, https://commons.wikimedia.org/wiki/File:Pizza-3007395.jpg. Licensed under: CC-0; Two drops of blood..., Rogeriopfm, 2009,https://commons.wikimedia.org/wiki/File:NIK_3232-Drops_of_blood_medium.JPG. Licensdd under: CC-BY-3.0; Jack Daniels & Coke..., Sam Howzit, 2013, https://commons.wikimedia.org/wiki/File:Jack_%26_Coke_at_Encounter_Restaurant_%26_Bar_at_Los_Angeles_International_Airport_(11300470795).jpg. Licensed under: CC-BY-2.0; Close-up of hands holding soil, USDA NRCS MOntana, 2011, https://commons.wikimedia.org/wiki/File:Soil_Survey09_(39041768842).jpg. Licensed under: PD USDA

2.3.2 Properties of Mixtures

- A mixture is usually heterogeneous; for example, iron filings and sulfur powder.
- A mixture shows the properties of all the constituents present in it.
- A mixture can be separated into its constituents by physical processes.
- The composition of a mixture is variable; the constituents can be present in any proportion by mass.
- A mixture does not have a definite melting point and boiling point.
- As energy is usually neither given out nor absorbed in the preparation of a mixture, the formation of a mixture involves only a physical change.

Table 2.3 lists the differences between mixtures and compounds

Table 2.3 Differences between mixtures and compounds

Mixture	Compound
1. A mixture can be separated into its constituents by using many physical processes like filtration, sublimation, evaporation, distillation, solvents, magnet, etc.	A compound cannot be separated into its constituents by physical processes; it can only be separated into its constituents by chemical processes.
2. A mixture shows the properties of its constituents.	The properties of a compound are totally different from those of its constituents.
3. A mixture does not have a fixed melting point, boiling point, etc.	A compound has a fixed melting point, boiling point, etc.
4. A mixture does not have a definite formula as its composition is variable and the constituents can be present in any proportion by mass.	A compound has a definite formula as its composition is fixed and the constituents are present in fixed proportion by mass.
5. During the preparation of a mixture, energy in the form of heat, light, etc. is usually neither given out nor absorbed.	During the preparation of a compound, energy in the form of heat, light, etc. is usually given out or absorbed.

In a nutshell, we can say that a compound is always homogeneous while a mixture can be homogeneous or heterogeneous. We cannot decide whether a substance is a mixture or a compound by it homogeneous or heterogeneous nature only. For this purpose, we have to consider these points:

(i) If the given substance is separable into its constituents by physical methods, it is a mixture and if not, it is a compound.

(ii) If the given substance shows the properties of its constituents, it is a mixture and if not, it is a compound.

(iii) If the given substance does not have a fixed value melting and boiling point it is a mixture, but if it has a fixed value melting and boiling point, it is a solution.

(iv) If the given substance has variable composition it is a mixture, but if the composition is fixed it is a compound.

(v) If the given substance's preparation involves no heat change it is a mixture, but if heat change is involved, it is a compound.

TEST YOUR KNOWLEDGE

1. Classify the given substances into compounds and mixtures:
   Cloud, iodised table salt, steam, sucrose, steel, aerated drink, dry ice, milk, 22-carat gold, marsh gas, petrol, smoke

Solution: Compounds: Cloud, steam, sucrose, dry ice, marsh gas

Mixture: Iodised table salt, steel, aerated drink, petrol, smoke, milk, 22-carat gold

2. Classify the following into elements, compounds and mixtures:
   Potassium, Soil, Salt solution, Gold, Magnesium carbonate, Tin, Silicon, Coal, Air, Soap, Methane, Carbon dioxide, Blood

Solution: We can classify the given materials into elements, compounds and mixtures as follows:

Elements	Compounds	Mixtures
Potassium	Magnesium carbonate	Soil
Gold	Soap	Salt solution
Tin	Methane	Coal
Silicon	Carbon dioxide	Air
		Blood

3. Give the names of the elements present in the following compounds:
   (a) Quicklime,
   (b) Hydrogen iodide,
   (c) Baking soda,
   (d) Sodium sulfate

Solution:
(a) Quicklime is calcium oxide, CaO, and the elements present in it are calcium (Ca) and oxygen (O).
(b) Hydrogen iodide is HI and the elements present in it are hydrogen (H) and iodine (I).
(c) Baking soda is sodium hydrogen carbonate, $NaHCO_3$, and the elements present in it are sodium (Na), hydrogen (H), carbon (C) and oxygen (O).
(d) Sodium sulfate is Na_2SO_4 and the elements present in it are sodium (Na), sulfur (S) and oxygen (O).

Quick Review

- A substance is a type of matter that cannot be separated into other kinds of matter by any physical process. It is a pure form of matter.
- A pure substance comprises a single type of particle and is always homogeneous.
- An element is a substance that cannot be split into two or more simpler substances by the usual chemical methods of applying heat, light and electric energy.
- Elements can be divided into three types: non-metals, metalloids and metals.
- A non-metal is an element that is neither malleable nor ductile and cannot conduct electricity [with the exception graphite (C)]. All non-metals are gases or solids, except bromine.
- The elements which show some properties of metals and some other properties of non-metals are known as metalloids.
- A metal is an element that is malleable and ductile and can conduct electricity. All metals are solids except mercury, which is a liquid.

- *The bonding or strong interaction that holds the metal atoms firmly together, due to forces of attraction between metal ions and mobile electrons, is known as metallic bonding.
- A compound is a substance made up of two or more elements chemically combined in a fixed proportion by mass. It is formed as a result of chemical reactions, between the constituent elements.
- Mixtures are constituted of more than one kind of pure form of matter. That is, a mixture is a substance which consists of two or more elements or compounds not chemically combined together.
- Those mixtures in which the substances are completely mixed, are indistinguishable from one another and form a single phase are known as homogeneous mixtures (solutions). These can be of three types: solid–liquid mixture (mixture of sugar in water), solid–solid mixture (alloys) and liquid–liquid mixture [rectified spirit (alcohol + gasoline)].
- The mixtures in which all the substances remain separate and one substance is spread throughout the other substance as small particles, droplets or bubbles are known as heterogeneous mixtures.

Exercise 2.1

All questions marked * are practical based.

Section A: Multiple Choice Questions
(1 Mark)

1. What is a pure substance made up of two or more elements that are chemically combined?
   - (a) Mixture
   - (b) Solution
   - (c) Element
   - (d) Compound

2. Which of the following statements describes elements?
   - (a) Elements have unique sets of properties
   - (b) Elements cannot be joined together in chemical reactions
   - (c) All of the particles in the same element are different
   - (d) Elements can be broken down into simpler substances

3. A group of two or more substances that are physically combined is:
   - (a) Compound
   - (b) Molecule
   - (c) Mixture
   - (d) Element

4. The following can be separated by chemical means:
   - (a) Compound
   - (b) Mixture
   - (c) Element
   - (d) All of the above

5. The following can be separated by physical means:
   - (a) Compound
   - (b) Mixture
   - (c) Element
   - (d) All of the above

6. Which of the following substances can be separated into pure substances by physical means?
   - (a) Water
   - (b) Gold
   - (c) Saltwater
   - (d) Calcium

7. Which one of the following is a mixture?
   - (a) Sugar
   - (b) Air
   - (c) Salt
   - (d) Iron

8. Which of the following is not true about compounds?
   - (a) Elements are chemically bonded
   - (b) The properties of the compound are the same as the properties of the constituent elements
   - (c) Composition of elements is always the same
   - (d) Can be separated by chemical means

9. Which of the following is not true about mixtures?
   - (a) Substances are chemically combined
   - (b) Can be separated by physical means
   - (c) Properties of mixture are the same as the properties of individual substances
   - (d) Composition of substances can vary

10. Antimony and arsenic can be classified as:
    - (a) Metalloids
    - (b) Non-metals
    - (c) Metals
    - (d) Cannot be said

11. Which pair of elements may be liquid?
    - (a) Iodine, sulfur
    - (b) Sodium, mercury
    - (c) Bromine, mercury
    - (d) Mercury, chlorine

12. The least malleable metal here is:
    - (a) Aluminium
    - (b) Iron
    - (c) Gold
    - (d) Silver

13. One of the following substances is neither a good conductor of electricity nor an insulator. This substance is:
    (a) Manganese   (b) Germanium
    (c) Gallium   (d) Sodium

14. Which of the following is a compound?
    (a) Graphite
    (b) Stainless steel
    (c) Bronze
    (d) Hydrogen chloride

15. In which of the following may the constituents be present in any ratio?
    (a) Solution   (b) Colloid
    (c) Mixture   (d) Compound

16. *Which one of the following will result in the formation of a mixture?
    (a) Adding sodium metal to water
    (b) Crushing of a marble tile into small particles
    (c) Adding milk to water
    (d) Breaking of ice cubes into small pieces

17. The purity of a solid substance can be checked by its:
    (a) Solubility in water   (b) Melting point
    (c) Solubility in alcohol (d) Boiling point

18. Which of the following is an example of a heterogeneous substance?
    (a) Pieces of silver   (b) Candle
    (c) Bottled water   (d) NaBr

19. Assume air contains only nitrogen, oxygen and inert gas. Then air will be a:
    (a) Heterogeneous mixture
    (b) Colloids
    (c) Compound
    (d) Homogeneous mixture

20. Which set represents correct statements:
    (i)   Metals have lustre with high tensile strength
    (ii) Metals are mostly silvery or grey in colour
    (iii) Both graphite and diamond are soft
    (iv) Both graphite and diamond can conduct electricity
    (a) (i), (ii)   (b) (i), (ii), (iv)
    (c) (i), (ii), (iii)   (d) (i), (ii), (iii), (iv)

Assertion–Reason Questions

Direction: In the following question two statements (Assertion) A and Reason (R) are given Mark.
(a) if A and R both are correct and R is the correct explanation of A;
(b) if A and R both are correct but R is not the correct explanation of A;
(c) A is true but R is false;
(d) A is false but R is true

Assertion	Reason
1. An element or a compound are pure substances.	1. Both are made up of only one type of particle.
2. Blood is an example of a homogeneous mixture.	2. In a homogeneous mixture, the composition is uniform.
3. A mixture cannot have a definite boiling or melting point.	3. The composition of a mixture is variable without any definite formulae.
4. Formation of iron sulfide is a chemical change.	4. Heat is evolved here.
5. A solid compound is pure, if it has a sharp melting point.	5. A liquid compound is pure, if it has a fixed boiling point.

Section B: Very Short Answer Questions   (2 Marks)

1. Name a method to check the purity of a liquid.

2. What is the difference between a compound and a mixture?

3. Name one non-metal and one metal which exist as liquids at room temperature.

4. Name a non-metal which is a good conductor of electricity.

5. Is water free from suspended impurities a pure substance: 'yes or no'?

6. Name a liquid which can be classified as a pure substance and which conducts electricity.

7. By measuring which property can you distinguish between pure water and brine solution without any chemical test?

8. If you were given a sample having a fixed melting point, will you regard it as a mixture or a pure substance?

9. What is the main difference between a solution and an ordinary mixture?

10. Give one example each of a homogeneous and a heterogeneous mixture.

11. Is 22-karat gold a pure substance or a mixture?

12. *When we heat iron filings and sulfur till red hot, do we get a compound or a mixture?

13. If we take sea water or tap water, which will boil at a higher temperature?

14. Classify the following into elements and compounds:
(i) H_2O, (ii) He, (iii) Cl_2, (iv) CO_2, (v) Co, (vi) Ar, (vii) NH_3

15. What is the meaning of the subscript 2 to the right of the H and the lack of a subscript to the right of the O in the formula for water?

Section C: Short Answer Questions
(3 Marks)

1. What is the difference between an atom and a molecule?

2. Define each of the following terms:
(i) Atom, (ii) Molecule, (iii) Ion

3. What is a formula unit? How is it used in chemistry?

4. If compounds contain more than one element, why do we classify them as pure substances rather than as mixtures?

5. List four characteristics by which compounds can be distinguished from mixtures.

6. When two elements come together and chemically combine:
   (i) What happens to their properties?
   (ii) What happens to their identities?
   (iii) What happens to their atoms?

7. What is meant by saying that non-metals are brittle?

8. Mention two physical properties on the basis of which metals can be distinguished from non-metals.

9. Explain why a solution of salt in water is considered a mixture and not a compound.

10. Explain whether air is a mixture or a compound. Give at least three reasons for your answer.

11. Give reasons why:
    (i) Graphite is used for making electrodes in a dry cell.
    (ii) Copper is used for making electric wires.

12. Compare the properties of metals and non-metals with respect to (i) hardness, (ii) ductility and (iii) electrical conductivity.

Section D: Long Answer Questions
(5 Marks)

1. State whether each of the following represents an atom, molecule or ion:
(i) Se, (ii) CCl_4, (iii) HCO_3^-, (iv) C_6H_6, (v) Pb, (vi) Br_2

2. Name a non-metal:
   (i) which is required for combustion
   (ii) which is lustrous
   (iii) which is filled in electric bulbs
   (iv) other than carbon which shows allotropy
   (v) which is known to form the largest number of compounds
   (vi) whose allotrope form is a good conductor of electricity; also name the allotrope

3. At room temperature and pressure, which of the substances listed below is:
   (i) a solid element
   (ii) a liquid non-metal element
   (iii) a gaseous mixture
   (iv) a solid mixture
   (v) a liquid compound
   (vi) a solid compound
   (vii) a liquid metal
   (viii) a liquid mixture
   (ix) gaseous element
   (x) a gaseous compound
   Bromine, carbon dioxide, helium, steel, air, oil, marble, copper, water, sand, tin, bronze, mercury, common salt.

4. Name a metal:
   (i) which can be easily cut with a knife

   (ii) which forms amalgams
 (iii) which is used to wrap cigarettes
 (iv) which has no fixed shape
  (v) which is yellow in colour
 (vi) which is used to make the filament of electric bulbs
(vii) which is used in making mobiles batteries

5. State whether each of the following is an element, a compound or a mixture:
(i) Water, (ii) A multicoloured rock, (c) A sample of stainless steel, (d) CO_2, (v) Bromine, (vi) A solution of sugar in water, (vii) Fe, (viii) Sand and water mixed together in a container

6. For each of the following formulae of compounds, indicate what elements (their names, not symbols) are present in the compound and also state how many atoms of each element are present in one formula unit or molecule of the compound:
Example: $NaNO_3$: sodium, 1 atom; nitrogen, 1 atom; oxygen, 3 atoms
(i) Na_3PO_4, (ii) CCl_4, (iii) $Al_2(SO_4)_3$, (iv) $(NH_4)_2CO_3$, (v) $Ba(OH)_2$

Section E: Case Study or Passage-Based Questions   (4 Marks)

1. On the basis of physical properties, elements can be further divided into metals, non-metals and metalloids. Metal and non-metals differ in many properties like physical structure, brittleness, hardness, density, conductivity and tensile strength. Most metals are solid with brittleness and high values of density, boiling point and tensile strength. Non-metals are mostly gaseous, but may be liquid and solid also. Non-metals have low boiling point and melting point. Among metals, mercury shows many exceptions in properties.
   (i) Which type of elements, metals or non-metals, show the property of brittleness?
  (ii) Why is mercury a poor conductor of electricity?
 (iii) Name two metals with low tensile strength.
 (iv) Name two metals which become liquid at slightly above room temperature.
  (v) Why are metals used in making sheets and wires?

2. Consider the following jumbled up list of substances which includes elements, compounds and mixtures and answer the given questions:

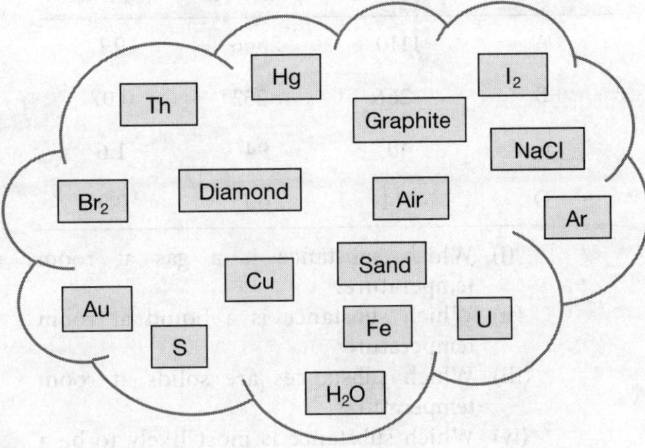

(i) Here, the number of allotropes is:
(a) 1   (b) 2
(c) 3   (d) 4

(ii) Which pair represents a mixture?
(a) $NaCl$, H_2O (b) $NaCl$, Air
(c) Air, Sand   (d) Sand, H_2O

(iii) Which pair represents a compound?
(a) Air, H_2O (b) $NaCl$, H_2O
(c) H_2O, Diamond   (d) Sand, H_2O

(iv) Which pair is liquid in nature?
(a) Br_2, I_2 (b) Br_2, Hg
(c) Hg, S   (d) Br_2, Ar

(v) Which pair can show maximum electric conductance?
(a) Fe, S   (b) Hg, S
(c) Cu, Au   (d) Fe, Cu

Higher Order Thinking Skills (HOTS) Questions

1. Which of the substances listed below are:
   (i) metallic elements
  (ii) non-metallic elements
 (iii) compounds
 (iv) mixtures
Silicon, sea water, calcium, argon, water, air, carbon monoxide, iron, sodium chloride, diamond, brass, copper, dilute sulfuric acid, sulfur, oil, nitrogen gas, ammonia.

2. The table below shows the melting points, boiling points and densities of substances A to D.

Substance	Melting point/°C	Boiling point/°C	Density/g cm^{-3}
A	1110	2606	9.1
B	−266	−252	0.07
C	40	94	1.6
D	−14	60	0.9

   (i) Which substance is a gas at room temperature?

   (ii) Which substance is a liquid at room temperature?

   (iii) Which substances are solids at room temperature?

   (iv) Which substance is most likely to be a metal?

   (v) What is the melting point of the least dense non-metal?

   (vi) Which substances are gases at 72°C?

3. How many atoms of the different elements are present in the formulae of the compounds given below?

   (i) Nitric acid, HNO_3

   (ii) Methane, CH_4

   (iii) Copper nitrate, $Cu(NO_3)_2$

   (iv) Ethanoic acid, CH_3COOH

   (v) Sugar, $C_{12}H_{22}O_{11}$

   (vi) Phenol, C_6H_5OH

   (vii) Ammonium sulfate, $(NH_4)_2SO_4$

4. *A student heated a mixture of iron filings and sulfur. He saw a red glow spread through the mixture as the reaction continued. At the end of the experiment, a black solid had been formed.

   (i) Explain what the red glow indicates.

   (ii) Give the chemical name of the black solid.

   (iii) Write a word equation and a balanced chemical equation to represent the reaction which has taken place.

   (iv) The black solid is a compound. Explain the difference between the mixture of sulfur and iron and the compound formed by the chemical reaction between them.

Answers

Section A: Multiple Choice Questions

1. (d)	**2.** (a)	**3.** (c)	**4.** (a)
5. (b)	**6.** (c)	**7.** (b)	**8.** (b)
9. (a)	**10.** (a)	**11.** (c)	**12.** (b)
13. (b)	**14.** (d)	**15.** (c)	**16.** (c)
17. (b)	**18.** (b)	**19.** (d)	**20.** (a)

Assertion–Reason Questions

1. (a)	**2.** (d)	**3.** (a)	**4.** (b)
5. (b)			

Section B: Very Short Answer Questions

1. By determining the boiling point, the purity of a liquid can be checked.

2. In a compound, elements are chemically combined. In a mixture, the components are not chemically combined, only physically mixed.

3. One non-metal is bromine and one metal is mercury which is liquid.

4. Graphite (Carbon).

5. No, as it may have dissolved impurities.

6. Mercury

7. By measuring the boiling point.

8. Pure substance.

9. Solution is homogeneous while mixture is heterogeneous.

10. Homogeneous mixture is salt in water and heterogeneous mixture is salt and sulfur.

11. It is a mixture as it contains 22 parts by weight of pure gold and 2 parts by weight of copper or silver.

12. When we heat iron filings and sulfur till red hot, we get a compound.

13. Due to impurities, sea water will boil at a higher temperature.

14. Elements are He, Ar, Cl_2 and Co and compounds are H_2O, CO_2 and NH_3.

15. Two atoms of hydrogen and one atom of oxygen combine to form a molecule of water.

16. Metal.

17. Due to resistance to the flow of electric current, mercury is a poor conductor.

18. Sodium and potassium.

19. Gallium and cesium.

20. As metals are malleable and ductile, they can be used in making sheets and wires.

Section C: Short Answer Questions

1. An atom is the fundamental unit of an element. A molecule is the fundamental unit of a molecular compound. A molecule is composed of atoms in chemical combination.

2. (i) An atom is the smallest particle of an element that can exist and still be that element.
   (ii) A molecule is the smallest particle of a molecular (or covalent) compound that can exist and still be that compound.
   (iii) An ion is an atom or grouping of atoms that has developed an electrical charge.

3. A formula unit is the simplest particle of an ionic compound that can exist and still be the compound. The term is used to differentiate between those compounds that are ionic from those that are molecular or covalent.

4. Because in a compound, the elements are chemically combined, meaning that it is a different substance with properties different from the elements before they were combined. In a mixture, the components are simply physically mixed, meaning that the properties are retained.

5. Refer to Section 2.3.2, Properties of Mixtures.

6. (i) Entirely new properties.
   (ii) New substance formed.
   (iii) Atoms become chemically combined with other atoms, forming molecules or formula units of the new substance.

7. Refer to Section 2.2.1, Elements.

8. Refer to Table 2.2.

9. Refer to Section 2.3.1, Types of Mixtures.

10. Refer to Section 2.3.1, Types of Mixtures.

11. Refer to Section 2.2.1, Elements.

12. Refer to Table 2.2.

Section D: Long Answer Questions

1. (i) Atom, (ii) Molecule, (iii) Ion, (iv) Molecule, (v) Atom, (vi) Molecule.

2. (i) Oxygen, (ii) Iodine, (iii) Argon, (iv) Sulfur, (v) Carbon, (vi) Carbon, graphite.

3. (i) Copper, tin, (ii) Bromine, (iii) Air, (iv) Bronze, steel, sand, (v) Water, (vi) Common salt, (vi) Mercury, (viii) Oil, (ix) Helium, (x) Carbon dioxide.

4. (i) Sodium, (ii) Mercury, (iii) Aluminium, (iv) Mercury, (v) Gold, (vi) Tungsten, (vii) Lithium.

5. (i) Compound, (ii) Heterogeneous mixture, (iii) Homogeneous mixture, (iv) Compound, (v) Element, (vi) Homogeneous mixture, (vii) Element, (viii) Heterogeneous mixture.

6. (i) Na_3PO_4: Sodium, 3 atoms; phosphorus, 1 atom; oxygen, 4 atoms
   (ii) CCl_4: Carbon, 1 atom; chlorine, 4 atoms
   (iii) $Al_2(SO_4)_3$: Aluminum, 2 atoms; sulfur, 3 atoms; oxygen, 12 atoms
   (iv) $(NH_4)_2CO_3$: Nitrogen, 2 atoms; hydrogen, 8 atoms; carbon,1 atom; oxygen, 3 atoms
   (v) $Ba(OH)_2$: Barium, 1 atom; oxygen, 2 atoms; hydrogen, 2 atoms

Section E: Case Study or Passage-Based Questions

1. (b)   2. (c)   3. (b)   4. (b)
5. (c)

Higher Order Thinking Skills (HOTS) Questions

1. (i) Metallic elements: calcium, iron, copper
   (ii) Non-metallic elements: silicon, argon, sulfur, diamond
   (iii) Compounds: water, carbon monoxide, sodium chloride, nitrogen gas, ammonia

(iv) Mixtures: sea water, air, brass, dilute sulfuric acid, oil

2.   (i) B, (ii) D, (iii) A, C, (iv) A, (v) −266°C, (vi) C, D

3.   (i) 5, (ii) 5, (iii) 9, (iv) 8, (v) 45, (vi) 13, (vii) 15.

4.   (i) Red glow indicates chemical change.
(ii) Black solid is iron sulfide (FeS).

(iii) The word equation to represent the reaction between iron and sulfur is:

$$\text{Iron + Sulfur} \xrightarrow{\text{Heat}} \text{Iron sulfide}$$

(iv) The mixture having sulfur and iron has a yellow colour due to sulfur and reddish colour due to iron, but when iron sulfide is formed, only black colour is present.

Hints

Section A: Multiple Choice Questions

14.   Here, hydrogen chloride (HCl) is a compound of hydrogen and chloride, stainless steel and bronze are alloys whereas graphite is the allotropic form of carbon.

17.   As every solid has a fixed melting point, it can be used.

18.   Candle is a heterogeneous mixture of wax and threads while silver is an element and bottled water and table salt (NaBr) are compounds.

20.   As diamond is hard and a non-conductor.

2.4 Physical and Chemical Changes

In some changes, new substances are formed, while in other changes, no new substances are formed. On the basis of the formation of new substances, we can categorise these changes as physical and chemical. In order to understand the difference between a pure substance and a mixture, let us understand the difference between a physical change and a chemical change.

2.4.1 Physical Changes

In the previous chapter, we learnt about a few physical properties of matter. The properties that can be observed and specified, such as colour, hardness, rigidity, fluidity, density, melting point and boiling point are the physical properties. The inter-conversion of states is a physical change because these changes occur without a change in composition and there is no change in the chemical nature of the substance. A change in which no new substances are formed, but the physical form of the substance changes, is known as a physical change. The product formed is chemically identical to the starting substance. A physical change is a temporary one as it can be easily reversed to form the original substance.

For example, when ice is heated, it changes into liquid water and on further heating, water changes into steam. However, water in solid form (ice), liquid form or gaseous form (steam) is chemically the same substance (H_2O). This transformation represents a physical change and such a change can be reversed easily. For example, steam on cooling forms liquid water, which on further cooling changes into ice (Fig. 2.9).

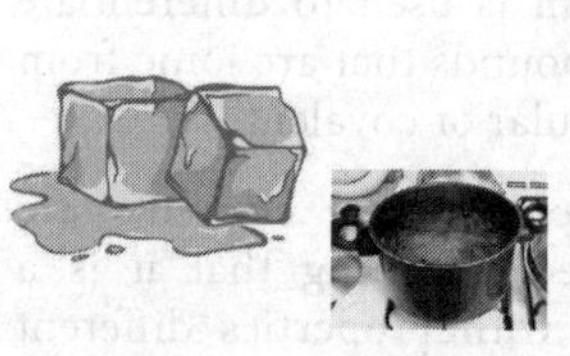

Fig. 2.9 Examples of physical changes

The glowing of an electric bulb is a physical change as when it is switched on, an electric current passes through its filament and as a result, the filament becomes hot and glows to give light. However, when the bulb is switched off, the glow stops as the filament retains its normal state.

The making of a solution, for example, common salt in water, is also a physical change.

2.4.2 Chemical Changes

A change in which one or more substances changes into a new substance(s) is known as a chemical change. It is usually irreversible or cannot be reversed easily. This process causes a change in the chemical composition which brings about a change in the chemical properties also. A chemical change is also called a chemical reaction.

For example, both water and cooking oil are liquids, but their chemical characteristics are different. They differ in odour and inflammability. Oil burns in air whereas water extinguishes fire. So, it is the chemical property of oil that makes it different from water (burning is a chemical change). When a magnesium wire is heated, it burns in air to give a white powder of magnesium oxide (a new substance); this is a chemical change.

When electricity is passed through water, it decomposes into two new substances, hydrogen and oxygen. Similarly, rusting of iron and calcination of limestone are examples of chemical changes (Fig. 2.10).

$$Fe + S \xrightarrow{\text{Heat}} FeS$$

Table 2.3 lists the differences between physical and chemical changes.

Table 2.3 Differences between physical and chemical changes

Physical change	Chemical change
No new substance is formed	A new substance is formed
It is a temporary change	It is a permanent change
It is easily reversible	It is usually irreversible
The mass of the substance does not alter	The mass of the substance does alter
Very little heat or light energy is generally absorbed or evolved	A lot of heat or light energy is absorbed or evolved

Burning of Mg wire

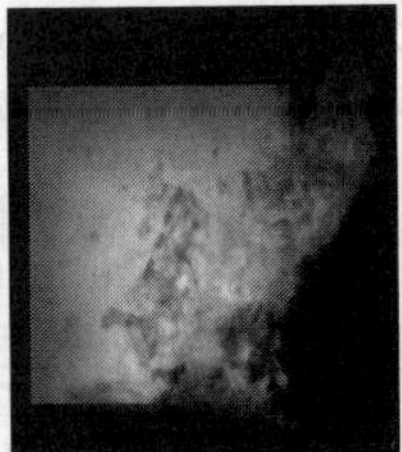
Burning of paper

Rusting of iron

Ripening of fruits

Fig. 2.10 Examples of chemical changes

Source: This "Burning Paper" was created..., UBhapE2, 2011, https://commons.wikimedia.org/wiki/File:Burning_Paper.jpg. Licensed under: CC-BY-SA-3.0; Rusted Bolt...,
Thester11, 2008, https://commons.wikimedia.org/wiki/File:Rust_Bolt.JPG. Licensed under: CC-BY-SA-3.0; Fruit ripening of..., Gmihail at Serbian Wikipedia, 2020, https://commons.wikimedia.org/wiki/File:Fruit_ripening_of_Ziziphus_jujuba.jpg. Licensed under: CC-BY-SA-3.0-RS

Competition Edge

The following processes involve both physical and chemical changes: sublimation of NH_4Cl, heating of zinc hydroxide, heating of sodium nitrate and burning of a candle.

TEST YOUR KNOWLEDGE

1. Classify the following as physical or chemical changes:
   (a) Dissolving of salt in water (b) Burning of magnesium ribbon in air (c) Crystallisation of copper sulfate (iv) Conversion of milk into curd (v) Burning of coal (vi) Electrolysis of sodium chloride solution by passing current (vii) Rusting of iron nails

Solution: (i) Physical change (ii) Chemical change (iii) Physical change (iv) Chemical change (v) Chemical change (vi) Chemical change (vii) Chemical change

> **2.** Classify each of the following as a physical or a chemical change. Give reasons.
>   (a) Drying of a shirt in the sun.
>   (b) Rising of hot air over a radiator.
>   (c) Burning of kerosene in a lantern.
>   (d) Change in the colour of black tea on adding lemon juice to it.
>   (e) Churning of milk cream to get butter.
>
> **Solution:** (a), (b), (e): Physical changes because there is no change in chemical composition.
> (c), (d): Chemical changes because new substances are formed.

2.5 SOLUTIONS, SUSPENSIONS AND COLLOIDS

In order to understand solutions, suspensions and colloids, first we must know the meaning of and difference between the terms solute and solvent.

A **solute** is a substance that is dissolved in a liquid to make a solution and the liquid used here is called the **solvent**. The solute particles are known as dispersed particles and the solvent is known as the dispersion medium. Normally, the solute is a smaller amount and the solvent is taken in bulk for preparing a solution. For example, in a sugar solution, we add sugar to water, so sugar is the solute and water is the solvent.

Solutes may be in any physical state, that is, solid, liquid or gas. Solvents are mostly in liquid state like water, alcohol and acetone, but can be in gas or solid state also.

2.5.1 Solutions

A solution is a homogeneous mixture of two or more pure substances. For example, lemonade and soda water are solutions. In a solution, there is homogeneity at the particle level. For example, lemonade tastes the same throughout. This shows that particles of sugar or salt are evenly distributed in the solution.

A solution consists of two components: a solvent and a solute. The component of the solution that dissolves the other component in it (usually the component present in larger amount) is known as the solvent while the component of the solution that is dissolved in the solvent (usually present in smaller quantity) is known as the solute.

For example, a solution of sugar in water is a solid-in-liquid solution. Here, sugar is the solute and water is

the solvent. A solution of iodine in alcohol, known as tincture of iodine, has iodine (solid) as the solute and alcohol (liquid) as the solvent (Fig. 2.11).

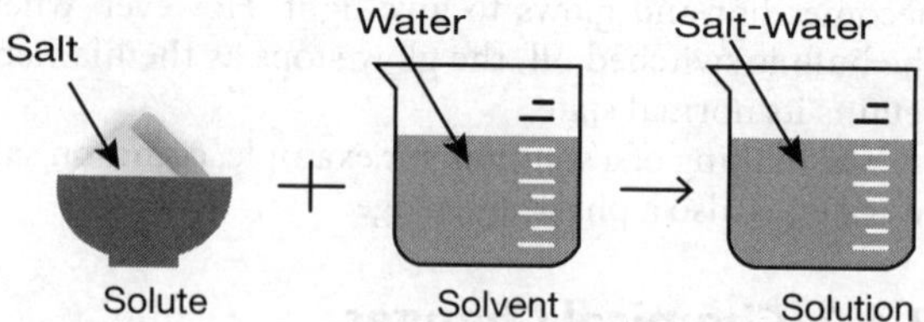

Fig. 2.11 Components and examples of solutions

Types of Solutions

We can divide solutions into different categories:
- **Based on the amount of solute in the solution**: When different amounts of solute are used to make solutions, but the quantity of the solvent is kept the same. We obtain different types of solutions depending upon the dissolution of the solute in the solvent. Solutions can be unsaturated, saturated or supersaturated (Fig. 2.12).

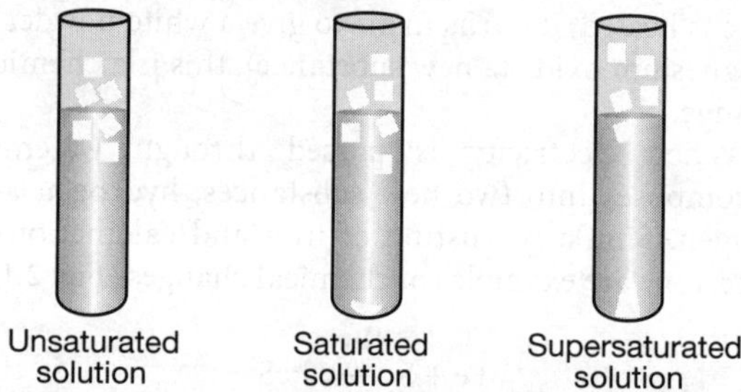

Fig. 2.12 Types of solutions based on saturation

- **Unsaturated solution:** An unsaturated solution is one in which a solvent is capable of dissolving more solute at a given temperature without any rise in temperature.
- **Saturated solution:** A saturated solution is one in which a solvent is not capable of dissolving any more solute at a given temperature. It means the solution has maximum amount of solute in the dissolved state.
- **Supersaturated solution:** A supersaturated solution comprises a large amount of solute at a particular temperature. When the temperature is reduced, the extra solute will crystallise quickly.
- **Effect of temperature:** It is important to note that a solution is saturated at a specific temperature only as the solubility of the solute varies with change in temperature (Fig. 2.13). If the temperature is lowered,

some of the dissolved solute will be separated out in the form of solid crystals as solubility decreases with decrease in temperature or on cooling.

For example, if we consider a common salt aqueous solution at 293 K or 20°C in which in 100 g of water, a maximum 36 g of common salt (NaCl) is dissolved, it is a saturated solution. If the temperature is below 20°C, the amount of NaCl will be less than 36 g as some amount of NaCl will crystalise out.

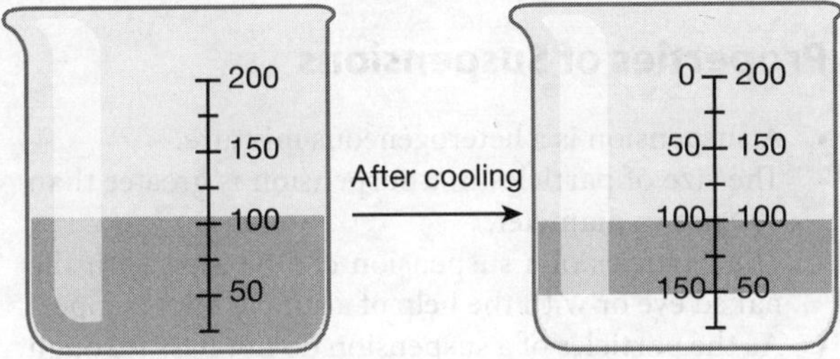

Fig. 2.13 Effect of temperature

- **Based on the nature of the solvent**: These solutions are of two types, depending on whether the solvent is water or not (Fig. 2.14).

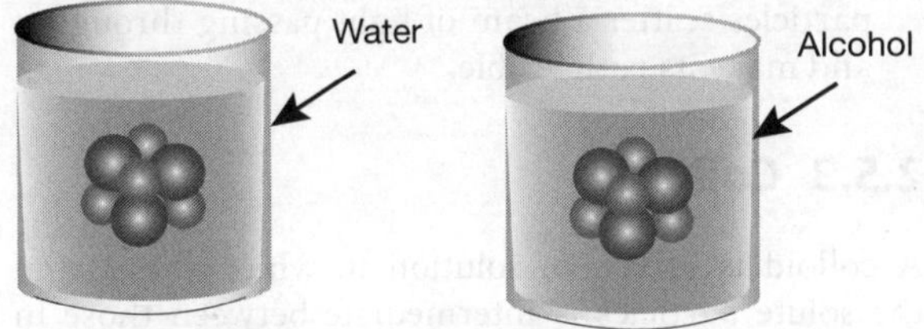

Fig. 2.14 Types of solutions based on the nature of the solvent

- **Aqueous solution:** When a solute is dissolved in water, the solution is known as an aqueous solution. For example, salt in water, sugar in water and copper sulfate in water.
- **Non-aqueous solution:** When a solute is dissolved in a solvent other than water, it is known as a non-aqueous solution. For example, iodine in carbon tetrachloride, sulfur in carbon disulfide, phosphorus in ethyl alcohol.
- **Based on the amount of solute**: Depending on the amount of solute added to the solvent, solutions can be dilute or concentrated (Fig. 2.15).
  - **Dilute solution:** A dilute solution contains a small amount of solute in a large amount of solvent.

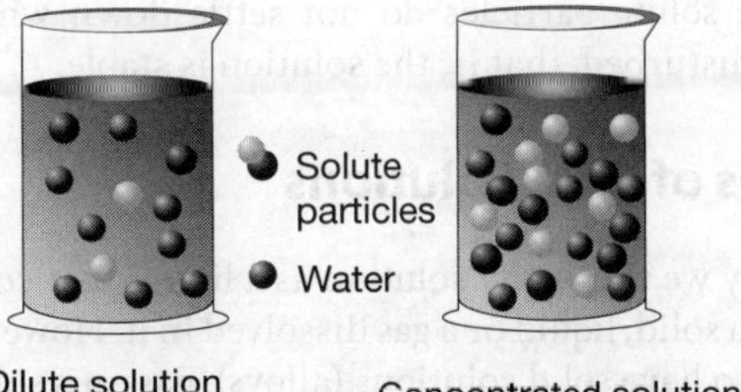

Fig. 2.15 Types of solutions based on the amount of solute

- **Concentrated solution:** A concentrated solution contains a large amount of solute dissolved in a small amount of solvent.
- **Based on solute particle size**: On the basis of the size of particles, we can further classify a solution or substance into three types: true solutions, suspensions and colloids.
  - **True solution:** A true solution is a solution in which the particles of the solute can be broken down to such a fine state that they cannot be seen even under a powerful microscope.
  - **Suspensions:** A suspension is a heterogeneous mixture in which the small particles of a solid are spread throughout a liquid without dissolving in it.
  - **Colloids:** A colloid is a type of solution in which the size of the solute particles is intermediate between those in true solutions and those in suspensions. The size of the solute particles in a colloid is larger than that of a true solution but smaller than that of a suspension.

Properties of Solutions

- A solution is always homogeneous in nature.
- The properties of the solute are retained in a true solution. For example, a sugar solution is sweet to taste and a solution of salt in water is saline to taste.
- The particles of a solution are smaller than 1 nm (10^{-9} m) in diameter. So they cannot be seen by the naked eye.
- A true solution does not show Tyndall effect because of very small particle size. (Tyndall effect is the scattering of a beam of light passing through a solution so that the path of light is visible in the solution). That is, true solutions are transparent to light.
- The solute particles in a true solution can easily pass through a filter paper. Hence the solute particles cannot be separated from the mixture by the process of filtration.

- The solute particles do not settle down when left undisturbed, that is, the solution is stable.

Types of True Solutions

Usually we think of a solution as a liquid that contains either a solid, liquid or a gas dissolved in it. However, we can also have solid solutions (alloys), gaseous solutions (air), and so on.

Solution of a solid in a solid: Metal alloys are solutions of a solid in a solid. For example, brass is a solution of zinc in copper. It is prepared by mixing molten zinc with molten copper and cooling the mixture.

Solution of a liquid in a liquid: Vinegar is a solution of acetic acid in water.

Solution of a gas in a liquid: Aerated drinks like soda water are gas in liquid solutions. These contain carbon dioxide (gas) as the solute and water (liquid) as the solvent.

Solution of a gas in a gas: Air is a mixture of a gas in a gas and it is a homogeneous mixture. Its two main constituents are oxygen (21%) and nitrogen (78%), and the other gases (CO_2, Ar) are present in very small quantities (Table 2.4).

Table 2.4   Types of solutions

Type of solution	Solute	Solvent	Common examples
Gaseous solutions	Gas	Gas	Mixture of oxygen and nitrogen gases
	Liquid	Gas	Chloroform mixed with nitrogen gas
	Solid	Gas	Camphor in nitrogen gas
Liquid solutions	Gas	Liquid	Oxygen dissolved in water
	Liquid	Liquid	Ethanol dissolved in water
	Solid	Liquid	Glucose dissolved in water
Solid solutions	Gas	Solid	Solution of hydrogen in palladium
	Liquid	Solid	Amalgam of mercury with sodium
	Solid	Solid	Copper dissolved in gold

2.5.2 Suspensions

A suspension is a heterogeneous mixture in which the small particles of a solid are spread throughout a liquid without dissolving in it. In a suspension, the solute particles do not dissolve but remain suspended throughout the bulk of the medium. For example, sand particles suspended in water, muddy water, chalk–water mixture, flour in water and milk of magnesia. Most medicines are suspensions.

Properties of Suspensions

- A suspension is a heterogeneous mixture.
- The size of particles in a suspension is greater than 100 nm in diameter.
- The particles of a suspension can be seen with the naked eye or with the help of a simple microscope.
- As the particles of a suspension do not pass through filter paper, it is possible to separate them by ordinary filtration.
- As the particles of a suspension settle down when the suspension is left undisturbed, the suspension is unstable.
- A suspension is not transparent to light as the particles scatter a beam of light passing through it and make its path visible.

2.5.3 Colloids

A colloid is a type of solution in which the size of the solute particles is intermediate between those in true solutions and those in suspensions. The size of the solute particles in a colloid is larger than that of a true solution but smaller than that of a suspension. The particles of a colloid are uniformly spread throughout the solution. Due to the relatively smaller size of particles as compared to that of a suspension, the mixture appears to be homogeneous, but actually, a colloidal solution is a heterogeneous mixture. For example, milk. A colloid consists of dispersed particles and the dispersion medium.

Dispersed particles: The solute particles are known as dispersed particles.

Dispersion medium: Solvents are known as the dispersion medium.

Classification of Colloids

Colloids are classified according to the physical state of the dispersed phase (solute) and the dispersion medium (solvent). Most colloids can be classified into eight groups (Table 2.5, Fig. 2.16).

Table 2.5 Types of colloids

Dispersed phase	Dispersion medium	Colloidal system	Examples
Gas	Liquid	Foam or froth	Soap solutions, lemonade froth, whipped cream
Gas	Solid	Solid foam	Pumice stone, styrene foam, foam rubber, dried sea foam
Liquid	Gas	Aerosols of liquids	Fog, clouds, mist, fine insecticide sprays
Liquid	Liquid	Emulsions	Milk, cold cream, tonics
Liquid	Solid	Gels	Cheese, butter, boot polish, table jellies
Solid	Gas	Aerosols of solids	Smoke, dust, haze
Solid	Liquid	Solutions	Paints, starch dispersed in water, gold sol, muddy water, ink
Solid	Solid	Solid solutions	Ruby glass, minerals, gem stones

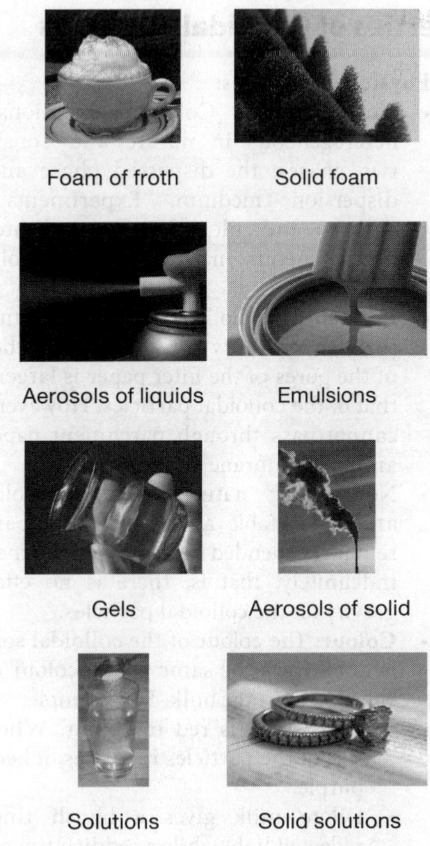

Fig. 2.16 Examples of different types of solutions

Source: Cup of Coffee with foam, Lotus Head, 2008, https://commons.wikimedia.org/wiki/File:Cup_of_Coffee_with_foam_cropped.jpg. Licensed under: CC-BY-SA-3.0-migrated; Detail of a solid polymeric foam..., Giovanna Canu, Eva Santini, 2019, https://commons.wikimedia.org/wiki/File:Solid-polymeric-foam.jpg. Licensed under: CC-BY-SA-4.0; Aerosol Spray Can, PiccoloNamek, 2005, https://commons.wikimedia.org/wiki/File:Aerosol.png. Licensed under: CC-BY-SA-3.0-migrated-with-disclaimers; If used, credit must be given..., United Soybean Board, 2009, https://commons.wikimedia.org/wiki/File:Open_Soy_Paint_Can_(10481948313).jpg. Licensed under: CC-BY-2.0; A highly viscous, ternary..., Olli Niemitalo, 2012, https://commons.wikimedia.org/wiki/File:Cubic_phase_gel.jpg. Licensed under: CC-0; A smoke plume from the..., Walter Baxter, 2013, https://commons.wikimedia.org/wiki/File:A_smoke_plume_from_the_Dunbar_Cement_Works_chimney_-_geograph.org.uk_-_3765299.jpg. Licensed under: CC-BY-SA-2.0; Solution of Salt in Water, Chris 73, 2004, https://commons.wikimedia.org/wiki/File:SaltInWaterSolutionLiquid.jpg. Licensed under: CC-BY-SA-3.0; I don't think I ever..., Jennifer Dickert, 2008, https://commons.wikimedia.org/wiki/File:Diamond_rings_photo_by_Jennifer_Dickert_(2).jpg. Licensed under: CC-BY-2.0

Properties of Colloidal Solutions

(i) Physical Properties:

- **Heterogeneity**: Colloidal solutions are heterogeneous in nature and consist of two phases: the dispersed phase and the dispersion medium. Experiments like dialysis and ultrafiltration indicate the heterogeneous nature of the colloidal system.
- **Filterability**: Colloidal particles can pass through ordinary filter papers as the size of the pores of the filter paper is larger than that of the colloidal particles. However, they cannot pass through parchment paper, an animal membrane or an ultrafilter.
- **Non-setting nature:** Colloidal solutions are quite stable as the colloidal particles remain suspended in the dispersion medium indefinitely, that is, there is no effect of gravity on the colloidal particles.
- **Colour:** The colour of the colloidal solution is not always the same as the colour of the substances in the bulk. For example:
  - Finest gold is red in colour. When the size of the particles increases, it becomes purple.
  - Dilute milk gives a bluish tinge in reflected light while a reddish tinge is the transmitted light.
- **Stability:** Colloidal solutions are quite stable.
- **Visibility:** Due to the small size of colloidal particles, we cannot see them with the naked eye. As these particles can easily scatter a beam of visible light, they become visible when viewed through an ultramicroscope.

(ii) Optical Properties (Tyndall Effect): When a strong beam of light is passed through a colloidal solution, its path becomes visible (bluish light) when viewed at right angles to the beam of light. The Tyndall effect (Fig. 2.17) was used by Zsigmondy and Siedentop in devising the ultramicroscope. It is due to the scattering of light by colloid particles that the sky looks blue, dust particles can be seen in the light coming from a ventilator into a dark room, the tail of a comet is visible, and so on.

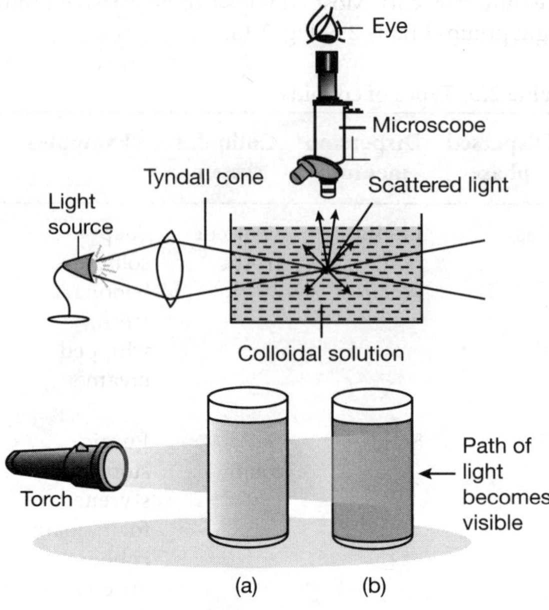

Fig. 2.17 Tyndall effect

(iii) Mechanical Properties:

- **Brownian Movement:** Colloidal particles show ceaseless, random and swarming zig-zag motion in the dispersal medium known as Brownian movement (shown for the first time by Robert Brown; Fig. 2.18).

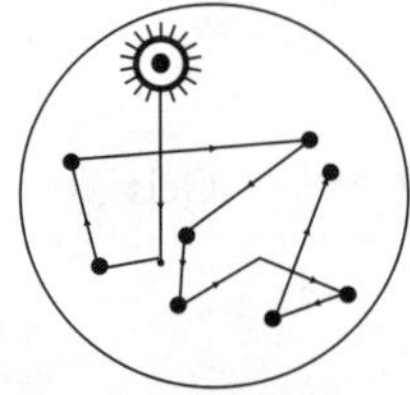

Fig. 2.18 Brownian movement

- **Sedimentation:** The heavier solution particles tend to settle down very slowly under the influence of gravity. This is called sedimentation (Fig. 2.19).

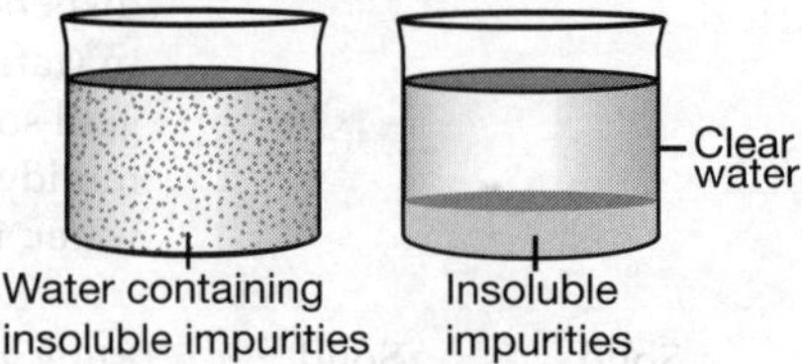

Fig. 2.19 Sedimentation

- **Diffusion:** The colloidal particles have a tendency to diffuse from the high concentration side to the low concentration side. However, the rate of diffusion is less than that of true solutions.

Concentration Terms of Solutions

As mentioned earlier, depending on the amount of solute added to the solvent, solutions can be of two types: dilute and concentrated.

Although the concentration of a solution can be expressed in multiple ways, the most common way is the 'percentage method' which refers to the 'percentage of solute' present in a solution. The percentage of solute can be 'by mass' or 'by volume'.

Concentration: The concentration of a solution is the amount of solute present in a given mass or volume of solution. It can also be defined as the amount of solute dissolved in a given mass or volume of solvent.

Concentration of solution = Amount of solute/Amount of solution, or, Amount of solute/Amount of solvent

$$C = \frac{w\,(g)}{V\,(L)} = \frac{w\,(g)}{V\,(mL)} \times 1000$$

Mass percentage (w/W%): When a solid solute is taken in a liquid solvent, mass percentage of the solute is considered to find the concentration of the solution. w% by mass% indicates that w grams of solute is taken in 100 grams of solution. For example, a 5% by mass solution of glucose means that 5 grams of glucose is present in 100 grams of the solution. Here, mass of solvent can be found out by subtracting the mass of the solute from the mass of the solvent.

The concentration of a solution in terms of mass percentage of solute is given by the following formula:

$$\text{Concentration of solution} = \frac{\text{Mass of solute}}{\text{Mass of solution}} \times 100$$

Mass of solution = Mass of solute + Mass of solvent
Mass of solvent = Mass of solution – Mass of solute

So, we can obtain the mass of the solvent by subtracting the mass of the solute from it.
In the above example,
Mass of solute (salt) = 5 g
Mass of solvent (water) = 95 g
So, Mass of solution = Mass of solute + Mass of solvent
= 5 + 95 = 100 g

$$\text{Concentration of solution} = \frac{5}{100} \times 100 = 5\% \text{ (by mass)}$$

We can simply write 5% by mass as 5% solution also as glucose is a solid. So it is clear that percentage by mass as volume of solid is not considered in making the solution.

Percentage strength (w/V%): It is known as percentage by strength or mass by volume percentage w/V%. This signifies that w grams of solute is taken in 100 mL of solution. For example, a 10% solution of glucose by strength signifies that 10 grams of glucose is present in 100 mL of the solution.

The concentration of a solution in terms of percentage by strength of solute is given by the following formula:

$$\text{Concentration of solution} = \frac{\text{Mass of solute}}{\text{Volume of solution}} \times 100$$

Volume strength (v/V%): This is used for a liquid solute dissolved in a liquid solvent. The concentration of a solution is defined as volume strength. v/V% by strength means v mL of the solute is present in 100 mL of the solution. For example, a 42% v/V solution of ethyl alcohol means that 42 mL of ethyl alcohol is present in 100 mL of the solution.

$$\text{Concentration of solution} = \frac{\text{Volume of solute}}{\text{Volume of solution}} \times 100$$

Competition Edge

If a very small amount of a solute is taken in a large amount of solvent, the value of mass percentage or mass by volume percentage will be very low. In such cases, the concentration is expressed in parts per million (ppm) and part per billion (ppb).

Concentration is ppm

$$= \frac{\text{Mass of solute}}{\text{Mass of solution}} \times 10^6 \text{ or } \frac{\text{Volume of solute}}{\text{Volume of solution}} \times 10^6$$

Concentration is ppb

$$= \frac{\text{Mass of solute}}{\text{Mass of solution}} \times 10^9 \text{ or } \frac{\text{Volume of solute}}{\text{Volume of solution}} \times 10^9$$

TEST YOUR KNOWLEDGE

1. A syrup is prepared by dissolving 125 g of sucrose in 175 g of water. Find the mass% of sucrose in this syrup.

Solution: Mass of solute (sucrose) = 125 g
Mass of solvent = 175 g
So, Mass of solution = Mass of solute + Mass of solvent = 125 + 175 = 300 g

$$\text{Concentration of solution} = \frac{\text{Mass of solute}}{\text{Mass of solution}} \times 100$$

$$\text{Concentration of solution} = \frac{125}{300} \times 100 = 41.66\%$$
by mass

2. A solution is prepared by dissolving 90 g of glucose in 200 mL of water. Find the percentage strength of the glucose solution.

Solution: Mass of solute (glucose) = 90 g
Volume of solvent = 200 mL

$$\text{Concentration of solution} = \frac{\text{Mass of solute}}{\text{Volume of solution}} \times 100$$

$$\text{Concentration of solution} = \frac{90}{200} \times 100 = 45 \text{ by}$$
strength

Solubility

The maximum amount of a solute that can be dissolved in 1 litre of a solution at a specified temperature is known as the solubility of that solute in that solvent (mostly water or may be another). It can be given in g/L or mol/L. It is important to note that if we take water as a solvent, solubility is always stated as 'mass of solute per 100 grams of water'. For example, at 20°C or 293 K (Fig. 2.20), sodium chloride (NaCl) has maximum solubility in water, that is, 36 g of NaCl per 100 g of water; similarly for potassium nitrate (KNO_3), solubility at this temperature is 32 g of potassium nitrate per 100 g of water.

Substance (or Solute)	Solubility
1. Ammonium chloride	37 g
2. Copper sulfate	21 g
3. Potassium chloride	34 g
4. Potassium nitrate	32 g
5. Sugar	204 g
6. Sodium chloride	36 g

Fig. 2.20 Solubility in water (at 20°C)

Factors Affecting Solubility

Effect of temperature: The solubility of solids in liquids generally increases when temperature increases and decreases when temperature decreases. The solubility of gases in liquids generally decreases on increase in temperature and increases on decrease in temperature as the dissolution of a gas in a solvent is exothermic (heat is evolved).

Let us consider the example of water having some air. On heating it, the air departs as bubbles (Fig. 2.21); so the solubility of a gas decreases in a liquid.

Fig. 2.21 Effect of temperature on solubility of air in water

Effect of pressure: The solubility of solids in liquids remains unaffected by change in pressure. The solubility of gases in liquids increases on increase in pressure and decreases on decrease in pressure as the solubility of a gas is directly proportional to the partial pressure of the gas.

Let us consider the example of carbonated water or soda water having dissolved CO_2. On opening the bottle, due to decrease in pressure, CO_2 is released (Fig. 2.22).

Fig. 2.22 Carbonated water

Source: Glass of carbonated water..., Egoite, 2009, https://commons.wikimedia.org/wiki/File:Verre-eau-gazeuse-au-jardin.jpg. Licensed under: CC-BY-SA-3.0

In short, we can say that the solubility of a gas in a liquid is directly proportional to pressure and inversely proportional to temperature.

Competition Edge

- Gases having more intermolecular forces of attraction are more soluble. For example, CO_2 is more soluble than oxygen in water.
- Solutes having a smaller size of particles are more soluble. For example, powdered sugar dissolves more than granules in water.

TEST YOUR KNOWLEDGE

1. 18 grams of sodium sulfate dissolves in 90 grams of water at 60°C. What is its solubility in water at that temperature.

Solution: Here, 90 grams of water dissolves 18 grams of sodium sulfate. So we have to find how much sodium sulfate will be dissolved in 100 grams of water, as follows:

90 g of water dissolves = 18 g of sodium sulfate

So, 100 g of water will dissolve = $\dfrac{18}{90} \times 100$g of sodium sulfate

$$= 20 \text{ g of sodium sulfate}$$

Hence the solubility of sodium sulfate in water is 20 g at 60°C.

2. 2.7 g of a solute is dissolved in 27 g of water to form a saturated solution at 298 K. Find out the solubility of solute at this temperature?

Solution: Mass of solute = 2.7 g

Mass of solvent = 27 g

$$\text{Solubility} = \dfrac{\text{Mass of solute}}{\text{Mass of solvent}} \times 100$$

$$= \dfrac{2.7}{27} \times 100 = 10 \text{ g}$$

3. A student determined the solubility of four substances, potassium nitrate, sodium chloride, potassium chloride and ammonium chloride, in water at five different temperatures of 10°C, 20°C, 40°C, 60°C and 80°C and obtained the following data:

Substance	Solubility				
	10°C	20°C	40°C	60°C	80°C
1. Potassium nitrate	21 g	32 g	62 g	106 g	167 g
2. Sodium chloride	36 g	36 g	36 g	37 g	37 g
3. Potassium chloride	35 g	35 g	40 g	46 g	54 g
4. Ammonium chloride	24 g	37 g	41 g	55 g	66 g

(a) What mass of potassium nitrate would be needed to make a saturated solution of potassium nitrate in 50 grams of water at 40°C?

(b) What would be observed if a saturated solution of potassium chloride at 80°C is left to cool to room temperature?

(c) What is the solubility of each salt at 20°C? Which salt has the highest solubility at this temperature?

(d) What is the effect of change in temperature on the solubility of a salt as shown by the above data?

Solution: (a) Here, the solubility of potassium nitrate at 40°C is 62 grams, that is, 62 grams of potassium nitrate is needed to make a saturated solution of potassium nitrate in 100 grams of water at 40°C. So, to make a saturated solution in 50 grams of water, we will need half of 62 grams of potassium nitrate, that is, 31 grams of potassium nitrate.

(b) When a saturated solution of potassium chloride at 80°C is left to cool, crystals of solid potassium chloride will gradually separate out from the solution as the solubility decreases on cooling.

(c) The solubility of various salts at 20°C is as follows:

Potassium nitrate 32 g; Sodium chloride 36 g; Potassium chloride 35 g and Ammonium chloride 37 g. Hence, ammonium chloride has the highest solubility here.

(d) The given data shows that the solubility of a salt increases on increase in temperature or decreases with decrease in temperature.

Quick Review

- A change in which no new substances are formed but the physical form of the substance changes is known as a physical change.
- A change in which one or more substances change into new substance(s) is known as a chemical change. It cannot be reversed easily.
- A solution is a homogeneous mixture of two or more pure substances.
- A suspension is a heterogeneous mixture in which the small particles of a solid are spread throughout a liquid without dissolving in it.
- A colloid is a type of solution in which the size of the solute particles is intermediate between those in true solutions and those in suspensions. Colloidal solutions are heterogeneous in nature and consist of two phases: the dispersed phase and the dispersion medium.
- Based on the dispersed phase and dispersion medium, colloids are of eight types. Lyophillic colloids are solvent-loving while lyophobic colloids are solvent-hating.
- Colloid particles show ceaseless random and swarming zig-zag motion in the dispersal medium, known as Brownian movement.
- The heavier particles tend to settle down very slowly under the influence of gravity. This is called sedimentation.
- Tyndall effect: When a strong and converging beam of light is passed through a colloidal solution, its path becomes visible (bluish light) when viewed at right angles to the beam of light.
- The concentration of a solution is the amount of solute present in a given amount of solution (mass or volume) or the amount of solute dissolved in a given mass or volume of solvent.
- The maximum amount of a solute that can be dissolved in 1 litre of a solution at a specified temperature is known as the solubility of that solute in that solvent. It can be given in g/L or mol/L.

Exercise 2.2

All questions marked * are practical based.

Section A: Multiple Choice Questions
(1 Mark)

1. Which of the following processes involves a physical change?
   - (a) Boiling of water into steam
   - (b) Burning of a piece of paper
   - (b) Glowing of an electric bulb
   - (d) Both (a) and (c)

2. Which of the following processes involves a chemical change?
   - (a) Burning of incense stick
   - (b) Electrolysis of water
   - (c) Cooking of food
   - (d) All of these

3. The physical properties of a mixture:
   (a) Depend on the organisation of the substance
   (b) Vary depending upon its components
   (c) Vary with the amount of substance
   (d) Depend on the volume of the substance

4. Which set represents chemical changes:
   (i) Breaking a glass tumbler, (ii) Burning of Mg ribbon, (iii) Condensation of steam into water, (iv) Rusting of iron.
   (a) (i), (ii)
   (b) (ii), (iii)
   (c) (ii), (iv)
   (d) (ii), (iii), (iv)

5. Which of the following is an example of a liquid solution?
   (a) Brine solution or aqueous NaCl
   (b) Iodine in air
   (b) Water drops in air
   (d) Brass

6. Which is an example of a liquid in gas solution?
   (a) Vinegar in water
   (b) Camphor in air
   (c) Clouds
   (d) CO_2 in water

7. Solutions with low concentrations of solutes are known as:
   (a) Dilute
   (b) Solvents
   (c) Concentrated
   (d) None of these

8. Tincture of iodine has antiseptic properties. This solution is made by dissolving:
   (a) Iodine in water
   (b) Iodine in potassium iodide
   (c) Iodine in alcohol
   (d) Iodine in vaseline

9. Substances whose solutions can readily diffuse through a parchment membrane are:
   (a) Non-electrolytes
   (b) Electrolytes
   (c) Crystalloids
   (d) Colloids

10. In colloidal state, particles size ranges from:
    (a) 10 to 1000 Å
    (b) 1 to 280 Å
    (c) 1 to 10 Å
    (d) 20 to 50 Å

11. CO_2 is more soluble in water when:
    (a) Temperature is high and pressure is low
    (b) Temperature is low and pressure is high
    (b) Temperature is low and pressure is low
    (d) Temperature is high and pressure is high

12. The mass of sulfuric acid present in 100 mL of 18% mass by mass solution of sulfuric acid with a density of 1.1 g per mL is:
    (a) 18 g
    (b) 19.8 g
    (c) 39.6 g
    (d) 1.98 g

13. 2.5 g of a solute is dissolved in 25 g of water to form a saturated solution at 298 K. What is the solubility of the solute at this temperature?
    (a) 5 g
    (b) 10 g
    (c) 15 g
    (d) 25 g

14. *An iron nail was dipped in a salt solution of 'A'. After some time a brown substance was deposited on the iron nail. The salt A could be:
    (a) Iron sulfate
    (b) Silver chloride
    (c) Copper sulfate
    (d) Silver nitrate

15. *A few pieces of zinc were treated with dilute H_2SO_4 in a test tube. Which of the following observations are correct?
    (a) It is a chemical change, zinc dissolves.
    (b) It is a chemical change and zinc dissolves with evolution of H_2 gas.
    (c) It is a physical change and zinc dissolves.
    (d) It is a physical change and H_2 gas is evolved.

16. *The magnesium oxide obtained on burning magnesium in air appears to be like:
    (a) Common salt
    (b) Powdered chalk
    (c) Wood ash
    (d) Powdered sugar

17. *A student was asked to prepare a true solution of sugar in water. By chance, he added sugar in excess. He stirred for quite some time but some of it settled down. He filtered the contents. The filtrate will be:
    (a) True solution
    (b) Colloidal solution
    (c) Suspension
    (d) Can be true solution or colloidal solution

18. *Which component of the mixture (Fe + S) reacts with dilute HCl and gives hydrogen gas?
    (a) Iron
    (b) Sulfur
    (c) Both
    (d) None of these

Assertion–Reason Questions

Direction: In the following question two statements (Assertion) A and Reason (R) are given Mark.
(a) if A and R both are correct and R is the correct explanation of A;
(b) if A and R both are correct but R is not the correct explanation of A;
(c) A is true but R is false;
(d) A is false but R is true

Assertion	Reason
1. On heating, a saturated solution becomes unsaturated.	1. On heating, solubility increases.
2. Rusting of iron is a chemical change.	2. Red rust of iron is $Fe_2O_3 \times H_2O$.
3. Colloid solutions show Tyndall effect.	3. Colloidal particles cannot scatter light.
4. Smog is an example of solid aerosol.	4. Smog is dust particles in air.
5. Concentration of a solution can be expressed in mass by mass % or mass by volume % mainly.	5. On the basis of amount of solute, a solution is further divided into dilute and concentrated solutions.

Section B: Very Short Answer Questions   (2 Marks)

1. Breaking of a glass tumbler is what kind of change?

2. Write the formula of red rust of iron.

3. A solution which contains the maximum amount of solute possible at a particular temperature is called?

4. *When magnesium is burnt in air, two compounds are formed. Name them.

5. The size of colloid particles in a nanometre is in the range of?

6. Name any two non-aqueous solvents.

7. In tincture of iodine, name the solute and solvent.

8. Why does tincture of iodine have antiseptic property?

9. Is a new compound formed when a mixture is formed?

10. *You are provided with a mixture of iron filings and sulfur powder. When you add carbon disulfide to the mixture, what would you observe?

Section C: Short Answer Questions   (3 Marks)

1. Define the terms solution, suspension and colloid. Make a comparison of the size of their particles.

2. What is meant in chemistry by the term physical change? Give some examples of physical changes.

3. What is meant in chemistry by the term chemical change? Give some examples of chemical changes.

4. How will you justify that rusting of iron is a chemical change?

5. A solution contains 27 g of sugar dissolved in 263 g of water. What is the concentration of the sugar solution?

6. Explain what happens when a beam of light is passed through a colloidal solution.

7. A solution has 18 g of urea in 130 g of the solution. What is the mass by mass percentage of the solution?

8. When will you say that a given solution of a solute is saturated? What would you observe when a hot saturated solution of the same solute is allowed to cool?

9. Explain why particles of colloidal solutions do not settle down when left undisturbed, while in the case of a suspension, they do settle.

10. *(i) Name the compound formed on heating a mixture of iron filings and sulfur.
    (ii) If dilute HCl is added to the above compound, name the gas evolved and write down two of its properties.

11. Vinegar which is used as a preservative and also added in Chinese foods is an example of a liquid-in-liquid solution. It is acetic acid dissolved in water.
    (i) What type of mixture is vinegar?
    (ii) Mention its two uses.
    (iii) What type of solution is lemonade and why does it taste sour?

12. *Four students were asked to add water to glucose powder, milk, sand and soil separately in four beakers. Classify the mixtures as true solution, colloid and suspension.

13. *Consider the change that occurs in each of the following. State whether it is a physical change or a chemical change and why.

(i) Sodium, a silvery metal, changes to a white solid in the presence of chlorine, a yellow-green gas.

(ii) A few crystals of sugar dissolve in the water in a laboratory beaker.

(iii) Alcohol, a liquid, changes to a vapour when heated.

14. (i) Define volume percentage.

(ii) 80 mL of alcohol was dissolved in 420 mL of water. Find the volume percentage of the solution.

15. (i) Mass of water must be added to 9 g of sodium chloride to make 9% mass by mass percentage of solution.

(ii) A solution has 80 g of common salt in 320 g of water. Find the concentration of the solution in terms of mass by mass percentage of solution.

Section D: Long Answer Questions
(5 Marks)

1. Which of the following represents a chemical change and which represents a physical change?

(i) Water is converted to steam by boiling.

(ii) Carbon, a black powder, is converted to a colourless gas when it reacts with oxygen.

(iii) Salt completely dissolves in water to form a solution that looks like ordinary water.

(iv) When the chemical mercury oxide is heated in a test tube, a silvery substance condenses near the top of the test tube.

(v) When acetic acid is cooled below 16.7°C (62°F), it is changed to a solid.

2. Define the following terms:

(i) Dispersed phase

(ii) Dispersion medium

(iii) Brownian movement

(iv) Solvent

3. To make a saturated solution, 45 g of sodium chloride is dissolved in 100 g of water at 293 K. Find its concentration at this temperature.

4. Calculate the mass of sodium sulfate required to prepare its 30% (mass percent) solution in 100 g of water.

5. During an experiment, the students were asked to prepare a 15% (mass/mass) solution of sugar in water. Shashwat dissolved 15 g of sugar in 100 g of water while Shanvi prepared it by dissolving 15 g of sugar in water to make 100 g of the solution.

(i) Are the two solutions of the same concentration?

(ii) Compare the mass % of the two solutions.

(iii) Who was right in making the mass by mass % solution.

Section E: Case Study or Passage-Based Question
(4 Marks)

1. A solution is a homogeneous mixture having two or more components. Solutions can be further divided into nine types, on the basis of solute (smaller amount component) and solvent (greater amount component). Solution and colloid differ in the fact that colloid is heterogeneous by nature. The difference between a solution and a colloid can be noticed by the size of particles scattering light. The concentration of a solution represents the amount of solute in a solution.

(i) In which pair of compounds is the solvent a solid?

(a) Brass, Soda water

(b) Brass, Hydrated copper sulfate

(c) Soda water, Iodine in air

(d) Camphor in air, Hydrogen on palladium

(ii) Which of the following comprises colloids only?

(a) Iodine in air, Air

(b) Aerated drink, Mist

(c) Smoke, Foam

(d) Milk, Sugar solution

(iii) With increase in temperature from 283 K to 353 K, whose solubility increases to the maximum extent?

(a) Sodium chloride
(b) Potassium chloride
(c) Potassium nitrate
(d) Ammonium chloride

(iv) Which pair is example of suspension?
(a) Paint, lime water
(b) Lime water, milk of magnesia
(c) Paints, milk
(d) Both (a) and (b)

(v) Which of the following sets represent correct statements about a solution?
(i) Solution is always homogeneous
(ii) It is mostly transparent in nature
(iii) A solution can never be coloured
(iv) Solute particles in solution cannot pass through filter paper

(a) (i), (ii)
(b) (i), (ii), (iii)
(c) (i), (ii), (iv)
(d) (i), (ii), (iii), (iv)

2. *When a mixture of iron filings and sulfur powder in a fixed ratio of mass 7 : 4 is heated strongly, a new substance iron sulfide (FeS) is formed. When we see this compound with a magnifying glass, iron particles and sulfur particles can be seen. Considering it answer the questions here:

(i) To prepare iron sulphide in the school laboratory, we heat a mixture of iron filings and sulfur powder in a:
(a) Copper dish
(b) Petri dish
(c) China dish
(d) Waatch glass

(ii) To test the properties of the iron sulfide formed, it should be taken in the form of:
(a) Lumps
(b) Small pieces
(c) Powder
(d) Solution

(iii) Observe the following diagrams and choose the correct option:

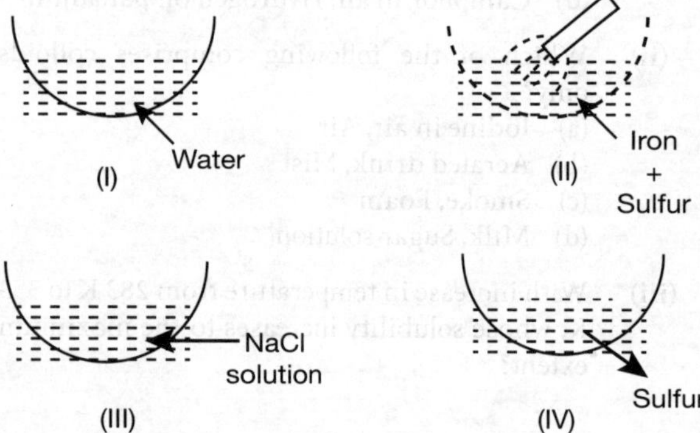

(a) III is an element
(b) II is a mixture
(c) IV is a compound
(d) I is an element

(iv) When iron filings and sulfur powder are taken in a china dish, mixed properly and heated strongly, then:
(a) A heterogeneous mixture is formed
(b) A compound is formed
(c) A homogeneous mixture is formed
(d) An element is formed

(v) A 50-mL saturated copper sulfate solution was taken in a beaker and a small amount of iron filings was added to it. The beaker was left undisturbed for some time. The mixture was then filtered to obtain the residue. The colour of residue was found to be:
(a) Greenish blue
(b) Yellow
(c) Steel grey
(d) Reddish brown

3. A solution is a homogeneous mixture while a colloid is a heterogeneous mixture. A solution can be further divided into many types like solid solution, liquid solution, gas solution, etc. Colloids can be further divided into eight types depending upon the medium used. Solution and colloid differ in many properties like size, Tyndall effect, etc. by which we can distinguish them easily.

(i) Are colloid particles visible to the naked eye?
(ii) The aqueous solution of common salt in water is known as?
(iii) Fog and mist are what type of solutions?
(iv) Which of the two will scatter light, soap solution or sugar solution?
(v) Name the solution which shows Tyndall effect.

Value-Based Questions

1. Shashwat went to watch a movie with his parents. In the dark, he found a small beam of light within which million of tiny particles were dancing. He thought, when there was light, no such beam appears. He was very surprised. He asked his science teacher Mr. Sharma about this. Mr. Sharma told him it was due to scattering of light.
(i) What values were exhibited by Shashwat?
(ii) What were the tiny particles?
(iii) What single term can be used for the scattering of light of such particles ?

(iv) Is scattering of light possible in a sugar solution?

Higher Order Thinking Skills (HOTS) Questions

1. Shanvi took 60 mL of water in two beakers at room temperature and added sugar to one beaker and sodium chloride to the other, till no more solute could be dissolved. Then she heated the contents of the beakers and added more solutes in them.
   (i) Will the amount of salt and sugar that can be dissolved in water at a given temperature be the same?
   (ii) What would you expect to happen if she cools the contents of the beakers? Justify your answer.

2. A teacher instructed three students Khushi, Shanvi and Shashwat to prepare a 40% (mass by volume) solution of sodium hydroxide (NaOH). Khushi dissolved 40 g of NaOH in 100 mL of water. Shanvi dissolved 40 g of NaOH in 100 g of water. Shashwat dissolved 40 g of NaOH in water to make 100 mL of solution. Which one of them has made the desired solution and why?

3. A teacher told three students X, Y and Z to prepare a 25% solution (mass by volume) of KOH. Student X dissolved 25 g of KOH in 100 g of water. Student Y dissolved 25 g of KOH in 100 mL of water and student Z dissolved 25 g KOH in water and made the volume 100 mL. Which one of them has made required 25% solution? Give your answer with reasons.

4. Name the phenomenon due to which sunlight that passes through the canopy of a dense forest becomes visible. Explain the cause of this phenomenon in the forest. Name one more substance through which this phenomenon can be observed.

5. *Four students A, B, C and D are asked to prepare colloidal solutions. The following diagrams show the preparations created by them. Name the student who will be able to prepare the colloidal solutions. Write two properties of colloidal solutions.

Answers

Section A: Multiple Choice Questions

1. (d)	2. (d)	3. (b)	4. (c)
5. (a)	6. (c)	7. (a)	8. (c)
9. (c)	10. (a)	11. (b)	12. (b)
13. (b)	14. (c)	15. (b)	16. (c)
17. (a)	18. (a)		

Assertion-Reason Questions

1. (a)	2. (b)	3. (c)	4. (a)
5. (b)			

Section B: Very Short Answer Questions

1. Physical change.

2. $Fe_2O_3.XH_2O$.

3. Supersaturated solution.

4. Magnesium oxide and magnesium nitride.

5. In the range of 1–100 nanometres.

6. Alcohol and carbon tetrachloride.

7. Solute is iodine and solvent is alcohol.

8. No. Elements or compounds only mix together to form a mixture and no new compound is formed.

9. We observe that sulfur powder dissolves and the solution turns yellow.

10. Tincture of iodine is prepared by dissolving iodine in alcohol.

Section C: Short Answer Questions

1. Refer to Section 2.5.2, Suspensions.

2. Physical change is a change that occurs without any transformation of substance. The substance is the same before and after the change. Examples include change of state (evaporation, condensation, freezing, melting) and change of texture (crushing of a granular solid to a powdery solid).

3. Chemical change is a transformation of one substance into one or more substances. Examples include the rusting of iron and the combination of hydrogen and oxygen to make water.

4. Rust is totally different from iron as iron is an element while rust is hydrated oxide of iron $[Fe_2O_3 \times H_2O]$; hence formation of rust from iron is a chemical change.

5. Mass of sugar = 27 g
    Mass of water = 263 g
    So, Mass of solution = Mass of solute + Mass of solvent = 27 + 263 = 300 g

    $$\text{Concentration of solution} = \frac{\text{Mass of solute}}{\text{Mass of solution}} \times 100$$

    $$\text{Concentration of solution} = \frac{27}{300} \times 100 = 9\% \text{ by mass}$$

6. Refer to Optical Properties, under Properties of Colloidal Solutions.

7. $$\text{Mass \% of solution} = \frac{\text{Mass of solute}}{\text{Mass of solution}} \times 100$$
    $$= \frac{18}{130} \times 100 = \frac{180}{13} = 13.84\%$$

8. When no more solute can be dissolved in the solution at a given temperature, the given solution is saturated. On cooling, crystals of solute separate out from the solution.

9. Refer to Mechanical Properties, under Properties of Colloidal Solutions.

10. (i) Iron sulfide
    (ii) The gas is hydrogen sulfide and its properties are: It is colourless. It has the smell of rotten eggs.

11. (i) It is a homogeneous mixture.
    (ii) It is used as a preservative and in the preparation of Chinese food.
    (iii) It is a true solution and it tastes sour as it has citric acid.

12. True solution: glucose powder with water
    Colloidal solution: milk with water
    Suspension: sand with water, soil with water

13. (i) Chemical change—there has been a transformation of substance.
    (ii) Physical change—sugar both before and after dissolving.
    (iii) Physical change—alcohol both before and after evaporation.

14. (i) Refer to Volume Strength under Concentration Terms of Solutions.
    (iii) Volume percentage
    $$= \frac{\text{Volume of solute}}{\text{Volume of solute} + \text{Volume of solvent}} \times 100$$
    $$= \frac{80}{80 + 420} \times 100 = 16\%$$

15. (i) As, Mass of solution = Mass of solute + Mass of solvent
    Mass of solvent = Mass of solution – Mass of solute
    $$100 - 9 = 91 \text{ g}$$
    (ii) Mass percentage
    $$= \frac{\text{Mass of solute}}{\text{Mass of solute} + \text{Mass of solvent}} \times 100$$
    $$= \frac{80}{80 + 320} \times 100 = 20\%$$

Section D: Long Answer Questions

1. (i) Physical change—water both before and after evaporation.
    (ii) Chemical change—there has been a transformation of substance.
    (iii) Physical change—salt both before and after dissolving.
    Chemical change—there has been a transformation of substance.
    Physical change—acetic acid before and after freezing.

2. Refer to Section 2.5.3, Colloids.

3. Mass of sodium chloride (solute) = 45 g
    Mass of water (solvent) = 100 g
    We know that, Mass of solution = Mass of solute + Mass of solvent = 45 g + 100 g = 145 g
    Concentration (mass percentage) of the solution

$$= \frac{\text{Mass of Solute}}{\text{Mass of Solution}} \times 100$$

$$= \frac{45\,\text{g}}{145\,\text{g}} \times 100 = 31.03\%$$

4. Let the mass of sodium sulfate required be = x g
   The mass of solution would be = (x + 100) g x g
   of solute in (x + 100) g of solution

$$30\% = \frac{x}{x+100} \times 100$$

$$30x + 3000 = 100x$$

$$70x = 3000$$

$$x = \frac{3000}{70} = 42.8\,\text{g}$$

5. No, Shanvi has a higher mass percentage.
   Solution made by Shashwat,

$$\text{Mass \%} = \frac{15}{(15+100)} \times 100$$

$$= \frac{15}{115} \times 100 = 13.04\%$$

   Solution made by Shanvi,

$$\text{Mass\%} = \frac{15}{100} \times 100 = 15\%$$

   Shanvi was correct in making a mass by mass solution.

Section E: Case Study or Passage-Based Question

1. (i) (b)   (ii) (c)   (iii) (c)   (iv) (d)   (v) (c)

2. (i) (c)   (ii) (c)   (iii) (b)   (iv) (c)   (v) (d)

3. (i) No
   (ii) Brine solution.
   (iii) Liquid in gas.
   (iv) Soap solution can scatter light.
   (v) A colloidal solution shows Tyndall effect.

Value-Based Questions

1. (i) Curiosity and good analytical skills.
   (ii) The tiny particles were particles of dust and smoke.
   (iii) Tyndall effect.
   (iv) No, as sugar particles cannot scatter light.

Higher Order Thinking Skills (HOTS) Questions

1. (i) No.
   (ii) Crystals of salt and sugar will appear as the solubility of a solid decreases with decrease in temperature.

2. Shashwat has prepared the desired solution/mass by volume (%) as he has taken the correct volume of the solution.
$$= \frac{\text{Mass of solute} \times 100}{\text{Volume of solution}} = \frac{40 \times 100}{100} = 40\%$$

3. Student Z made the required 25% solution.
   25 g of KOH in water and the volume of the solution is 100 mL.
   Mass by volume percentage of a solution
$$= \frac{\text{Mass of solute} \times 100}{\text{Volume of solution}} = \frac{25 \times 100}{100} = 25\%$$

4. (i) Tyndall effect.
   (ii) In the forest, the mist contains tiny droplets of water which acts as particles of colloid (dispersed phase) suspended in air (medium).
   (iii) Milk.

5. Student D will be able to prepare the colloidal solution as egg is not completely miscible in water.
   Two properties of colloidal solution are:
   (i) Particles of colloidal solutions cannot be separated.
   (ii) A colloidal solution appears to be homogeneous but actually it is a heterogeneous mixture of solute and solvent.

Hints

Section A: Multiple Choice Questions

3. Physical properties of mixtures are the same as of its components.

8. A solution of "Tincture of Iodine" is made by dissolving 2–7% iodine in alcohol. Tincture solutions are characterised by the presence of alcohol since alcohol is a good solvent. Iodine

does not dissolve in water. Tincture of iodine is used as an antiseptic.

10. Size of particles in colloidal state ranges from 10 to 1000 Å.

12. As 100/1.1 mL of solution contain sulfuric acid = 18 g
So 100 mL of the solution will contain sulfuric

$$\text{acid} = \frac{18 \times 1.1}{100} \times 100 \text{ g} = 19.8 \text{ g}$$

13. Mass of the solute = 2.5 g
Mass of the solvent = 25 g
So solubility of the solute

$$= \frac{\text{Mass of the solute}}{\text{Mass of the solvent}} \times 100 = \frac{2.5}{25} \times 100 = 10 \text{ g}$$

Section E: Case Study or Passage-Based Question

1. (v) As a solution can also be coloured.

2.6 SEPARATION OF MIXTURES

Most of the materials around us are mixtures and not pure substances. Mixtures have two or more substances mixed in them. We might require only one or two separate constituents of a mixture for our use. So, such mixtures need to be separated into their individual constituents to make them useful.

Various methods of separation are used to obtain individual components from a mixture. Separation makes it possible to study and use the individual components of a mixture. Heterogeneous mixtures can be separated into their respective constituents by simple physical methods like handpicking, sieving and filtering. Some special techniques can also be used for separating the components of a mixture (Fig. 2.23).

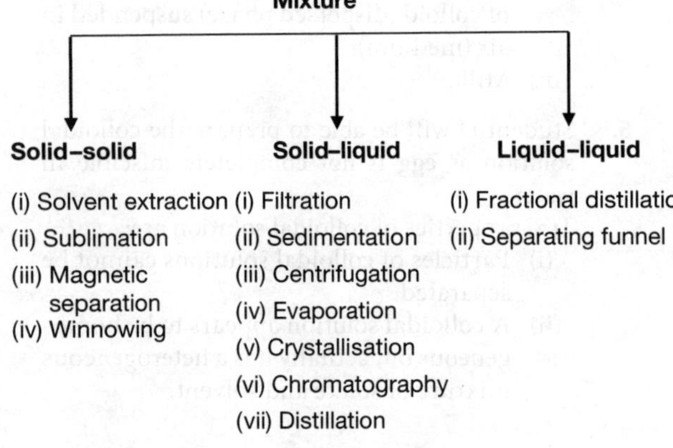

Fig. 2.23 Separation of mixtures

2.6.1 Separation of a Mixture of Two Solids

All mixtures having two solid substances can be separated by one of the following methods:

Separation by a Suitable Solvent: In some cases, one of the constituents of a mixture is soluble in a specific liquid solvent while the other is insoluble. This difference in the solubility of the constituents of a mixture can be used to separate them. Here are a few examples of this method:

- **Separation of sugar and sand:** Sugar is soluble in water while sand is insoluble. So a mixture of sugar and sand can be separated by using water as the solvent. Here, the mixture is taken in a beaker and water is added to it. It is then stirred to dissolve the sugar in water (Fig. 2.24).

Fig. 2.24 Sugar is soluble in water but sand is insoluble in water.

- **Separation of salt and sand:** As common salt (NaCl) is soluble in water while sand is insoluble, a mixture of salt and sand can be separated by using water as the solvent (Fig. 2.25). Here, the salt forms the solution while sand remains insoluble.

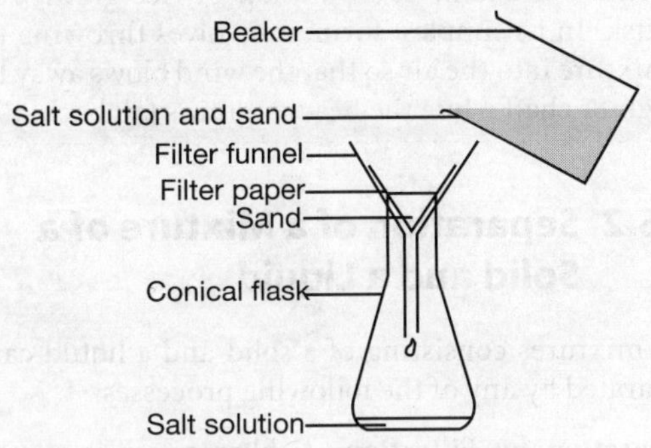

Fig. 2.25 Mixture of common salt and sand

*Ethanol is used as a solvent to separate a mixture of sugar and salt as salt is soluble in it but sugar is insoluble.

- **Separation of sulfur and sand:** As both sulfur and sand are insoluble in water, water cannot be used as a solvent to separate this mixture. However, sulfur is soluble in carbon disulfide while sand is insoluble in carbon disulfide, so this mixture of sulphur and sand can be separated by using carbon disulfide as solvent. This mixture is shaken with carbon disulfide where sulfur dissolves in carbon disulfide while sand remains undissolved. The sulfur solution having sand is filtered and sand is obtained as a residue on the filter paper and sulfur solution is obtained as a filtrate. On evaporating the filtrate, carbon disulfide solvent is eliminated and solid sulfur remains as residue.

As carbon, sugar and salt are also insoluble in carbon disulfide, these can also be separated from sulfur by using carbon disulfide as solvent.

Separation by Sublimation: The change of a solid substance into vapour by heating and the conversion of vapour into solid by cooling is known as sublimation. The process of sublimation can be used to separate those substances which sublime on heating, from a mixture. The solid substance obtained by cooling the vapour is called the sublimate (Fig. 2.26).

$$\text{Solid} \xrightarrow[\text{Cool, sublimation}]{\text{Heat, sublimation}} \text{Vapours}$$

Fig. 2.26 Sublimation

For example, iodine, naphthalene, anthracene, camphor and ammonium chloride can be separated from a mixture by sublimation. Sublimation is used to separate such substances from those which do not undergo sublimation like common salt, sand, sulfur, chalk, etc.

- **Separation of common salt and ammonium chloride:** As ammonium chloride sublimes on heating while common salt does not, the sublimation process can be used to separate ammonium chloride from a mixture of common salt and ammonium chloride (Fig. 2.27).

Here, the mixture is taken in a china dish and placed on a tripod stand. The china dish is covered with an inverted glass funnel and a loose cotton plug is placed in the upper, open end of this funnel to prevent ammonium chloride vapours from escaping into the atmosphere. On heating the mixture, ammonium chloride changes into white vapour which rises up and is converted into solid ammonium chloride on coming in contact with the cold, inner walls of the funnel. This can be removed while common salt does not change into vapour on heating, so it remains behind in the china dish.

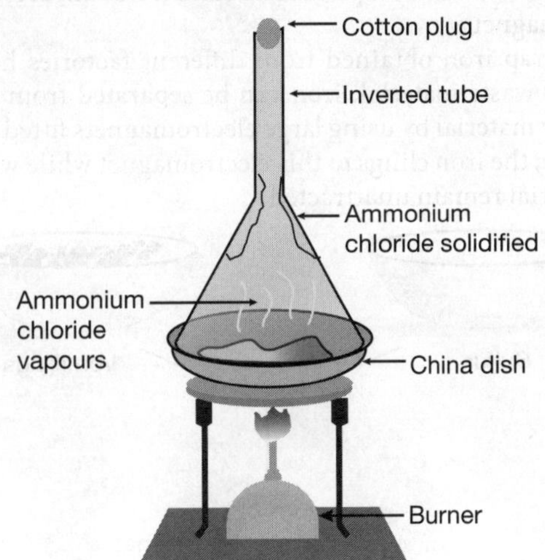

Fig. 2.27 Separation of ammonium chloride by sublimation.

- **Separation of iodine and common salt:** As iodine undergoes sublimation while common salt does not, they can be easily separated by the sublimation method, as shown in Fig. 2.28.

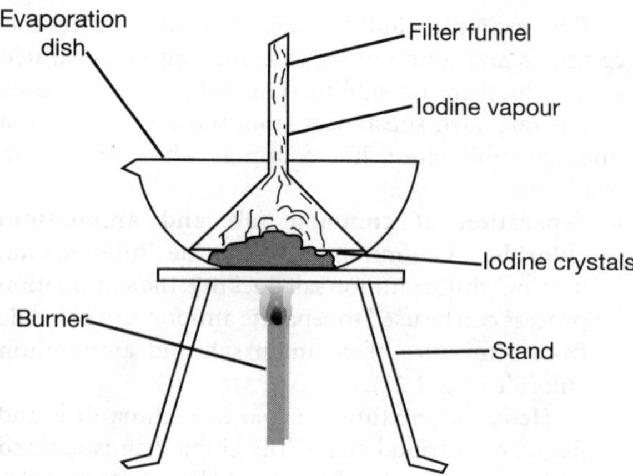

Fig. 2.28 Separation of iodine by sublimation

Separation by a Magnet: Iron is attracted by a magnet; this tendency of iron is used to separate it from a mixture having iron as one of the constituents. For example, a mixture of iron filings and sulfur powder can be separated by using a magnet as iron filings are attracted by a magnet while sulfur is not (Fig. 2.29). Similarly iron fillings from carbon powder or sand can also be separated as carbon powder or sand is not attracted by the magnet.

Scrap iron obtained from different factories has a lot of waste material. Iron can be separated from this waste material by using large electromagnets fitted to a crane; the iron clings to this electromagnet while waste material remain unattracted.

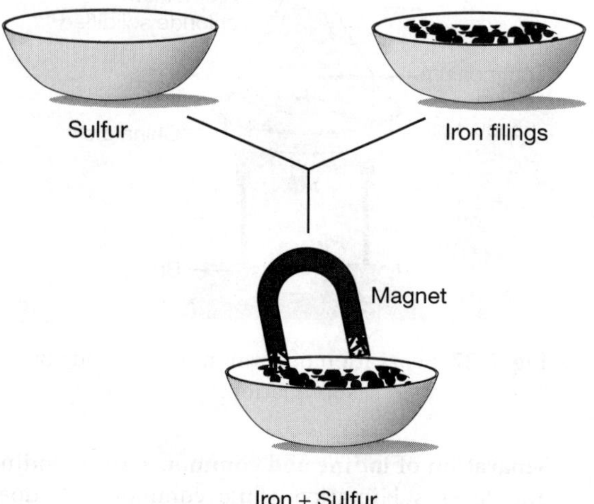

Fig. 2.29 Separation of iron from sulfur and waste materials

2.6.2 Separation of a Mixture of a Solid and a Liquid

All mixtures consisting of a solid and a liquid can be separated by any of the following processes:

Separation by Filtration: A filter paper is a special round-shaped paper having millions of tiny holes; only very minute particles can pass through these holes. For example, particles of chalk cannot pass through the holes due to their larger size. However, particles of water can.

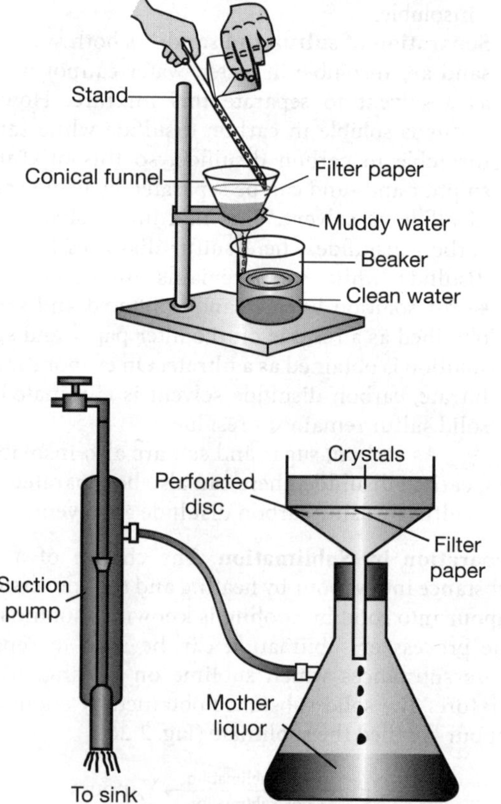

Fig. 2.30 Simple filter paper and Buchner's funnel (Quick filtration)

The process of removing insoluble solids from a liquid by using a filter paper is known as filtration. Here, the liquid passes through the filter paper and is collected in the beaker kept below the funnel. Therefore, the solid substance is left behind on the filter paper as residue and the clear liquid obtained is the filtrate. It must be noted that we cannot remove a solid substance dissolved in a liquid using this method. Filtration is a slow process so Buchner's funnel and water section pump are used for quick filtration (Fig. 2.30).

Sedimentation or Decantation: This method can be used for a mixture having one solid and one liquid component where the solid is insoluble in the liquid. In the sedimentation process, heavier components of the mixture settle at the bottom due to gravity. For example, settling of mud particles in water. Decantation follows sedimentation and this process involves pouring the clear, upper liquid out of the container, without disturbing the sediment. For example, after tea leaves have settled down, the clear liquid tea from the top can be poured into a cup. Hence, the transfer of clear tea is a case of decantation.

In cities, drinking water is supplied from water works and the source of the water supply is a nearby river or a lake (known as reservoir). This river and lake water usually contains many suspended solid substances, clay particles, harmful bacteria, etc. It means purification of water is a must and for that methods like sedimentation, decantation, loading, filtration and chlorination are used.

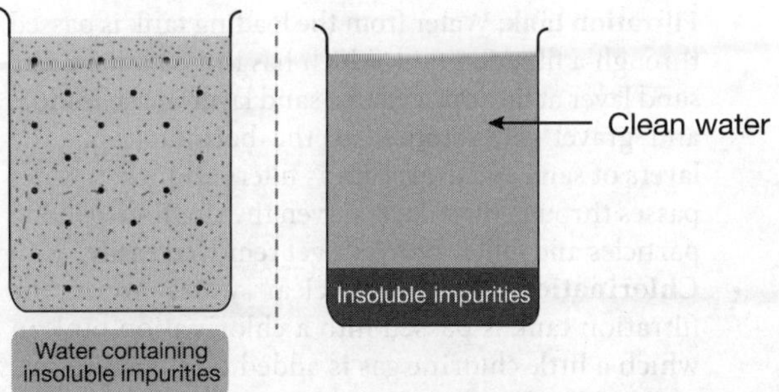

Fig. 2.31 Sedimentation

- **Sedimentation tank:** The water from a river or lake is pumped by the pumping station into a large reservoir known as the sedimentation tank. This water is allowed to stand in the sedimentation tank for some time so that many of the insoluble substances present in the water settle down at the bottom of the tank (Fig. 2.31).
- **Loading tank:** From the sedimentation tank, water is passed to a 'loading tank' in which some potash alum (phitkari) is mixed in water. Here, the heavy particles of dissolved alum deposit on the suspended clay particles in water, that is, the suspended clay particles 'loaded' with alum particles, become heavier and settle at the bottom of the tank (Fig. 2.32). It means the process of loading by using alum is to remove the suspended clay particles from water.

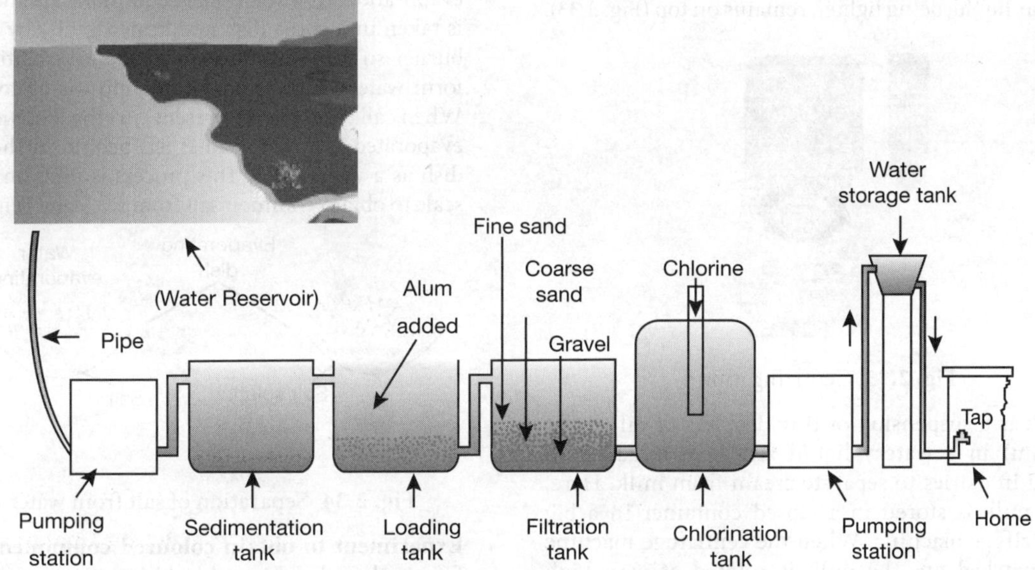

Fig. 2.32 Supply of drinking water in a city

- **Filtration tank:** Water from the loading tank is passed through a filtration tank which has three layers: a fine sand layer at the top, a coarse sand layer in the middle and gravel (tiny stones) as the bottom layer. The layers of sand and gravel act as filters and when water passes through these layers, even the small suspended particles and other materials get removed easily.
- **Chlorination tank:** The clear water from the filtration tank is passed into a chlorination tank in which a little chlorine gas is added to kill the germs present in the water. It is known as disinfecting or sterilising water. The water now becomes fit for drinking as it is free from bacteria.
- **High storage tanks:** The clean and disinfected water is now pumped by a pumping station into high storage tanks from which water is supplied to homes and factories in cities using a network of big and small pipes.

Separation by Centrifugation: This method is based on the principle that when a mixture is rotated at very high speed, the lighter particles stay on the surface of the liquid while the heavier particles are forced to the bottom of the liquid. The suspended particles of a substance in a liquid can be very rapidly separated by the centrifugation process with the help of a machine called a centrifuge. Here, the mixture of fine suspended particles in the liquid is taken in a test tube and the test tube is placed in a centrifuge machine and rotated rapidly for some time. As the mixture rotates, a centrifugal force acts on the heavier suspended particles and brings them down to the bottom of the test tube. The clear liquid, being lighter, remains on top (Fig. 2.33).

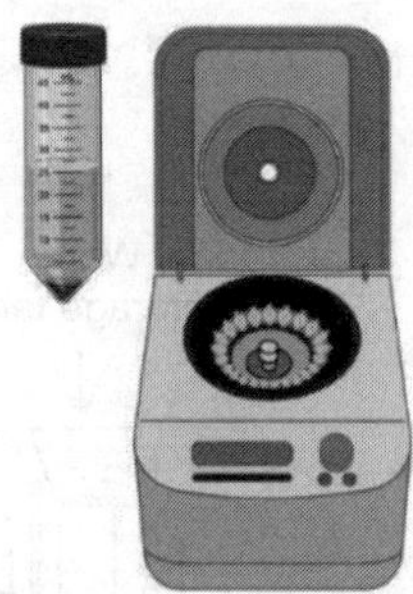

Fig. 2.33 Centrifugation

- Milk is a suspension of tiny droplets of oil (fat or cream) in a watery liquid and centrifugation is used in dairies to separate cream from milk. Here, the milk is stored in a closed container in a big centrifuge machine. When the centrifuge machine is switched on, the milk is rotated at very high speed in its container so it separates into 'cream' and 'skimmed milk'. The cream, being lighter, floats over the skimmed milk and can be removed easily.
- Clay particles suspended in water can be rapidly separated by centrifugation; the clay particles settle down at the bottom of the test tube and clear water remains at the top.
- The process of centrifugation is also used in washing machines to squeeze out water from wet clothes and make them dry.
- Centrifugation can be used in diagnostic laboratories for blood and urine tests, in dairies and at home to separate butter from cream.

Competition Edge

A mixture of common salt and water cannot separated by filtration or centrifugation as common salt dissolves completely in water.

Separation by Evaporation: The process of conversion of a liquid into vapour is called evaporation. It is used to separate a solid substance that has been dissolved in water or any other liquid solvent. The dissolved substance is left as a solid residue when all the water or liquid has evaporated. This method is based on the fact that liquids vaporise easily whereas solids do not vaporise easily. Evaporation of a liquid can take place slowly even at room temperature, but the solution is heated in order to hasten the process.

- **Separation of common salt from water:** Common salt (NaCl) dissolved in water can be separated by evaporation. The solution of common salt and water is taken in a china dish and heated gently by using a burner so the water present in the salt solution will form water vapour and escape into the atmosphere. When all the water present in the solution has evaporated, common salt is left behind in the china dish as a white solid. This process is used on a large scale to obtain common salt from sea water (Fig. 2.34).

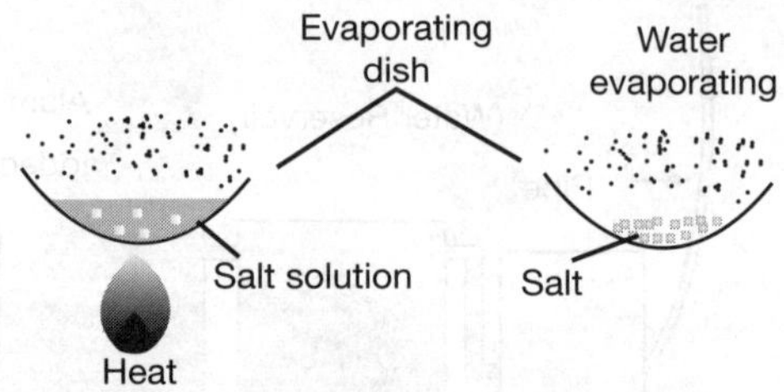

Fig. 2.34 Separation of salt from water

- **Experiment to obtain coloured component (dye) from ink:** Ink (blue or black) is a mixture of a dye in water and the dye can be separated from water by evaporation. Here, we put a few drops of ink on a watch glass and place it on a beaker half-full of water. When the beaker is heated, the water turns into steam which also heats up the ink; as a result, the water present in the ink evaporates and a blue or black residue in left on the watch glass (Fig. 2.35).

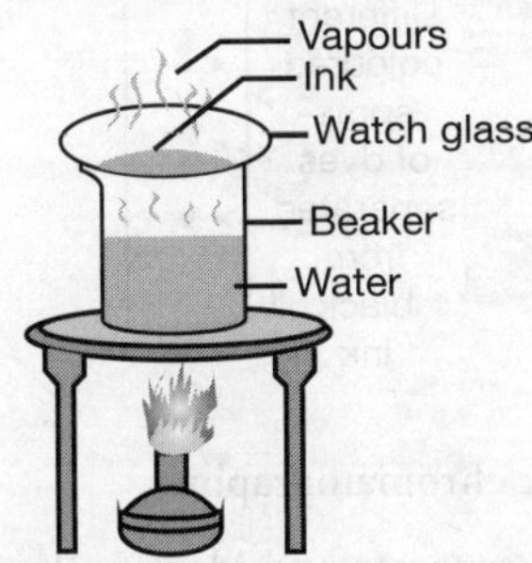

Fig. 2.35 Separation of dye from ink

This experiment proves that ink is not a single or pure substance but a mixture (mixture of dye in water).

Purification by Crystallisation: Crystals are the purest form of a substance having definite geometrical shape. The process of converting an impure compound into pure crystals is called crystallisation. Here, the hot, concentrated solution of a substance is cooled to obtain crystals.

Advantages of crystallisation over evaporation: In evaporation, we need to heat the solution to dryness, so evaporation gives us the solid substance in the form of a powder. If any impurity is present in the dissolved solid substance, the same will also be present when it is recovered by evaporation. This, however, is not the case with crystallisation. Crystallisation gives us proper crystals of solid particles with flat sides, and impurities are also removed; it is a method of purifying solid substances in which crystals of the pure substance are formed. Crystallisation begins by evaporating the liquid mixture by heating, but it is not continued up to dryness. Rather, when the solution becomes sufficiently concentrated or saturated, heating is stopped and the hot concentrated solution is allowed to cool slowly. After some time, crystals of pure solid substance appear in the solution.

Process: The process of crystallisation takes place as follows (Fig. 2.36):

1. The impure solid substance is dissolved in a minimum amount of water to form a solution.

2. The solution is filtered to remove insoluble impurities.

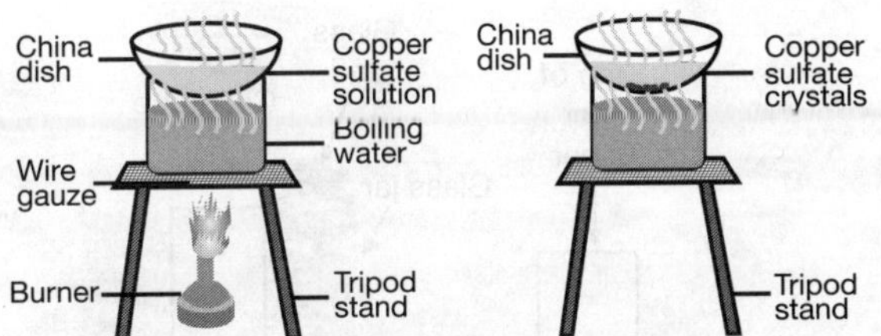

Fig. 2.36 Crystallisation

3. The clear solution is heated gently on a water bath till a concentrated or saturated solution is obtained. It can be tested by dipping a glass rod in hot solution again and again. When small crystals form on the glass rod, the solution is saturated.

4. Allow the hot, saturated solution to cool down slowly. Crystals of pure solid are formed while impurities remain dissolved in the solution.

5. Separate the crystals of pure solid by filtration and dry them.

Applications: (i) To remove impurities from the salt obtained from sea water. (ii) To obtain pure alum, pure potassium nitrate, etc.

Separation by Chromatography: This was introduced by Tswett; 'chromatography' means 'colour writing'. Chromatography is the most modern and versatile technique of separating two or more dissolved solids which are present in a solution in extremely small quantities or those that cannot be separated by other methods (Fig. 2.37). Chromatography is of following types: Thin Layer Chromatography (TLC), Gas-Liquid Chromatography (GLC) and Paper Chromatography.

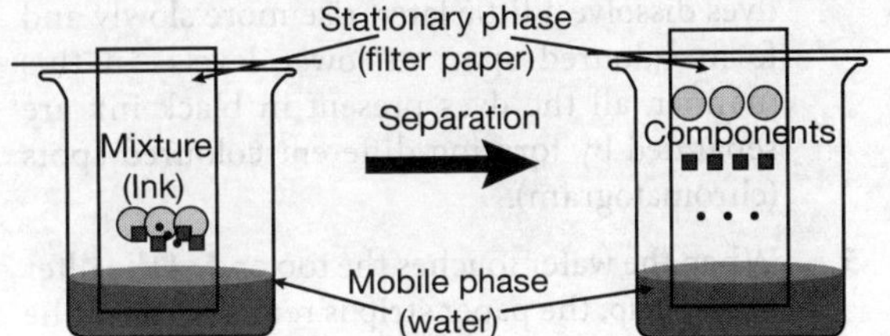

Fig. 2.37 Separation by chromatography

In paper chromatography, we can separate two or more different substances present in the same solution. It is based on the fact that though two or more substances are soluble in the same solvent, their solubility must be different.

- **To separate the dyes present in black ink:** The different coloured dyes present in black ink can be separated by performing a paper chromatography experiment, as shown in Fig. 2.38.

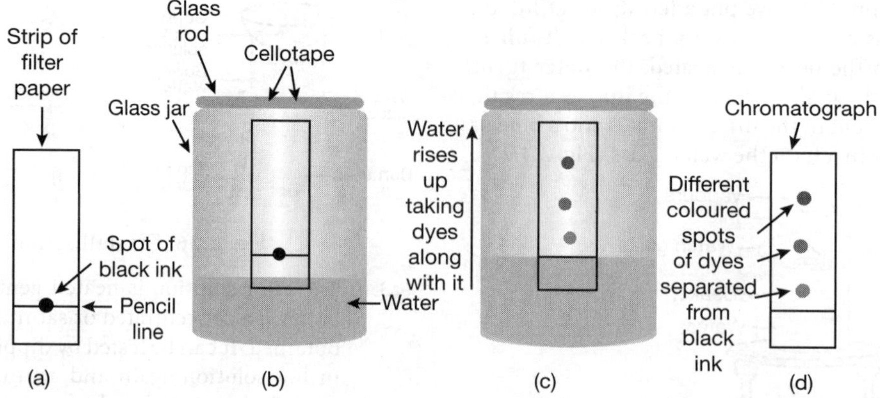

Fig. 2.38 Separation of the dyes in black ink by paper chromatography

1. Take a thin and long strip of filter paper. Draw a pencil line on it, nearly 3 centimetres from one end.

2. Put a small drop of black ink on the filter paper strip at the centre of the pencil line and let the ink dry.

3. When the drop of ink has dried, the filter paper strip is lowered into a tall glass jar having water in its lower part. The filter paper strip is held vertically by attaching its upper end to a glass rod with cellotape.

4. The water gradually rises up the filter paper strip and as it moves up the paper strip, it takes the dyes present in ink. The dye which is more soluble in water dissolves first, rises faster and produces a coloured spot on the paper at a higher position. On the other hand, less soluble dyes dissolve a little later, rise more slowly and form coloured spots at lower levels. In this manner, all the dyes present in black ink are separated by forming different coloured spots (chromatogram).

5. When the water touches the top end of the filter paper strip, the paper strip is removed from the jar and dried.

Applications: Some of the important applications of chromatography are as follows:

- It is used to separate solutions of coloured substances such as dyes and pigments like chlorophyll (green colour) and xanthophyl (yellow colour).
- It is used in forensic science to detect and identify trace amounts of substances like poisons and drugs in the contents of the bladder or stomach.
- It is used to separate small amounts of products of chemical reactions.
- It is used to separate and identify the amino acids obtained by hydrolysis of proteins.
- It is also used to monitor the progress of a reaction.

Competition Edge

Retardation (R_f) value is the ratio of the solute distance travelled to the solvent distance travelled.

$$R_f = \frac{\text{Distance moved by spot}}{\text{Distance moved by solvent}} = \frac{a}{b}$$

When the R_f value is calculated, the component of the mixture can be identified by comparison with R_f values in a data book. R_f values can be determined for all the components of a mixture.

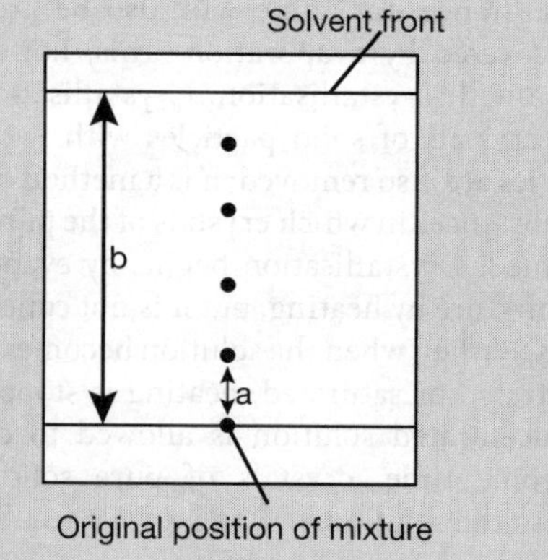

Separation by Distillation: Distillation is the process of heating a liquid to form vapour and then cooling the vapour to get back the liquid. Distillation can be shown as:

$$\text{Liquid} \underset{\text{cooling}}{\overset{\text{heating}}{\rightleftharpoons}} \text{Vapour (or gas)}$$

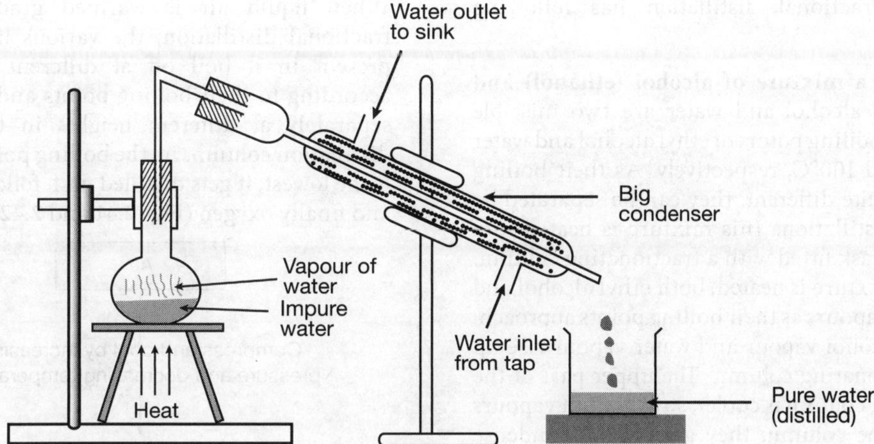

Fig. 2.39 Separation by distillation

The liquid obtained by condensing the vapour is known as the distillate. It is based on the principle that the more volatile component with the lower boiling point distills first while the less volatile component with the higher boiling point distills later. When the homogeneous mixture of solid and liquid is heated in a closed distillation flask, the liquid, being volatile, forms vapour. The vapours of the liquid are passed through a condenser where they are cooled and condense to form the pure liquid. This pure liquid is collected in a separate vessel. The solid, being non-volatile, is left behind in the distillation flask (Fig. 2.39).

- Using distillation, salt solution can be separated into salt and water.
- Using distillation, benzene and toluene can be separated.
- Using distillation, chlorobenzene and bromo benzene can be separated.

2.6.3 Separation of a Mixture of Two or More Liquids

Any mixture having two or more liquids can be separated by two methods: fractional distillation and using a separating funnel.

The liquids which can be mixed in any proportion and form a single layer are known as **miscible liquids.** For example, alcohol and water are miscible liquids and form a single layer on mixing. Such a mixture of miscible liquids can be separated by fractional distillation.

The liquids which do not mix with one another and which form separate layers are known as **immiscible liquids.** For example, oil and water are immiscible liquids and form separate layers on mixing. Water being heavier, forms the lower layer and oil being lighter, forms the upper layer. Such a mixture of immiscible liquids can be separated using an apparatus known as a separating funnel.

Separation by Fractional Distillation: If the boiling points of two miscible liquids are very close or less than 25 K, simple distillation cannot be used as both the liquids will be distilled together. In such cases, fractional distillation is used which involves repeated distillation and condensation. Fractional distillation is the process of separating two or more miscible liquids by distillation, the distillate being collected in fractions boiling at different temperatures. A mixture of two miscible liquids can be separated by the process of fractional distillation. The separation of two liquids by fractional distillation depends on the difference in their boiling points (10–25°C). Fractional distillation is carried out by using a fractionating column which provides hurdles to the ascending vapour and descending liquid and to increase the cooling surface area (Fig. 2.40).

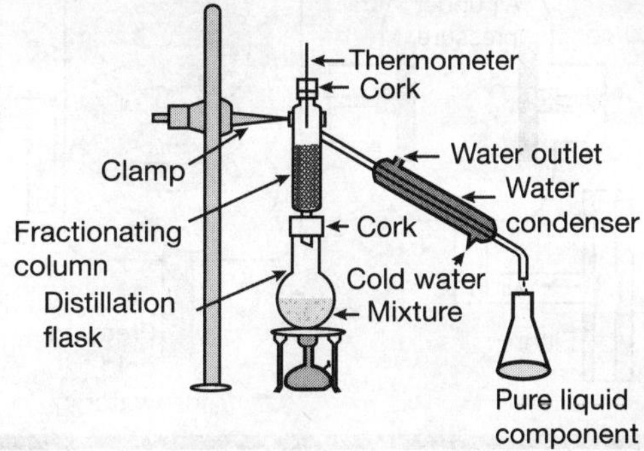

Fig. 2.40 Separation by fractional distillation

Applications: Fractional distillation has following applications:

- **To separate a mixture of alcohol (ethanol) and water:** Ethyl alcohol and water are two miscible liquids. The boiling points of ethyl alcohol and water are 78°C and 100°C, respectively. As their boiling points are quite different, they can be separated by fractional distillation. This mixture is heated in a distillation flask fitted with a fractionating column. When the mixture is heated, both ethyl alcohol and water form vapours as their boiling points approach. The ethyl alcohol vapour and water vapour rise up in the fractionating column. The upper part of the fractionating column is cooler, so as the hot vapours rise up in the column, they get cooled, condense and trickle back into the distillation flask. The more volatile liquid (lower boiling point) distils first and the less volatile liquid (higher boiling point) distils later. That is, ethanol distils first.

- **Separation of the gases of the air:** The main components of air are nitrogen (78.03%), oxygen (20.99%), argon (0.93%) and all the other gases (0.05%). All these gases can be separated from one another by the fractional distillation of liquid air. This separation is based on the fact that these gases have different boiling points in liquid form. The boiling points of nitrogen (−196°C), oxygen (−183°C) and argon (−186°C) are considered for fractional distillation.

  Liquid air is an extremely cold liquid and contains all these component gases in liquid form.

When liquid air is warmed gradually during fractional distillation, the various liquefied gases present in it boil off at different temperatures according to their boiling points and are collected separately at different heights in the fractional distillation column. As the boiling point of nitrogen is the lowest, it gets distilled first, followed by argon and finally oxygen (Figs 2.41 and 2.42).

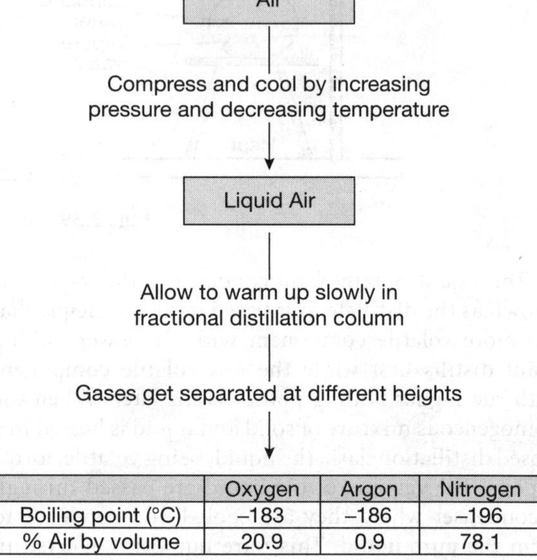

	Oxygen	Argon	Nitrogen
Boiling point (°C)	−183	−186	−196
% Air by volume	20.9	0.9	78.1

Fig. 2.41 Flow diagram for fractional distillation of air

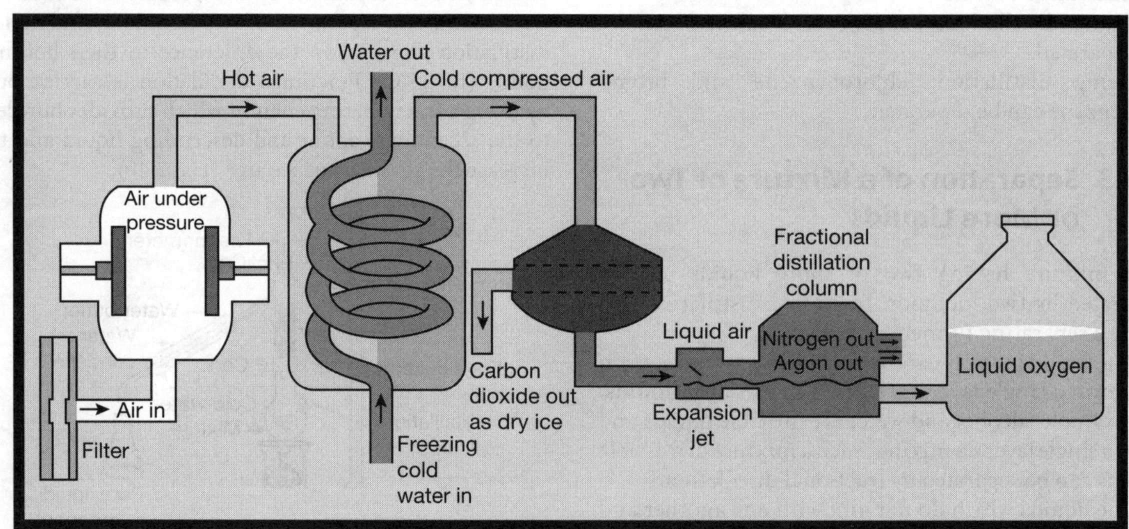

Fig. 2.42 Separation of gases of air

- Fractional distillation is also used in the separation of gasoline, kerosene, diesel, and so on, from crude petroleum.

Competition Edge

Fractional distillation cannot be used for a mixture in which both the components boil at the same temperature, that is, an azeotrope mixture. For example, rectified spirit (95% alcohol and 5% water).

Separation Using a Separating Funnel: A mixture of two immiscible liquids can be separated by using a separating funnel. A separating funnel is a special type of funnel having a stopcock in its stem to allow or to stop the flow of liquid from it. The separation of two immiscible liquids using a separating funnel depends on the difference in their densities. When such a mixture is put into a separating funnel and allowed to stand for some time, the mixture separates into two layers according to the densities of the liquids. The heavier (denser) liquid forms the lower layer while the lighter liquid forms the upper layer. On opening the stopcock of the separating funnel, the lower layer (heavier liquid) comes out first and is collected in a beaker. When the lower layer of heavier liquid has completely run off, the stopcock is closed. The lighter liquid in the upper layer is collected in a separate beaker by opening the stopcock again. For example, water and kerosene oil are two immiscible liquids. So they can be separated by using a separating funnel (Fig. 2.43).

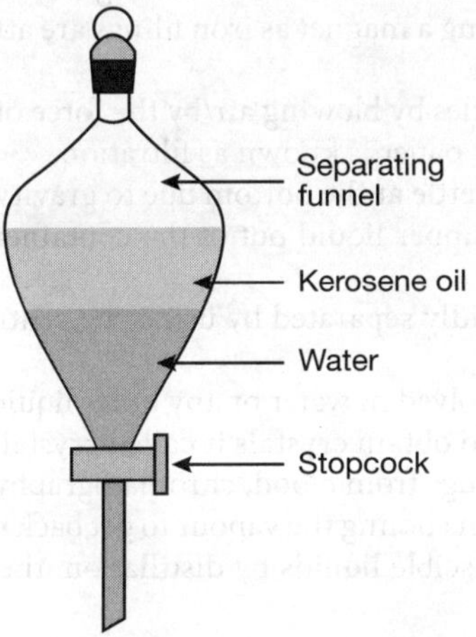

Fig. 2.43 Separation using a separating funnel

TEST YOUR KNOWLEDGE

1. How will you separate iron filings, ammonium chloride and sand from their mixture?

Solution: Whenever we want to separate the components of a mixture, we should first find out some properties which are different for different components. In this question, we have three components to be separated: iron filings, ammonium chloride and sand. Now, iron is attracted by a magnet whereas ammonium chloride undergoes sublimation.

However, sand does not have any of these properties. So, this difference in the properties of iron filings, ammonium chloride and sand will be used to separate them from their mixture.

The mixture containing iron filings, ammonium chloride and sand is separated as follows:

(i) Iron filings are attracted by a magnet so they are removed by magnetic separation. When a magnet is moved in this mixture, iron filings cling to it and get separated. We are then left with ammonium chloride and sand.

(ii) Ammonium chloride sublimes on heating whereas sand does not. So, ammonium chloride is separated from sand by sublimation. When the mixture containing ammonium chloride and sand is heated, ammonium chloride forms vapour easily. These vapours on cooling give pure ammonium chloride. Sand is left behind.

Please note that we have given many details in answering this question just to make you understand clearly. There is no need for students to write so many details while writing their answers.

2. Describe a method of separating common salt from a mixture of common salt and chalk powder.

Solution: This mixture has two constituents: common salt and chalk powder. Now, common salt is soluble in water whereas chalk powder is insoluble in water. So, this difference in their solubility will be used to separate them. This is performed as follows:

(i) Some water is added to the mixture of common salt and chalk powder, and stirred. Common salt dissolves in water to form a salt solution whereas chalk powder remains undissolved.

(ii) On filtering, chalk powder is obtained as a residue on the filter paper and the salt solution is obtained as filtrate.

(iii) The salt solution (filtrate) is evaporated to dryness when common salt is left behind.

3. You are given a mixture of sand, water and mustard oil. How will you separate the components of this mixture?

Solution: This mixture contains three components: sand, water and mustard oil. Now, sand is a solid which is insoluble in water as well as mustard oil. Water and mustard oil are immiscible liquids.

(i) The mixture of sand, water and mustard oil is filtered. Sand is left on the filter paper as residue. Water and mustard oil collect as filtrate.

(ii) The filtrate containing water and mustard oil is put in a separating funnel. Water forms the lower layer and mustard oil forms the upper layer. The lower layer of water is run out first by opening the stopcock of the separating funnel. Mustard oil is left behind in the separating funnel and can be removed separately.

4. Name the process used to separate the given mixtures:

(i) A mixture of methanol (338 K) and acetone (329 K)

(ii) Camphor having traces of common salt

(iii) Kerosene having water

Solution:

(i) By fractional distillation as the boiling points are close

(ii) Sublimation

(iii) Solvent extraction using a a separating funnel

5. Name the process used to separate the given mixtures:

(i) A mixture of liquid X (366 K) and liquid Y (353 K)

(ii) A mixture of liquid P (355 K) and liquid Q (415 K)

(iii) 0.1 g of a liquid having 3 components in very minute amounts

Solution:

(i) By fractional distillation as the boiling points are close (differ by only 12 K)

(ii) By simple distillation as the boiling points differ by a large difference of 60 K

(iii) Column chromatography

Quick Review

- Separation of mixtures: Separation makes it possible to study and use the individual components of a mixture.
- The process of sublimation can be used to separate those substances which sublime on heating from a mixture.
- A mixture of iron filings and sulfur powder can be separated by using a magnet as iron filings are attracted by the magnet while sulfur is not.
- Winnowing is used to separate lighter particles from heavier particles by blowing air/by the force of wind.
- The process of removing insoluble solids from a liquid using a filter paper is known as filtration.
- In the sedimentation process, heavier components of the mixture settle at the bottom due to gravity.
- Decantation follows sedimentation and it involves pouring clear, upper liquid out of the container, without disturbing the sediment.
- The suspended particles of a substance in a liquid can be very rapidly separated by using the centrifugation process with the help of a machine called the centrifuge.
- Evaporation is used to separate a solid substance that has been dissolved in water or any other liquid solvent.
- The process of cooling a hot, concentrated solution of a substance to obtain crystals is called crystallisation.
- To separate colours in a dye, pigments from natural colours and drugs from blood, chromatography is used.
- Distillation is the process of heating a liquid to form vapour and then cooling the vapour to get back the liquid.
- Fractional distillation is the process of separating two or more miscible liquids by distillation, the distillate being collected in fractions boiling at different temperatures.
- A mixture of two immiscible liquids can be separated by using a separating funnel.

Exercise 2.3

All questions marked * are practical based.

Section A: Multiple Choice Questions
(1 Mark)

1. The process of converting a solid directly into gaseous form is known as:
   (a) Evaporation   (b) Centrifugation
   (c) Sublimation   (d) Distillation

2. A mixture of $ZnCl_2$ and $PbCl_2$ can be separated by:
   (a) Sublimation
   (b) Fractional distillation
   (c) Distillation
   (d) Crystallisation

3. Which of these methods can separate sand and water?
   (a) Filtration
   (b) Chromatography
   (c) Crystallisation
   (d) Distillation

4. Which of the following is based on difference in size?
   (a) Fractional distillation
   (b) Sublimation
   (c) Filtration
   (d) Solvent extraction

5. Salt is separated from seawater by:
   (a) Filtration   (b) Distillation
   (c) Crystallisation   (d) Evaporation

6. Camphor can be purified by:
   (a) Sedimentation   (b) Sublimation
   (c) Distillation   (d) Filtration

7. Which of the following physical properties is a criterion when using filtration?
   (a) Substances with different boiling points
   (b) Substances with different solubility
   (c) Solid substance insoluble in the solvent
   (d) Liquid substance immiscible with the solvent

8. Distillation is a good separation technique for:
   (a) Solid alloys   (b) Gases
   (c) Solids   (d) Liquids

9. Magnetism can be beneficial for separating:
   (a) Non-metallic solids and solids such as sulfur
   (b) Non-magnetic solids from non-magnetic liquids
   (c) Gases and non-metallic liquids
   (d) Magnetic solids and solids like sulfur

10. Which is incorrect about centrifugation?
    (a) In it, lighter particles stay on the surface of the liquid.
    (b) Heavier particles go to the bottom of the liquid.
    (c) It is used in separating cream from milk.
    (d) It is used in removing water from blue ink.

11. A mixture of hexane and toluene can be separated by which method?
    (a) Sublimation   (b) Centrifugation
    (c) Distillation   (d) All of these

12. The process that can used to separate oil and water is:
    (a) Separating funnel
    (b) Chromatography
    (c) Distillation
    (d) Sublimation

13. Fractional distillation is based on the difference in ___________ of two liquids.
    (a) Solubility   (b) Melting point
    (c) Boiling point   (d) None of these

14. A mixture contains two substances. What property must the two substances have for them to be separated by paper chromatography?
    (a) Both substances must be coloured
    (b) Both substances must be solid at room temperature
    (c) Both substances must be soluble in the solvent
    (d) Both substances must have a low boiling point

15. Which of the following is not present in the water purification system in waterworks?
    (a) Sedimentation tank
    (b) Filtration tank
    (c) Chlorine tank
    (d) Oxidising tank

16. *A mixture containing (i) sodium chloride (ii) camphor and (iii) ammonium chloride was heated in a china dish. The substance left in the china dish was:
    (a) (ii) and (iii)   (b) (i) and (ii)
    (c) (i) only   (d) (iii) only

17. *Fractional distillation can be used to separate alcohol from water. This method of separation uses the fact that alcohol and water:
    - (a) Contain different pigments
    - (b) Burn at different temperatures
    - (c) Contain different sized molecules
    - (d) Boil at different temperatures

18. *Select the set with the correct statement:
    - (i) Chromatography can be used for any type of salt.
    - (ii) It is based on the difference in solubility of substances.
    - (iii) Here colour bends are formed.
    - (iv) It is applicable only when a substance is in bulk.
    - (a) (i), (ii)
    - (b) (i), (ii), (iii)
    - (c) (i), (ii), (iv)
    - (d) (i), (ii), (iii), (iv)

19. *When a bar magnet is rolled over a mixture 'A' of iron and sulfur and over the compound 'B' iron sulfide, which of the following observations is incorrect ?
    - (a) Mixture 'A' is heterogeneous
    - (b) Compound 'B' is homogeneous
    - (c) Iron clings to the magnet from mixture 'A'
    - (d) Iron clings to the magnet from compound 'B'

20. *When a mixture of sand, sodium chloride and ammonium chloride is heated in a china dish, dense white fumes are evolved. On cooling these fumes on a glass plate, a white deposit is obtained. The white deposit may be:
    - (a) Sodium chloride and ammonium chloride
    - (b) Sand
    - (c) Sodium chloride
    - (d) Ammonium chloride

Assertion–Reason Questions

Direction: In the following question two statements (Assertion) A and Reason (R) are given Mark.
- (a) if A and R both are correct and R is the correct explanation of A;
- (b) if A and R both are correct but R is not the correct explanation of A;
- (c) A is true but R is false;
- (d) A is false but R is true

Assertion	Reason
1. Blue or black ink is a homogeneous mixture of blue-black dye in water.	1. Blue or black ink can be separated from water by centrifugation.
2. Kerosene oil and water are separated by using a separating funnel.	2. Kerosene oil and water are immiscible liquids.
3. Fractional distillation is used to separate methanol and acetone.	3. Methanol and acetone forms an azeotropic mixture.
4. In crystallisation, impure compounds are converted into crystals.	4. Crystallisation is better than evaporation.
5. In paper chromatography, different colours are formed on a chromatogram.	5. Natural pigment can be identified by using chromatography.

Section B: Very Short Answer Questions (2 Marks)

1. What is filtration?

2. How can we separate a mixture of two immiscible liquids?

3. Name the process used to separate a mixture of salt and ammonium chloride.

4. How can we separate a mixture of two miscible liquids?

5. Name the apparatus used to separate oil from water.

6. How can we obtain different gases from air?

7. Name the process which is used in milk dairies to separate cream from milk.

8. How can we obtain pure copper sulfate from an impure sample?

9. For the separation of what kind of solutes is the process of chromatography used?

Section C: Short Answer Questions
(3 Marks)

1. Use research techniques (including the Internet) to obtain as many examples as you can in which a centrifuge is used.

2. State the principle of each of the following methods of separation of mixtures:
   (i) Separation using separating funnel, (ii) Centrifugation method.

3. What is the difference between simple distillation and fractional distillation?

4. Describe how you would use chromatography to show whether blue ink contains a single pure dye or a mixture of dyes.

5. Devise a method for obtaining salt (sodium chloride) from sea water in the school laboratory.

6. Can we separate alcohol dissolved in water by using a separating funnel? If yes, then describe the procedure. If not, explain.

7. Name the separation techniques which you will apply for the separation of the following mixtures:
   (i) Small pieces of metal in the engine oil of a car. (ii) Fine mud particles suspended in water. (iii) Oil from water. (iv) Sodium chloride from its solution in water. (v) Camphor from salt. (vi) Wheat grains from husk.

8. You are provided with a mixture of ammonium chloride, common salt, sand and iron filings. Which of these substances can be separated by sublimation and which by sedimentation?

9. While diluting a solution of salt in water, a student by mistake added acetone (boiling point 56°C). What technique can be employed to get back the acetone? Justify your choice?

10. *(i) When a mixture of sand, sodium chloride and ammonium chloride is heated in a china dish, dense white fumes are evolved. On cooling these fumes on a glass plate, a white deposit is obtained. The white deposit may be of which substance?
    (ii) The process used to separate ammonium chloride from a mixture of sand and common salt is known as?

Section D: Long Answer Questions
(5 Marks)

1. What separation techniques will you apply for the separation of the following:
   (i) Sodium chloride from its solution in water
   (ii) The different pigments from an extract of flower petals
   (iii) Butter from curd
   (iv) Oil from water
   (v) Tea leaves from tea
   (vi) Iron pins from sand
   (vii) Wheat grains from husk
   (viii) Fine mud particles floating in water

2. Explain the following terms used for separation of mixtures:
   (i) Filtration, (ii) Crystallisation,
   (iii) Evaporation, (iv) Sublimation

3. (a) Explain what is meant by a pure substance.
   (b) (i) Many people know that the boiling point of water is 100°C. However, a student of Chemistry was surprised to find that water he heated boiled at 105°C. State two possibilities that could have caused an increase in the boiling point of water.
      (ii) Would you expect the water sample which boiled at 105°C to freeze below or above 0°C at the same pressure?
   (c) Why is the purity of substances important in the food industry?
   (d) Filtration is vital in everyday life. Give one example of the industrial application of filtration in everyday life.

4. Name the method which is most suitable for separating the following:
   (i) oxygen from liquid air
   (ii) red blood cells from plasma
   (iii) petrol and kerosene from crude oil
   (iv) coffee grains from coffee solution
   (v) pieces of steel from engine oil
   (vi) amino acids from fruit juice solution
   (vii) ethanol and water

5. You are given a mixture of mustard oil and water.
   (i) Name the process that can be used to obtain mustard oil from the above mixture.
   (ii) Draw a well-labelled diagram of the above process.

6. Draw a flowchart showing the separation of components of air. Also name this process.

7. (i) Describe an activity to separate the crystal of alum (phitkari) from its impure sample and name this techniques.
    (ii) Give to applications of alum in your day-to-day life.

8. *You are provided with a mixture of camphor, common salt and soil. Using various techniques, how will you separate the components of this mixture? Write the various steps involved.

9.

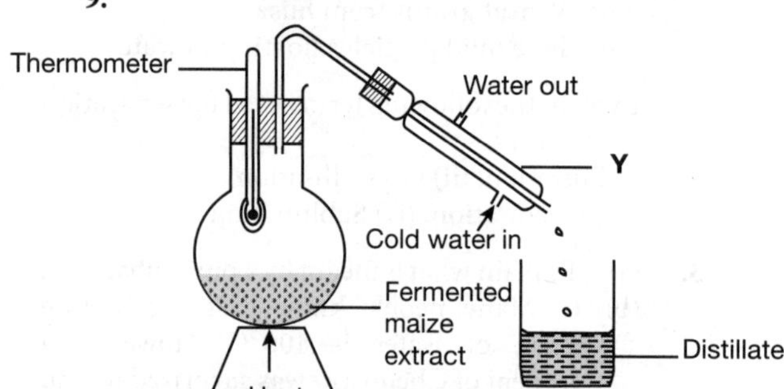

   *(i) Name the apparatus labelled Y.
   (ii) What improvement can be made to the above arrangement to obtain pure ethanol?
   (iii) State the purpose of cold water which passes through Y, and why it has to enter through the bottom.
   (iv) Name the separation technique being used above.
   (v) State the industrial application for the technique shown in the diagram above.

Section E: Case Study or Passage-Based Questions   (4 Marks)

1. Most natural substances are not chemically pure as they have various types of impurities. To remove these impurities or to separate the components of a heterogeneous mixture, we use many methods like sublimation, centrifugation, distillation and chromatography. These techniques are of great importance not only in chemical labs during practical work but also in our daily lives like in the dairy industry, supply of water, etc.

   (i) A mixture containing iron filings and sulfur powder is spread on white paper and a magnet is rolled in it. The particles which cling to the magnet are:
   (a) Iron sulfide
   (b) Iron particles
   (c) Sulfur
   (d) Mixture of iron and sulfur

   (ii) A mixture of sand and sulfur may best be separated by:
   (a) Fractional distillation
   (b) Fractional crystallisation from aqueous solution
   (c) Magnetic method
   (d) Dissolving in CS_2 and filtering

   (iii) A mixture of water and C_2H_5OH can be separated to a certain limit by:
   (a) Fractional distillation (b) Centrifugation
   (c) Sublimation   (d) Evaporation

   (iv) The correct sequence of steps taken for separating mixture of ammonium chloride, sand and common salt is:
   (a) Filtration, dissolving in water, sublimation and evaporation
   (b) Sublimation, dissolving in water, filtration and evaporation
   (c) Filtration, evaporation, sublimation and dissolving in water
   (d) Evaporation, dissolving in water, filtration and sublimation

   (v) Select the set having the correct statements:
   (i) Separating funnel is used to separate two immiscible liquids.
   (ii) Fractional distillation is used to separate gases of air.
   (iii) Evaporation is better than crystallisation.
   (iv) Fractional distillation is better than simple distillation.
   (a) (i), (ii)   (b) (i), (ii), (iii)
   (c) (i), (ii), (iv)   (d) (i), (ii), (iii), (iv)

2. Along with some physical methods like filtration and handpicking, many special techniques can be used to separate the components from a mixture. These techniques are evaporation, centrifugation, chromatography, etc. Evaporation is used to remove a coloured component from any liquid while filtration is used to remove solid particles. Centrifugation is used when filtration is not applicable.

   (i) Name the process by which coloured components can be obtained from blue ink?

   (ii) Give an application of crystallisation.

(iii) Can salt be separated by sedimentation? Justify your answer.

(iv) In the separation of gases from air, the last fraction contains which gas?

(v) Which gas is liquified first as the air is cooled and why?

(vi) Why is chlorine added to water during purification?

Value Based-Question

1. Aarushi's mother always squeezes water from wet clothes in the spinner of the washing machine and then uses them:
   (i) Write the principle of the technique used in the abovementioned process.
   (ii) Write one more application of this technique.
   (iii) What do you learn from Aarushi's mother?

2. Khushi's mother is a good cook. She uses natural colours such as turmeric for yellow colour, spinach for green colour and pomegranate for reddish colour. She always avoids synthetic food colours.
   Answer the following questions based on the above information:
   (i) Name the technique used to separate pigments from natural colours.
   (ii) Write the principle of that technique.
   (iii) Explain the values that are displayed by Khushi's mother.

High Order Thinking Skills (HOTS) Questions

1. The diagram below shows a chromatogram obtained during the analysis of some substances. A, B and C are standard substances whereas P, Q and R are unknown substances.
   (a) Which of substances P, Q and R
       (i) is a pure substance, (ii) is a mixture
   (b) Which of the standard substances can be used to make substance Q?
   (c) One spot of substance R is moved 20 cm from the baseline and the solvent front is 30 cm. Calculate the R_f value of that spot?

(d) Explain why the level of solvent is kept below the baseline at the start of the chromatographic process.

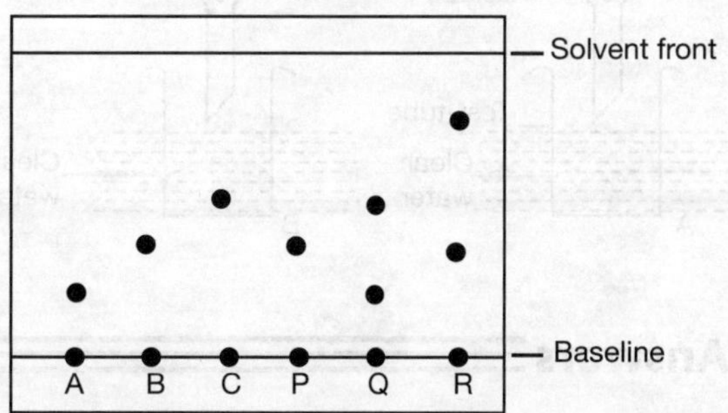

2. *Name the gas produced when a mixture of 7 g of iron filings and 4 g of sulfur powder is treated with dilute H_2SO_4 at room temperature. What gas will be produced if the same mixture is first heated strongly, cooled and then treated with dilute H_2SO_4? Write the cause of difference in behaviour.

3. *Observe the diagrams and answer the following questions:

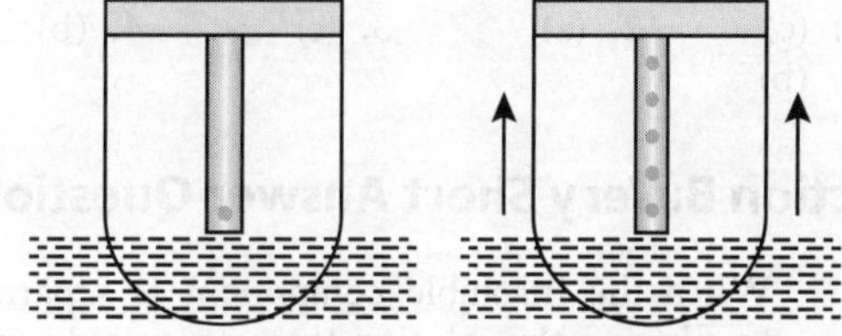

   (i) Identify the separation technique in the above diagram.
   (ii) What do you observe on the filter paper as the water rises?
   (iii) Give a reason for the rise of coloured spots on the paper strip.
   (iv) Give any two applications where you can use this technique.

4. *Four students A, B, C and D were given funnels, filter paper, test tubes, test tube stands, common salt, chalk powder, starch and glucose powder. They prepared a true solution, suspension and colloidal solution. Test tubes were arranged as shown in the figure. Observe the filtrate obtained in the test tubes and residue on filter paper. Conclude about filtrate, residue and type of solution.

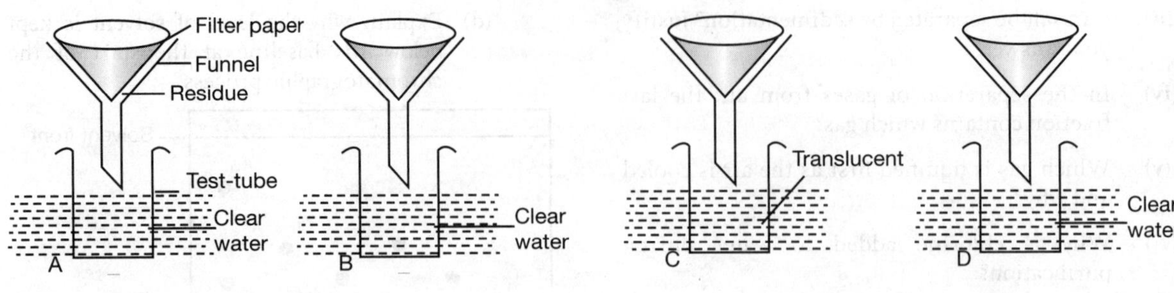

Answers

Section A: Multiple Choice Questions

1. (c)	**2.** (d)	**3.** (d)	**4.** (c)
5. (d)	**6.** (b)	**7.** (c)	**8.** (d)
9. (d)	**10.** (d)	**11.** (c)	**12.** (a)
13. (c)	**14.** (c)	**15.** (d)	**16.** (c)
17. (d)	**18.** (b)	**19.** (c)	**20.** (d)

Assertion–Reason Questions

1. (c)	**2.** (a)	**3.** (c)	**4.** (b)
5. (b)			

Section B: Very Short Answer Questions

1. When the insoluble component is separated by filtering the solution through a medium or membrane, it is known as filtration.

2. By using a separating funnel.

3. Sublimation is used to separate a mixture of salt and ammonium chloride.

4. By distillation.

5. Separating funnel is used to separate oil from water.

6. By fractional distillation.

7. Centrifugation is used in milk dairies to separate cream from milk.

8. By crystallisation.

9. Chromatography is used for the separation of those solutes that dissolve in the same solvent.

Section C: Short Answer Questions

1. Refer to Section 2.6.2, Separation of a Mixture of a Solid and a Liquid.

2. Refer to Section 2.6.2, Separation of a Mixture of a Solid and a Liquid.

3. Refer to Section 2.6.2, Separation of a Mixture of a Solid and a Liquid.

4. Refer to Section 2.6.2, Separation of a Mixture of a Solid and a Liquid.

5. Refer to Section 2.6.2, Separation of a Mixture of a Solid and a Liquid.

6. No, a mixture of water and alcohol cannot be separated since both are miscible and they form a solution.

7.   (i) Filtration,
    (ii) Sedimentation and Decantation,
   (iii) Separating funnel,
    (v) Evaporation,
    (v) Sublimation,
   (vi) Winnowing.

8. Ammonium chloride by sublimation. Sand and iron filings by sedimentation.

9. Acetone is soluble in water. So a homogeneous mixture is obtained and hence separation by separating funnel cannot be used as the boiling point of acetone (56°C) is much lower than that of water (100°C). So, acetone can be recovered by distillation. During distillation, acetone being volatile distils first, leaving water in the distillation flask.

10. (i) Ammonium chloride
    (ii) Sublimation

Section D: Long Answer Questions

1. Refer to Section 2.6, Separation of Mixtures.

2. Refer to Section 2.6.2, Separation of a Mixture of a Solid and a Liquid.

3. (a) Pure substance is one that has definite composition and properties which are constant throughout the sample.
   (b) (i) Presence of impurities in water, higher atmospheric pressure.
       (ii) Below 0°C (impurities lower melting point, raise boiling point)
   (c) To make food safe for consumption.
   (d) Filter tea from tea leaves, tap water is filtered at the distribution centre, air conditioners filter air.

4. Refer to Section 2.6.3, Separation of a Mixture of Two or More Liquids.

5. (i) Process: Using separating funnel.
   (ii) Refer to Section 2.6.3, Separation of a Mixture of Two or More Liquids.

6. Refer to Section 2.6.3, Separation of a Mixture of Two or More Liquids.

7. (i) Refer to Section 2.6.2, Separation of a Mixture of a Solid and a Liquid.
   (ii) Alum is used in purification of water and as an antiseptic (after shave).

8. (i) First, the mixture will be heated and as camphor is sublimable, it will vapourise. In this way it can be separated through sublimation.
   (ii) To separate mixture of common salt and soil, we will dissolve them in water. As salt is soluble in water, and soil is insoluble in water, soil can be separated through filtration.
   (iii) At the end, we get salt solution and salt can be separated from water by evaporation. Hence, sublimation, filtration and evaporation are used.

9. (i) Y is the condenser.
   (ii) By adding a fractionating column.
   (iv) Continuous supply of cold water provides a cooler environment in the condenser that changes vapour from the distillation flask into liquid. Water entering from the bottom aids to fill the entire volume of the condenser with cold water for effective cooling.
   (v) Simple distillation.
   (vi) Used for obtaining alcohol from fermented grains during alcohol production, obtaining distilled water from seawater for drinking and medical purposes.

Section E: Case Study or Passage-Based Questions

1. (i) (b) (ii) (d) (iii) (a) (iv) (b) (v) (c)
2. (i) By evaporation the coloured components can be obtained from blue ink.
   (ii) Purification of salt that we get from seawater.
   (iii) No, as common salt dissolves in water.
   (iv) Oxygen
   (v) Oxygen is liquified first as its boiling point is higher than that of the other components of the gas.
   (vi) It is added to kill bacteria.

Value-Based Questions

1. (i) Centrifugation principle: Denser particles are forced to the bottom and the lighter particles stay at the top when spun rapidly. Used in dairies to separate butter from cream. Multiple uses of the available resources to avoid wastage.

2. (i) Chromatography.
   (ii) Different colours get separated due to dissolution in the same solvent and different rates of rising due to different rates of absorption.
   (iii) Awareness, caring attitude for the family and a skillful person.

High Order Thinking Skills (HOTS) Questions

1. (a) (i) P, (ii) Q and R
   (b) A and C
   (c) $R_f = \dfrac{\text{Distance moved by spot}}{\text{Distance moved by solvent}}$

   $= \dfrac{20\ \text{cm}}{30\ \text{cm}} = \dfrac{2}{3}$

(d) So that the sample does not dissolve in the solvent and affect the results.

2. Here, hydrogen and hydrogen sulfide gases are produced, respectively.

    In case I: Mixture of iron filings and sulfur on reaction with dilute sulfuric acid gives hydrogen (H_2) gas.

    In case II: On heating, iron sulfide (FeS) is formed which on reaction with dilute sulfuric acid gives H_2S (hydrogen sulfide) gas.

3. (i) Chromatography.

    (ii) Different colours will appear at different heights.

(iii) Different colours have different rates of absorption.

(iv) Applications: To separate colours in a dye; to separate drugs from blood.

4. Test tube A has residue on the filter paper and the filtrate is a clear solution. It is a suspension. Test tube B has no residue on the filter paper and the filtrate is a clear solution. It is a true solution. Test tube C has no residue on the filter paper and the filtrate is translucent. It is a colloidal solution. Test tube D has no residue on the filter paper and the filtrate is a clear solution. It is a true solution.

Hints

Section A: Multiple Choice Questions

4. Filtration is based on size difference.

6. Camphor being a sublime substance can be purified by sublimation.

8. Distillation is a separation technique used for separation of miscible liquids having different boiling points.

9. Magnetism is useful for separation of magnetic and non-magnetic substances.

12. The mixture of oil and water can be separated by using a separating funnel as oil being less denser than water forms the upper layer.

13. Due to difference in boiling points and having different tendency to form vapours.

Section E: Case Study or Passage-Based Questions

1. (ii) Sulfur is dissolved in CS_2 and separates out as a filtrate. Sand gets collected on the filter paper as residue.

    (iii) Because both differ in boiling point.

Activities with Discussion and Conclusion

Activity 2.1 (To illustrate the concept of homogeneous and heterogeneous mixtures)
- Let us divide the class into groups A, B, C and D.
- Group A takes a beaker containing 50 mL of water and one spatula full of copper sulfate powder. Group B takes 50 mL of water and two spatula full of copper sulfate powder in a beaker.
- Groups C and D can take different amounts of copper sulfate and potassium permanganate or common salt (sodium chloride) and mix the given components to form a mixture.
- Report the observations on the uniformity in colour and texture.

Discussion:
(i) Both the groups A and B have obtained homogeneous mixtures as the composition of these mixtures or solutions is uniform throughout.
(ii) Although both the groups have obtained copper sulfate solutions, the intensity of colour of the two solutions is different. The intensity of blue colour in the solution obtained by group B which has two spatula full of copper sulfate is much higher than the solution obtained by group A which has one spatula full of copper sulfate.
(iii) Both the groups C and D have obtained heterogeneous mixtures as they not only have physically distinct boundaries but also their composition is not uniform. Mixtures of sodium chloride and iron filings, salt and sulfur, and oil and water are examples of heterogeneous mixtures.

Conclusion:
(i) Soluble substances such as copper sulfate, common salt and sugar when dissolved in water form homogeneous mixtures whose composition depend upon the amount of the substance dissolved.
(ii) When two or more solids which do not react chemically are mixed, they always form heterogeneous mixtures.

Activity 2.2 (To illustrate the difference between true solution, suspensions and colloidal solutions)
- Let us again divide the class into four groups – A, B, C and D.
- Distribute the following samples to each group:
  (i) Few crystals of copper sulfate to group A.
  (ii) One spatula full of copper sulfate to group B.
  (iii) Chalk powder or wheat flour to group C.
  (iv) Few drops of milk or ink to group D.

- Each group should add the given sample to water and stir properly using a glass rod. Are the particles in the mixture visible?
- Direct a beam of light from a torch through the beaker containing the mixture and observe from the front. Was the path of the beam of light visible?
- Leave the mixtures undisturbed for a few minutes (and set up the filtration apparatus in the meantime). Is the mixture stable or do the particles begin to settle after some time?
- Filter the mixture. Is there any residue on the filter paper?

Discussion: Group A and B
(i) Groups A and B have obtained true solutions. However, the intensity of the blue colour of the solution obtained by group A which has only a few crystals of copper sulfate is lower than that of the solution obtained by group B which has one spatula full of copper sulfate. Hence, the intensity of colour depends upon the amount of substance dissolved.
(ii) In both these solutions, particles are not visible to the naked eye.
(iii) When the beam of light is passed through the solution obtained by either group A or group B, the path of the beam of light was not visible. Hence, the particles of a true solution do not scatter light.
(iv) When the above solution is allowed to stand, the solution is stable and the particles of the solution do not settle down. Further, when the above solution is allowed to filter, the whole solution passes through the pores of the filter paper without leaving any residue on the filter paper.

Group C
(i) The group C has obtained a suspension.
(ii) As the particles of a suspension scatter light, they are visible to the naked eye.
(iii) When the above suspension is allowed to stand, it is not stable and the particles of the suspension settle down.
(iv) When the above suspension is filtered, the particles of chalk or wheat flour being bigger than the pores of the filter paper remain as residue on the filter paper.

Group D
(i) The group D has obtained a colloidal solution.
(ii) When the above solution is allowed to stand, the particles of the colloidal solution like those of the true solution do not settle down.

(iii) As the particles of a colloidal solution are nearly 100 times bigger than those of the true solution, when a beam of light is passed through a colloidal solution, the particles of the colloidal solution scatter light and hence the path of the beam becomes visible (Tyndall effect).

(iv) When the colloidal solution is passed through a filter paper, the particles of the colloidal solution like those of the true solution pass through the filter paper without leaving any residue on the filter paper.

Conclusion:

(i) The particles of a true solution are not visible even under a microscope and they do not settle down on standing and passing through the pores of the filter paper without leaving any residue.

(ii) The particles of a suspension are visible to the naked eye and they settle down on standing and also leave a residue on the filter paper.

(iii) As the particles of a colloidal solution scatter light, the path of light beam becomes visible when passed through it. A colloidal solution is stable. The particles of the colloidal solution do not immediately settle down on standing. They also pass through the pores of a filter paper.

Activity 2.3 (To demonstrate that different substances have different solubilities in the same solvent)

- Take approximately 50 mL of water each in two separate beakers.
- Add salt in one beaker and sugar or barium chloride in the second beaker with continuous stirring.
- When no more solute can be dissolved, heat the contents of the beaker.
- Start adding the solute again.
- Is the amount of salt and sugar or barium chloride, that can be dissolved in water at a given temperature the same?
- What would happen if you were to take a saturated solution at a certain temperature and cool it slowly?

Discussion:

(i) The amounts of common salt, sugar and barium chloride that can be dissolved in water (50 mL) at room temperature are different.

(ii) When a saturated solution at a certain temperature is cooled, the solubility decreases and the amount of the solute which exceeds the solubility at the lower temperature crystallises out of the solution.

Conclusion: Different substances have different solubilities in a given solvent at the same temperature and in general the solubility decreases as the solution is cooled and the extra amount of solute crystallises out.

Activity 2.4 (To separate the components of a mixture by evaporation)

- Fill half a beaker with water.
- Put a watch glass on the mouth of the beaker.
- Put a few drops of ink on the watch glass.

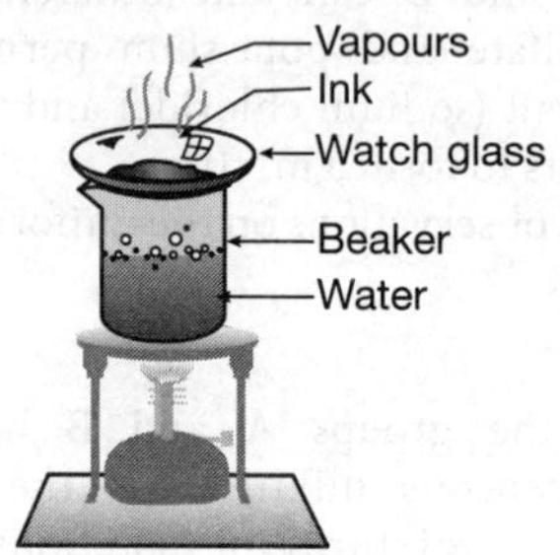

- Now start heating the beaker. We do not want to heat the ink directly. You will see that evaporation is taking place from the watch glass.
- Continue heating as evaporation proceeds and stop heating when you do not see any further change on the watch glass.
- Observe carefully and record your observations.

Now Answer

- What do you think has evaporated from the watch glass?
- Is there a residue on the watch glass?
- What is your interpretation? Is ink a single substance (pure) or is it a mixture?

Discussion: Ink (blue or black) is a mixture of dyes in water. When the ink is heated, the volatile component of the mixture, that is, water, gets evaporated while the non-volatile components of the mixture, the dyes, remain as the residue.

Conclusion: The method of evaporation can be used to separate the volatile component (solvent) of the mixture from its non-volatile solute.

Precaution: On direct heating, ink may undergo decomposition; therefore, it is heated by steam on a watch glass.

Activity 2.5 (To separate cream from milk by centrifugation)

- Take some full-cream milk in a test tube.
- Centrifuge it by using a centrifuging machine for two minutes. If a centrifuging machine is not

available in the school, you can do this activity at home by using a milk churner in the kitchen.

- If you have a milk dairy nearby, visit it and ask (i) how they separate cream from milk and (ii) how they make cheese from milk.

Now Answer

- What do you observe on churning the milk?
- Explain how separation of cream from milk takes place.

Discussion: When the milk is churned, the lighter fat particles collide with one another to form cream. The cream thus separated starts floating on the surface of milk and is continuously collected from the outlet provided near the top of the centrifuging machine.

Conclusion: Liquid mixtures which have solute particles which can easily pass through a filter paper (colloids) can be separated by using a centrifuging machine.

Activity 2.6 (To illustrate separation of two immiscible liquids using a separating funnel)

- Let us try to separate kerosene oil from water using a separating funnel.

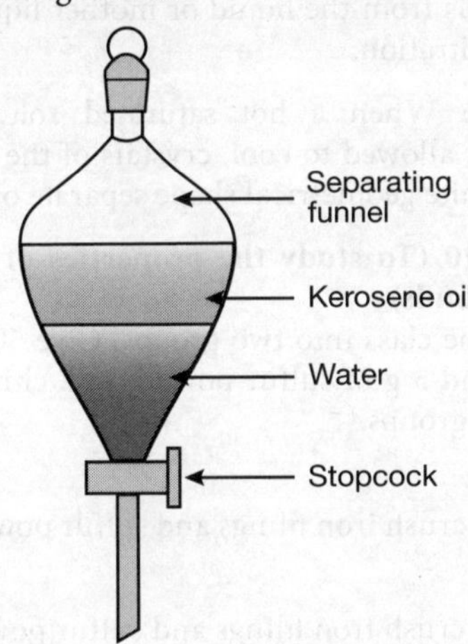

- Pour the mixture of kerosene oil and water in a separating funnel.
- Let it stand undisturbed for some time so that separate layers of oil and water are formed.
- Open the stopcock of the separating funnel and pour out the lower layer of water carefully.
- Close the stopcock of the separating funnel as the oil reaches the stopcock.

Discussion:

(i) Kerosene oil and water being insoluble form two separate layers. Kerosene oil being lighter than water forms the upper layer while water being heavier forms the lower layer.

(ii) Please note that as water and alcohol form a homogeneous mixture, they cannot be separated by using a separating funnel.

Conclusion: As immiscible liquids separate out in different layers depending upon their densities, they can be separated using a separating funnel.

Activity 2.7 (To separate the different coloured compounds of ink using chromatography)

- Take a thin strip of filter paper.
- Draw a line on it using a pencil, approximately 3 cm above the lower edge.

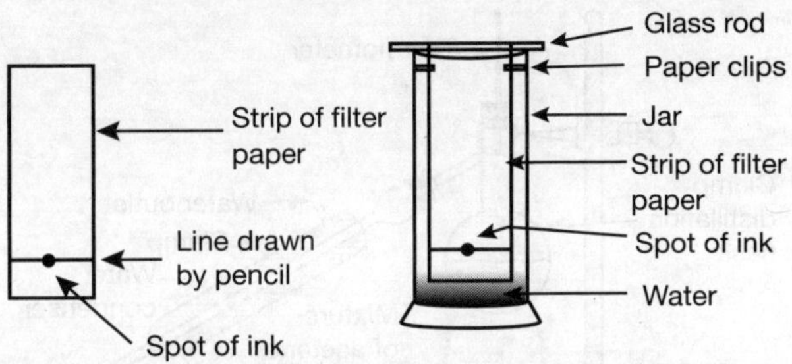

- Put a small drop of ink (water soluble, that is, from a sketch pen or fountain pen) at the centre of the line. Let it dry.
- Lower the filter paper into a jar/glass/beaker/test tube containing water so that the drop of ink on the paper is just above the water level, and leave it undisturbed.
- Watch carefully as the water rises up the filter paper. Record your observations.

Now Answer

- What do you observe on the filter paper as the water rises on it?
- Do you obtain different colours on the filter paper strip?
- What according to you, could be the reason for the rise of the coloured spot on the paper strip?

Discussion:

(i) As water rises on the filter paper, it takes along with it different components of the ink.

(ii) As a dye usually consists of two or more coloured substances, different colours are obtained on the filter paper strip as colour bends.

(iii) Further, due to different solubilities (in water) of coloured components present in the ink, they rise to different heights.

Conclusion: Separation of different coloured components present in ink occurs on a chromatographic paper due to their different solubilities in water.

Activity 2.8 (To separate a mixture of two miscible liquids using distillation)
- Let us try to separate acetone and water from their mixture.
- Take the mixture in a distillation flask. Fit it with a thermometer.
- Arrange the apparatus as shown.
- Heat the mixture slowly, keeping a close watch on the thermometer.
- The acetone vaporises, condenses in the condenser and can be collected from the condenser outlet.
- Water is left behind in the distillation flask.

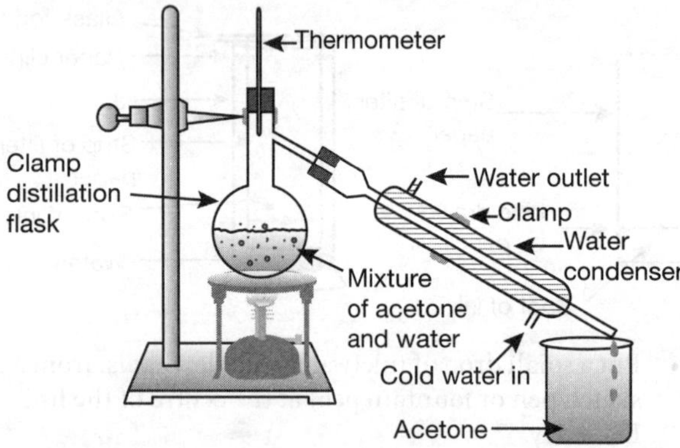

Now Answer
- What do you observe as you start heating the mixture?
- At what temperature does the thermometer reading become constant for some time?
- What is the boiling point of acetone?
- Why do the two components separate?

Discussion:
(i) When the mixture is heated, the vapours of the low boiling liquid which is acetone are first formed. These vapours move upwards and on passing through the condenser, condense to form liquid acetone that can be collected in the beaker.

(ii) The boiling point of acetone is 56°C or 329 K; hence, the thermometer reading becomes constant at 329 K when acetone is distilling.

(iii) Water is left behind in the flask.

Conclusion: Separation of the components of a mixture having two miscible liquids which boil without decomposition and have sufficient difference of 30–50 K in their boiling points can be carried out by simple distillation. This is because at the boiling point of each liquid, the vapours almost entirely consist of that liquid.

Activity 2.9 (To obtain pure copper sulfate from an impure sample using crystallisation)

- Take some (approximately 5 g) impure sample of copper sulfate in a china dish.
- Dissolve it in minimum amount of water.
- Filter the impurities out.
- Evaporate water from the copper sulfate solution so as to get a saturated solution.
- Cover the solution with a filter paper and leave it undisturbed at room temperature to cool slowly for a day.
- You will obtain crystals of copper sulfate in the china dish.
- This process is called crystallisation.

Now Answer
- What do you observe in the china dish?
- Do the crystals look alike?
- How will you separate the crystals from the liquid in the china dish?

Discussion: When a hot saturated solution of copper sulfate is allowed to cool, crystals of copper sulfate separate out. As all these crystals have a definite geometrical shape, they look alike. We can separate these crystals from the liquid or mother liquor by the process of filtration.

Conclusion: When a hot saturated solution of a substance is allowed to cool, crystals of the substance having definite geometrical shape separate out.

Activity 2.10 (To study the properties of mixtures and compounds)
- Divide the class into two groups. Give 50 g of iron filings and 3 g of sulfur powder in a china dish to both the groups.

Group I
- Mix and crush iron filings and sulfur powder.

Group II
- Mix and crush iron filings and sulfur powder. Heat this mixture strongly till red hot. Remove from the flame and let the mixture cool.

Groups I and II
- Check for magnetism in the material obtained. Bring a magnet near the material and check if the material is attracted to it.
- Compare the texture and colour of the material obtained by the groups.
- Add carbon disulphide to one part of the material obtained. Stir well and filter.
- Add dilute sulphuric acid or dilute hydrochloric acid to the other part of the material obtained.*(Note: teacher supervision is necessary for this activity).*

- Perform all the above steps with both the elements (iron and sulphur) separately.

Now Answer

- Did the material obtained by the two groups look the same?
- Which group obtained a material with magnetic properties?
- Can we separate the components of the material obtained?
- On adding dilute sulfuric acid or dilute hydrochloric acid, did both the groups obtain a gas? Did the gas in both the cases smell the same or different?

Discussion:

(i) The material obtained by students of group I is a heterogeneous mixture, that is, when we examine the mixture under a microscope, it is found that though the iron particles lie very close to the sulfur particles, at some places, there are more of iron particles whereas at others, there are more sulfur particles.

On the other hand, the material obtained by students of group II is a compound which has altogether different properties as compared to those of its constituents, that is, iron filings and sulfur.

$$\underset{\text{(Element)}}{\text{Iron}} + \underset{\text{(Element)}}{\text{Sulfur}} \xrightarrow[\text{Heat}]{\text{Strongly}} \underset{\text{(Compound)}}{\text{Iron sulfide}}$$

If you examine iron sulfide under a microscope, you will find that the composition, texture and colour of the compound (black) is the same throughout.

(ii) Material obtained by group I contains iron filings which has magnetic properties.

Alternatively, material obtained by group II is a compound which does not contain iron filings and hence it does not have magnetic properties.

(iii) As the material obtained by students of group I is a mixture, its components can be separated by simple physical methods like magnet or carbon disulfide.

If you put a magnet in the mixture of iron filings and sulfur powder, the iron particles are attracted by the magnet, they stick to the magnet and get separated from sulfur.

Alternatively, if you shake the mixture of iron filings and sulfur powder with carbon disulfide (CS_2), sulfur dissolves in it but iron filings do not. On filtration, iron filings are obtained as a residue on the filter paper while sulfur is recovered from the filtrate by evaporating carbon disulfide.

(iv) Material obtained by students of both group I and group II on treatment with dilute sulfuric acid or dilute hydrochloric acid evolve a gas but the gas evolved is different in each case. When dilute sulfuric acid is added to the material, that is, the mixture of iron filings and sulfur powder obtained by students of group I, the iron part reacts with dilute sulfuric acid to produce hydrogen gas (H_2) which is a colourless, odourless and combustible gas. When dilute sulfuric acid is added to the material, that is, iron sulfide obtained by students of group II, hydrogen sulfide (H_2S) is obtained. H_2S is a colourless gas having the smell of rotten eggs.

Conclusion: A mixture shows the properties of its constituents, but the properties of a compound are altogether different from those of its constituents.

Exemplar Problems

Short Answer Questions

1. Suggest separation technique(s) one would need to employ to separate the following mixtures:
   (a) Mercury and water
   (b) Potassium chloride and ammonium chloride
   (c) Common salt, water and sand
   (d) Kerosene oil, water and salt

2. Which of the tubes in figures (a) and (b) will be more effective as a condenser in the distillation apparatus?

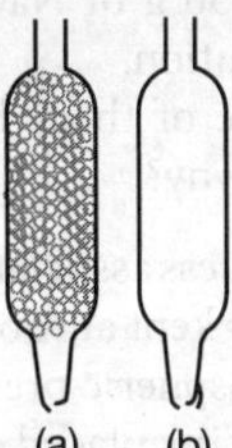

3. Salt can be recovered from its solution by evaporation. Suggest some other technique for the same.

4. Sea water can be classified as a homogeneous as well as heterogeneous mixture. Comment.

5. While diluting a solution of salt in water, a student by mistake added acetone (boiling point 56°C). What technique can be employed to get back the acetone? Justify your choice.

6. What would you observe when:
   (a) A saturated solution of potassium chloride prepared at 60°C is allowed to cool to room temperature
   (b) An aqueous sugar solution is heated to dryness
   (c) A mixture of iron filings and sulfur powder is heated strongly?

7. Explain why particles of a colloidal solution do not settle down when left undisturbed, while in the case of a suspension they do.

8. Smog and fog are what type of sols and how do they differ?

9. Classify the following as physical or chemical properties:
   (a) The composition of a sample of steel is 98% iron, 1.5% carbon and 0.5% other elements.
   (b) Zinc dissolves in hydrochloric acid with the evolution of hydrogen gas.
   (c) Metallic sodium is soft enough to be cut with a knife.
   (d) Most metal oxides form alkalis on interacting with water.

10. The teacher instructed three students 'A', 'B' and 'C' to prepare a 50% (mass by volume) solution of sodium hydroxide (NaOH). 'A' dissolved 50 g of NaOH in 100 mL of water, 'B' dissolved 50 g of NaOH in 100 g of water while 'C' dissolved 50 g of NaOH in water to make 100 mL of solution.

    Which one of them has made the desired solution and why?

11. Name the process associated with the following:
    (a) Dry ice is kept at room temperature and at one atmospheric pressure.
    (b) A drop of ink placed on the surface of water contained in a glass spreads throughout the water.
    (c) A potassium permanganate crystal is in a beaker and water is poured into the beaker with stirring.

   (d) An acetone bottle is left open and the bottle becomes empty.
   (e) Milk is churned to separate cream from it.
   (f) Settling of sand when a mixture of sand and water is left undisturbed for some time.
   (g) Fine beam of light entering through a small hole in a dark room, illuminates the particles in its paths.

12. You are given two samples of water labelled as 'A' and 'B'. Sample 'A' boils at 100°C and sample 'B' boils at 102°C. Which sample of water will not freeze at 0°C? Comment.

13. What are the favourable qualities given to gold when it is alloyed with copper or silver for the purpose of making ornaments?

14. An element is sonorous and highly ductile. Under which category would you classify this element? What other characteristics do you expect the element to possess?

15. Give an example each for the mixture having the following characteristics. Suggest a suitable method to separate the components of these mixtures.
    (a) A volatile and a non-volatile component.
    (b) Two volatile components with appreciable difference in boiling points.
    (c) Two immiscible liquids.
    (d) One of the components changes directly from solid to gaseous state.
    (e) Two or more coloured constituents soluble in some solvent.

16. Fill in the blanks.
    (a) A colloid is a _________ mixture and its components can be separated by the technique known as _________.
    (b) Ice, water and water vapour look different and display different _________ properties but they are _________ the same.
    (c) A mixture of chloroform and water taken in a separating funnel is mixed and left undisturbed for some time. The upper layer in the separating funnel will be of _________ and the lower layer will be that of _________.
    (d) A mixture of two or more miscible liquids, for which the difference in the boiling points is less than 25 K can be separated by the process called _________.

(e) When light is passed through water containing a few drops of milk, it shows a bluish tinge. This is due to the ________ of light by milk and the phenomenon is called ________. This indicates that milk is a ________ solution.

17. Sucrose (sugar) crystals obtained from sugarcane and beetroot are mixed together. Will it be a pure substance or a mixture? Give reasons for the same.

18. Give some examples of Tyndall effect observed in your surroundings?

19. Can we separate alcohol dissolved in water by using a separating funnel? If yes, then describe the procedure. If not, explain.

20. On heating, calcium carbonate is converted into calcium oxide and carbon dioxide.
    (a) Is this a physical or a chemical change?
    (b) Can you prepare one acidic and one basic solution by using the products formed in the above process? If so, write the chemical equation involved.

21. Non-metals are usually poor conductors of heat and electricity. They are non-lustrous, non-sonorous, non-malleable and are coloured.
    (a) Name a lustrous non-metal.
    (b) Name a non-metal which exists as a liquid at room temperature.
    (c) The allotropic form of a non-metal is a good conductor of electricity. Name the allotrope.
    (d) Name a non-metal which is known to form the largest number of compounds.
    (e) Name a non-metal other than carbon which shows allotropy.
    (f) Name a non-metal which is required for combustion.

22. Classify the substances given in the figure into elements and compounds.

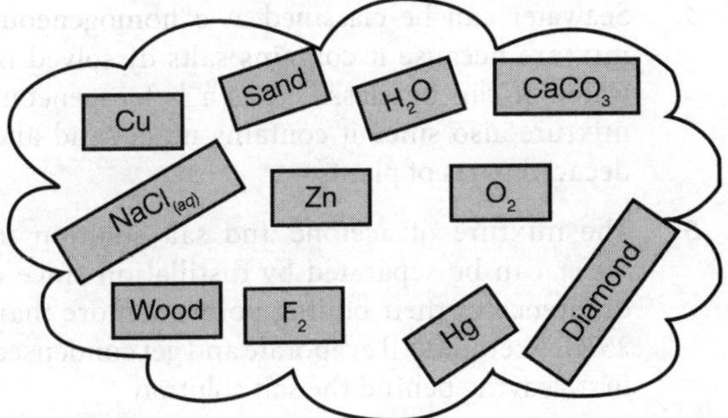

23. Which of the following are not compounds?
    (a) Chlorine gas
    (b) Potassium chloride
    (c) Iron
    (d) Iron sulphide
    (e) Aluminium
    (f) Iodine
    (g) Carbon
    (h) Carbon monoxide
    (i) Sulphur powder

Long Answer Questions

1. Fractional distillation is suitable for separation of miscible liquids with a boiling point difference of about 25 K or less. What part of the fractional distillation apparatus makes it efficient and possess an advantage over a simple distillation process? Explain using a diagram.

2. (a) Under which category of mixtures will you classify alloys and why?
   (b) A solution is always a liquid. Comment.
   (c) Can a solution be heterogeneous?

3. Iron filings and sulfur were mixed together and divided into two parts, 'A' and 'B'. Part 'A' was heated strongly while Part 'S' was not heated. Dilute hydrochloric acid was added to both the parts and evolution of gas was seen in both cases. How will you identify the gases evolved?

4. A child wanted to separate the mixture of dyes constituting a sample of ink. He marked a line by the ink on the filter paper and placed the filter paper in a glass containing water, as shown in the figure. The filter paper was removed when the water moved near the top of the filter paper.

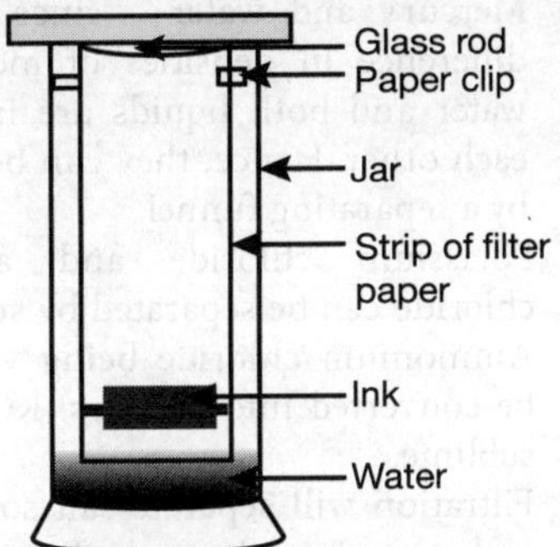

(i) What would you expect to see, if the ink contains three different coloured components?
(ii) Name the technique used by the child.

(iii) Suggest one more application of this technique.

5. A group of students took an old shoe box and covered it with black paper on all sides. They fixed a source of light (a torch) at one end of the box by making a hole in it and made another hole on the other side to view the light. They placed a milk sample contained in a beaker/tumbler in the box, as shown in the figure. They were amazed to see that milk taken in the tumbler was illuminated. They tried the same activity by taking a salt solution but found that light simply passed through it?

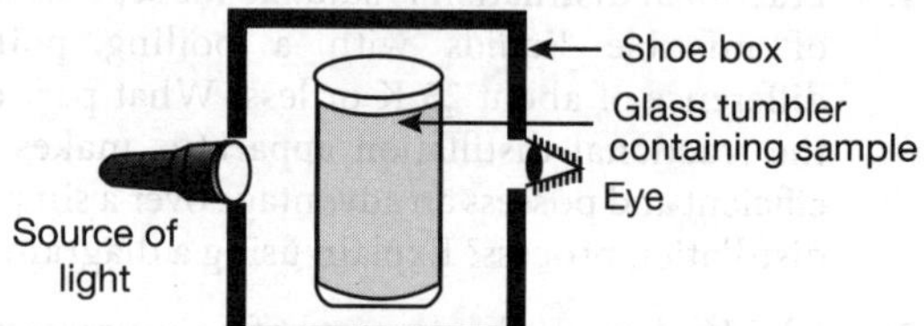

(a) Explain why the milk sample was illuminated. Name the phenomenon involved.
(b) The same results were not observed with a salt solution. Explain.
(c) Can you suggest two more solutions which would show the same effect as shown by the milk solution?

6. Classify each of the following as a physical or a chemical change. Give reasons.
(a) Drying of a shirt in the sun.

(b) Rising of hot air over a radiator.
(c) Burning of kerosene in a lantern.
(d) Change in the colour of black tea on adding lemon juice to it.
(e) Churning of milk cream to get butter.

7. During an experiment, the students were asked to prepare a 10% (mass/mass) solution of sugar in water. Ramesh dissolved 10 g of sugar in 100 g of water while Sarika prepared it by dissolving 10 g of sugar in water to make 100 g of the solution.
(a) Are the two solutions of the same concentration?
(b) Compare the mass % of the two solutions.

8. You are provided with a mixture containing sand, iron filings, ammonium chloride and sodium chloride. Describe the procedures you would use to separate these constituents from the mixture?

9. Arun has prepared 0.01% (by mass) solution of sodium chloride in water. Which of the following correctly represents the composition of the solutions?
(a) 1.00 g of NaCl + 100 g of water
(b) 0.11 g of NaCl + 100 g of water
(c) 0.01 g of NaCl + 99.99 g of water
(d) 0.10 g of NaCl + 99.90 g of water

10. Calculate the mass of sodium sulfate required to prepare its 20% (mass percent) solution in 100 g of water?

Answers

Short Answer Questions

1. (a) Mercury and water – since there is a difference in densities of mercury and water and both liquids are insoluble in each other. Hence, they can be separated by a separating funnel.
   (b) Potassium chloride and ammonium chloride can be separated by sublimation. Ammonium chloride being volatile will be converted into vapours. KCl does not sublime.
   (c) Filtration will separate salt solution (salt and water) and sand. Evaporation of salt solution will separate salt from the solution.
   (d) Separating funnel will separate kerosene oil and salt solution. Evaporation of salt solution will separate salt and water.

2. The tube (a) containing beads will be more effective as a condenser because the surface area of the tube increases.

3. Salt solution can be concentrated by heating to make a supersaturated solution. Crystallisation will occur when the solution is left for cooling and salt will separate out from the solution.

4. Seawater can be classified as a homogeneous mixture because it contains salts dissolved in water. It can be classified as a heterogeneous mixture also since it contains mud, sand and decayed parts of plants.

5. The mixture of acetone and salt solution in water can be separated by distillation since a difference in their boiling points is more than 25°C. Acetone will evaporate and get condensed first, leaving behind the salt solution.

6. (a) Crystals of potassium chloride will separate out.
   (b) On heating sugar solution, water will evaporate first. Once the solution dries, it would turn black and sugar will get charred.
   (c) Iron sulfide is formed when a mixture of iron filings and sulfur is heated strongly.

7. The size of colloidal particles in a colloidal solution is smaller than in a suspension. These particles are in a random motion and hence do not settle down when left undisturbed. The particles of suspension are bigger and they tend to settle down under the effect of gravity.

8. Both smog and fog have air as the dispersion medium, so they are aerosols. They differ in the disperse phases. Smog is made up of fine particles of carbon while fog contains particles of water.

9. (a) Physical property
   (b) Chemical property
   (c) Physical property
   (d) Chemical property

10. By definition, 50% mass by volume percent solution means 50 grams of a solute dissolved in 100 mL of solution. Hence, student C made the desired solution as he first dissolved 50 g of NaOH in small volume water and then he added more water to make the volume of solution 100 mL. Student A dissolved 50 g of NaOH in 100 mL of water. So the solution is diluted and it is not the desired solution. By definition, 50% mass by mass percent solution means 50 grams of a solute dissolved in 100 grams of solution. Student B dissolved 50 g of NaOH in 150 g of solution so, it is not the desired solution.

$$\text{Mass by volume \%} = \frac{\text{Mass of solute}}{\text{Volume of solution}} \times 100$$

$$= \frac{50}{100} \times 100 = 50\% \text{ mass by volume}$$

11. (a) Sublimation of dry ice (solid) to CO_2 (gas)
    (b) Diffusion of ink into water
    (c) Diffusion or dissolution of solid into liquid
    (d) Evaporation, diffusion of acetone in air
    (e) Centrifugation
    (f) Sedimentation
    (g) Tyndall effect – Scattering of light

12. Sample 'B' which boils at 102°C contains impurities. It will not freeze at 0°C. There will be a depression in freezing point.

13. Pure gold is highly malleable and soft. When it is alloyed with copper or silver, it becomes hard and strong and can be moulded into various shapes.

14. It is a metal. The element is expected to be lustrous, malleable and a good conductor of heat and electricity.

15. (a) Evaporation or distillation – Common salt solution
    (b) Distillation – Acetone–water mixture
    (c) Separation using separating funnel – Oil–water mixture
    (d) Sublimation – Mixture of common salt and ammonium chloride
    (e) Chromatography – Ink

16. (a) Heterogeneous, centrifugation
    (b) Physical, chemically
    (c) Water, chloroform
    (d) Fractional distillation
    (e) Scattering, Tyndall effect, colloidal

17. The composition of the sugar (sucrose) will remain constant irrespective of the source of its preparation. Hence, sugar or sucrose is a pure substance with fixed composition.

18. Examples of Tyndall effect:
    I. When light rays enter a dark room through a hole or a small window.
    II. Sunlight passing through a group of trees in the forest.
    III. Path of light rays seen in front of the projector in a cinema hall.

19. No, a mixture of water and alcohol cannot be separated since both are miscible and they form a solution. Only immiscible liquids can.

20. $$CaCO_3 \xrightarrow{\Delta} CaO + CO_2$$
    (a) It is a chemical change in which new substances are formed.
    (b) Calcium oxide when dissolved in water forms a basic solution.
    $$CaO + H_2O \rightarrow Ca(OH)_2$$
    Carbon dioxide when dissolved in water forms an acidic solution
    $$CO_2 + H_2O \rightarrow H_2CO_3$$

21. (a) Iodine   (b) Bromine
     (c) Graphite   (d) Carbon
     (e) Sulfur, phosphorus (f) Oxygen

22. Elements – Cu, Zn, F_2, O_2, diamond (C), Hg
     Compounds – $CaCO_3$, H_2O

23. Chlorine gas, iron, aluminium, iodine, carbon
     and sulfur powder are not compounds.

Long Answer Questions

1. In fractional distillation, a fractionating column
    is used which is packed with glass beads or
    small plates. It increases the surface area for the
    vapours and they quickly lose energy when they
    come in contact with beads or plates and can
    be quickly condensed. The length of the column
    would increase the efficiency of the process.

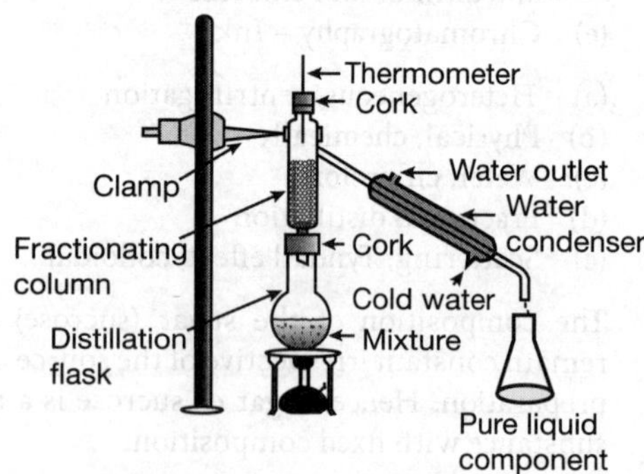

2. (a) Alloys are homogeneous mixtures
         because they have uniform composition
         throughout.
    (b) No, a solution can be solid (alloys) or
         gaseous (air) also.
    (c) No, a solution is a homogeneous mixture.

3. Part A – Iron sulfide is formed which gives out
    hydrogen sulfide gas with HCl.

    $$Fe + S \xrightarrow{\Delta} FeS$$
    $$FeS + 2HCl \rightarrow FeCl_2 + H_2S$$

    H_2S gas can be identified by its smell. It has
    a foul smell and it turns lead acetate solution
    black.
    Part B – Fe and S will not react, when HCl is
    added to this mixture, only Fe will react with
    HCl to give out H_2 gas.

    $$Fe + 2HCl \rightarrow FeCl_2 + H_2$$

    Hydrogen gas burns with a pop sound and
    hence can be identified by bringing a burning
    matchstick near it.

4. (i) The components of the ink will travel
         with water and we would see three bands
         on the filter paper at various lengths.

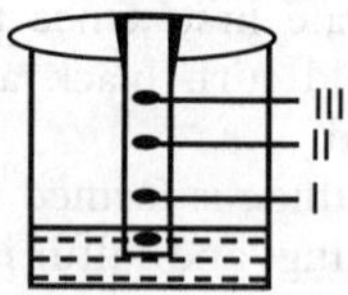

    (ii) The technique is called chromatography.
    (iii) Separation of pigments present in
          chlorophyll.

5. (a) Milk is a colloidal solution and hence
         shows Tyndall effect.
    (b) True solutions do not show Tyndall effect
         because they do not scatter light.
    (c) Detergent solution, sulfur solution.

6. (a, b, e): Physical changes because there is
    no change in chemical composition, (c), (d):
    Chemical changes because new substances are
    formed.

7. (a) No, Sarika has a higher mass percentage.
    (b) Solution make by Ramesh,

    $$\text{Mass \%} = \frac{10}{(10 + 100)} \times 100$$
    $$= \frac{10}{110} \times 100 = 9.09\%$$

    Solution made by Sarika,
    $$\text{Mass\%} = \frac{10}{100} \times 100 = 10\%$$

8. Main Mixture – Sand + Fe + NH_4Cl + NaCl
    Step I – Separate iron filings using magnet.
    Step II – Separate NH_4Cl by sublimation.
    Step III – Add, water, stir and filter to separate
    sand.
    Step IV – Evaporate filtrate to get salt (NaCl).

9. Mass%
    $$= \frac{\text{Mass of solute}}{\text{Mass of solute + Mass of solvent}} \times 100$$

    (a) Mass percentage $= \dfrac{1.00}{1.0 + 100} \times 100 = 0.99$

    (b) Mass percentage
         $$= \frac{0.11}{0.11 + 100} \times 100 = 0.1098$$

    (c) Mass percentage
         $$= \frac{0.01}{0.01 + 99.99} \times 100 = 0.01$$

(d) Mass percentage

$$= \frac{0.10}{0.10 + 99.90} \times 100 = 0.1$$

10. Let the mass of sodium sulfate required be $= x$ g

The mass of solution would be $= (x + 100)$ g x g of solute in $(x + 100)$ g of solution

$$20\% = \frac{x}{x + 100} \times 100$$

$$20x + 2000 = 100x$$

$$80x = 2000$$

$$x = \frac{2000}{80} = 25 \text{ g}$$

NCERT CORNER

In-text Questions

1. What is meant by a pure substance?

2. List the points of difference between homogeneous and heterogeneous mixtures.

3. Differentiate between homogeneous and heterogeneous mixtures with examples.

4. How are sol, solution and suspension different from each other?

5. To make a saturated solution, 36 g of sodium chloride is dissolved in 100 g of water at 293 K. Find its concentration at this temperature.

6. How will you separate a mixture containing kerosene and petrol (difference in their boiling points is more than 25°C), which are miscible with each other?

7. Name the technique to separate:

(i) Butter from curd, (ii) Salt from sea water, (iii) Camphor from salt

8. What types of mixtures are separated by the technique of crystallisation?

9. Classify the following as chemical or physical changes
    (i)   Cutting of trees,
    (ii) Melting of butter in a pan,
    (iii) Rusting of almirah,
    (iv) Boiling of water to form steam,
    (v)   Passing of electric current, through water and the water breaking down into hydrogen and oxygen gases,
    (vi) Dissolving common salt in water,
    (vii) Making a fruit salad with raw fruits, and
    (viii) Burning of paper and wood

10. Try segregating the things around you as pure substances or mixtures: (i) Wood, (ii) Coal, (iii) Milk, (iv) Sugar, (v) Common salt, (vi) Soap, (vii) Soil, (viii) Rubber.

Exercises

1. Which separation techniques will you apply for the separation of the following?
    (a) Sodium chloride from its solution in water.
    (b) Ammonium chloride from a mixture containing sodium chloride and ammonium chloride.
    (c) Small pieces of metal in the engine oil of a car.
    (d) Different pigments from an extract of flower petals.
    (e) Butter from curd. (f) Oil from water.
    (g) Tea leaves from tea.
    (h) Iron pins from sand.
    (i) Wheat grains from husk.
    (j) Fine mud particles suspended in water.

2. Write the steps you would use for making tea. Use the words solution, solvent, solute, dissolve, soluble, insoluble, filtrate and residue.

3. Pragya tested the solubility of three different substances at different temperatures and collected the data as given below (results are given in the following table, as grams of substance dissolved in 100 grams of water to form a saturated solution).

Substance dissolved	Temperature in K				
	283	293	313	333	353
	Solubility				
Potassium nitrate	21	32	62	106	167
Sodium chloride	36	36	36	37	37
Potassium chloride	35	35	40	46	54
Ammonium chloride	24	37	41	55	66

(a) What mass of potassium nitrate would be needed to produce a saturated solution of potassium nitrate in 50 grams of water at 313 K?

(b) Pragya makes a saturated solution of potassium chloride in water at 353 K and leaves the solution to cool at room temperature. What would she observe as the solution cools? Explain.

(c) Find the solubility of each salt at 293 K. Which salt has the highest solubility at this temperature?

(d) What is the effect of change in temperature on the solubility of a salt?

4. Explain the following giving examples.
(a) Saturated solution   (b) Pure substance
(c) Colloid   (d) Suspension

5. Classify each of the following as a homogeneous or heterogeneous mixture: Soda water, wood, air, soil, vinegar, filtered tea.

6. How would you confirm that a colourless liquid given to you is pure water?

7. Which of the following materials fall in the category of a 'pure substance'?

(a) Ice, (b) Milk, (c) Iron, (d) Hydrochloric acid, (e) Calcium oxide, (f) Mercury, (g) Brick, (h) Wood and (i) Air.

8. Identify the solutions among the following mixtures:
(a) Soil, (b) Sea water, (c) Air, (d) Coal, (e) Soda water.

9. Which of the following will show 'Tyndall effect'?
(a) Salt solution, (b) Milk, (c) Copper sulfate solution, (d) Starch solution.

10. Classify the following into elements, compounds and mixtures:
(a) Sodium, (b) Soil, (c) Sugar solution,
(d) Silver, (e) Calcium carbonate, (f) Tin,
(g) Silicon, (h) Coal, (i) Air, (j) Soap,
(k) Methane, (l) Carbon dioxide, (m) Blood.

11. Which of the following are chemical changes?
(a) Growth of a plant, (b) Rusting of iron, (c) Mixing of iron filings and sand, (d) Cooking of food, (e) Digestion of food, (f) Freezing of water, (g) Burning of a candle.

Answers

In-text Questions

1. Substances having only one type of particle is known as pure substance.
For example: Hydrogen.

2.

Homogeneous mixture	Heterogeneous mixture
1. Its constituent particles cannot be seen easily.	Its constituent particles can be seen easily.
2. There are no visible boundaries of separation.	There are visible boundaries of separation between the constituents.
3. Its constituents cannot be easily separated. Examples: Alloys, solution of salt in water, etc.	Its constituents can be separated by simple methods. Examples: Mixture of sand and common salt, mixture of sand and water, etc.

3. See the answer to Q2.

4. Refer to Section 2.5.2, Suspensions.

5. Mass of sodium chloride (solute) = 36 g
Mass of water (solvent) = 100 g
We know that, mass of solution = Mass of solute + Mass of solvent = 36 g + 100 g = 136 g
Concentration (mass percentage) of the solution

$$= \frac{\text{Mass of solute}}{\text{Mass of solution}} \times 100$$

$$= \frac{36\ \text{g}}{136\ \text{g}} \times 100 = 26.47\%$$

6. Simple distillation is the method which can separate the mixture of kerosene and petrol (boiling points differ by more than 25°C).
Refer to Section 2.6.3, Separation of a Mixture of Two or More Liquids.

7. (i) By using the centrifugation method, butter can be separated from curd.

(ii) By using the evaporation method, salt can be separated from sea water. Water vaporises on evaporation leaving behind the salt.

(iii) Camphor from salt can be separated by sublimation method. On subliming, camphor will be converted into vapour leaving behind the salt.

8. Crystallisation method can be used for the purification of those mixtures which contain insoluble and/or soluble impurities. They are crystalline in nature and cannot be separated by filtration as some impurities are soluble.

9. Physical Change: Cutting of trees, Melting of butter in a pan, Boiling of water to form steam, Dissolving common salt in water, Making a fruit salad with raw fruits.
Chemical Change: Rusting of almirah, Passing of electric current through water and the water breaking down into hydrogen and oxygen gases, Burning of paper and wood.

10. (i) Mixture, (ii) Mixture, (iii) Mixture, (iv) Pure substance, (v) Pure substance, (vi) Compound/mixture, (vii) Mixture, (viii) Pure substance.

Exercises

1. (a) Evaporation, (b) Sublimation, (c) Filtration, (d) Chromatography, (e) Centrifugal machine or churning the curd by hand. (f) Decantation, (g) Filtration, (h) Magnetic Separation, (i) Winnowing, (j) Coagulation and decantation.

2. Method of preparing tea
   (i) Take some water (solvent) in a pan and heat it.
   (ii) Add some sugar (solute) and boil to dissolve the sugar completely; obtained homogeneous mixture is called solution.
   (iii) Add tea leaves (or tea) in the solution and boil the mixture.
   (iv) Now add milk and boil again.
   (v) Filter the mixture through the tea Stainer and collect the filtrate or soluble substances, that is, tea in a cup. The insoluble tea leaves are left behind as residue in the strainer.

3. (a) Mass of potassium nitrate needed to produce its saturated solution in 100 g of water at 313 K = 62 g
   Mass of potassium nitrate needed to produce its saturated solution in 50 g of water at 313

$$K = \frac{62}{100} \times 50g = 31g$$

(b) Crystals of potassium chloride are formed. This happens as solubility of solid decreases with decreasing temperature.

(c) Solubility of each salt at 293 K
Potassium nitrate   32 g per 100 g water
Sodium chloride   36 g per 100 g water
Potassium chloride   35 g per 100 g water
Ammonium chloride   37 g per 100 g water
Note: Solubility of a solid is that amount in grams which can be dissolved in 100 g of water (solvent) to make a saturated solution at a particular temperature. Ammonium chloride has the maximum solubility (37 g per 100 g of water) at 293 K.

(d) Solubility of a (solid) salt decreases with decrease in temperature while it increases with rise in temperature.

4. (a) Saturated solution: A solution in which no more amount of solute can be dissolved at a particular temperature is called a saturated solution. For example, when sugar is dissolved repeatedly in a given amount of water, a condition is reached at which further dissolution of sugar is not possible in that amount of water at room temperature.

(b) Pure substance: A substance made up of one type of particles (atoms and/or molecules) is called a pure substance. All elements and compounds are said to be pure. For example, water, sugar, etc.

(c) Colloid: A heterogeneous mixture in which the solute particle is too small to be seen with the naked eye, but is big enough to scatter light is known as a colloid. There are two phases in a colloidal solution: dispersed phase and dispersion medium. For example, milk, clouds, etc.

(d) Suspension: A suspension is a heterogeneous mixture in which the solute particles do not dissolve but remain suspended throughout the bulk of the medium. Particles of suspension are visible to the naked eye. For example, mixture of sand, water and muddy water.

5. Homogeneous mixtures: Air, soda water, vinegar, filtered tea.
Heterogeneous mixtures: Wood, soil.

6. If the given colourless liquid boils at 100°C sharp, it is pure water, otherwise not.

7. Ice, iron, calcium oxide, mercury are pure substances as they have definite composition. Milk is a colloid, so it is a heterogeneous mixture. Hydrochloric acid is also a mixture of hydrogen chloride gas and water.

8. Sea water, air and soda water: Homogeneous mixture
    Coal, Soil: Heterogeneous solution.

9. Milk and starch solution will show 'Tyndall effect' as both are colloids.

10. Elements: Sodium, silver, tin and silicon
    Compounds: Calcium carbonate, methane, and carbon dioxide
    Mixtures: Soil, sugar solution, coal, air, soap and blood.

11. Growth of a plant, rusting of iron, cooking of food, digestion of food, burning of a candle are chemical changes because the chemical composition of the substance changes.

CHAPTER 3

Atoms and Molecules

Learning Objectives

After studying this unit, you will be able to:

- State the laws of chemical combination
- Explain Dalton's atomic theory
- Describe atoms, symbols of elements and their representation
- Explain atomic mass and its calculation
- Define molecules, ions and radicals
- Explain valency and write chemical formulae
- Describe molecular mass/formula mass and its calculation
- Explain the concept of mole

3.1 Introduction

Ancient Indian and Greek philosophers have always wondered about the unknown and unseen form of matter. The idea of divisibility of matter is believed to have been considered in India around 500 BCE. The Indian philosopher Maharishi Kanad was one of the first to propose that matter (*padarth* in Hindi) is made up of very small particles called *parmanu* (John Dalton called these particles **atoms**). The word 'atom' means 'indivisible' – something that cannot be divided further. Another Indian philosopher, Pakudha Katyayama, elaborated that the particles of matter (atoms) normally exist in a combined form known as **molecules**.

All matter is made up of small particles known as atoms and molecules. Different kinds of atoms and molecules have different properties due to which different kinds of matter also show different properties. By the end of the eighteenth century, scientists recognised the difference between elements and compounds and became interested in finding out how and why elements combine and what happens when they combine together.

3.2 Laws of Chemical Combination

Antoine L Lavoisier laid the foundations for the chemical sciences by establishing two important laws of chemical combination. These two laws played a significant role in the development of Dalton's atomic theory of matter.

3.2.1 Law of Conservation of Mass

In the eighteenth century, scientists observed that when a chemical reaction is carried out in a closed vessel, there is no change in mass. This led to the development of the law of conservation of mass, introduced by Lavoisier in 1774. According to this law, matter can neither be created nor destroyed in a chemical reaction. The substances which combine together in a chemical reaction are called **reactants** and the new substances formed as a result of the chemical reaction are called **products**. The law of conservation of mass states that in a chemical reaction, the total mass of products is equal to the total mass of reactants and that is there is no change in mass.

For example, Lavoisier observed that on heating mercuric oxide, it produced free mercury and oxygen. The sum of the masses of mercury and oxygen was found to be equal to the mass of mercuric oxide.

$$\text{Mercuric oxide} \rightarrow \text{Mercury} + \text{Oxygen}$$
$$100\,\text{g} \qquad 92.6\,\text{g} \qquad 7.4\,\text{g}$$

Mass of the reactant = 100 g
Mass of the products = 92.6 + 7.4 g = 100.0 g
= 100 g

Here, the total mass of the products formed is equal to the total mass of the reactants undergoing reaction, and hence, this data is in agreement with the law of conservation of mass.

When 100 g of calcium carbonate is heated, 56 g of calcium oxide and 44 g of carbon dioxide are formed. It means the total mass of the reactants (calcium carbonate) and products (calcium oxide and carbon dioxide) is the same, that is, 100 g.

$$\underset{\text{100 g}}{\text{Calcium carbonate}} \xrightarrow[\text{(chemical reaction)}]{\text{Heat}} \underset{\text{56 g}}{\text{Calcium oxide}} + \underset{\text{44 g}}{\text{Carbon dioxide}}$$

$$56 + 44 = 100 \text{ g}$$

A piece of ice (solid water) is taken in a small conical flask. It is well corked and weighed. The flask is heated gently to melt the ice (solid) into water (liquid).

$$\text{Ice} \xrightarrow{\Delta} \text{Water}$$

The flask is again weighed. It is found that there is no change in weight during this physical change.

Experimental Verification of the Law of Conservation of Mass

Let us take a clean conical flask fitted with a rubber cork, and a small test tube having a long thread tied to its neck, as shown in Fig. 3.1. All these items are weighed together on a sensitive balance to find the initial mass of this apparatus. Now let us perform the experiment.

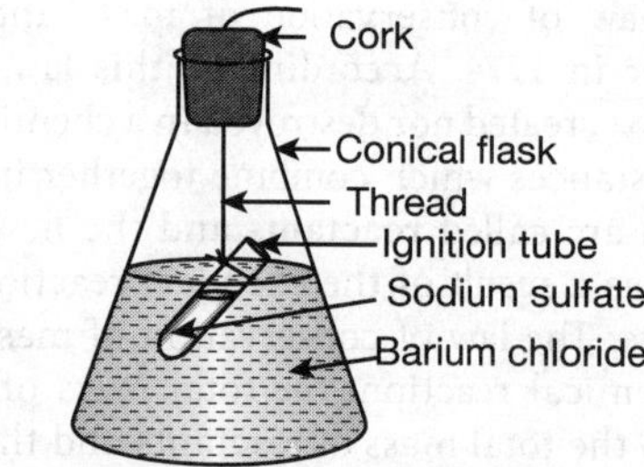

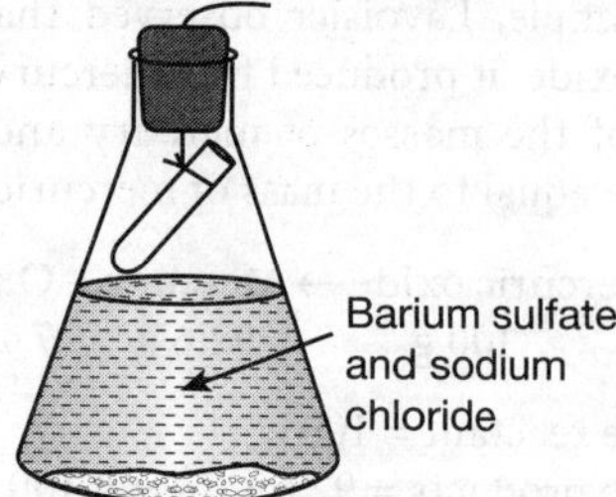

Fig. 3.1 Experimental verification of the law of conservation of mass

1. Take some barium chloride solution in the conical flask and put some sodium sulfate solution in a small test tube and lower it carefully into the conical flask by holding the free end of the thread tied to its neck.

2. Fix a rubber cork in the mouth of the flask so that it holds the thread firmly. The mouth of the small test tube should be kept above the liquid level in the flask so that the reactants cannot mix at this stage.

3. Find the mass of the apparatus along with the reactants by weighing on a balance. On subtracting the initial mass of the apparatus from this, we can obtain the mass of the reactants alone. Let this be 'w'.

4. Now remove the rubber cork from the mouth of the conical flask for a little while so that the thread may become loose. The small test tube having sodium sulfate solution drops into the flask, due to which the sodium sulfate solution mixes with the barium chloride solution. The solutions react to give a white precipitate of barium sulfate and sodium chloride solution.

5. Find the mass of the apparatus along with the products by weighing on a balance. On subtracting the initial mass of the apparatus from this mass, we can get the mass of the products (W). Now, if the mass of the products (W) is equal to the mass of the reactants (w), the experiment verifies the law of conservation of mass.

The chemical reaction taking place in this experiment is as follows:

$$\underset{\text{(solution)}}{\text{Barium chloride}} + \underset{\text{(solution)}}{\text{Sodium sulfate}} \rightarrow \underset{\text{(white solid)}}{\text{Barium sulfate}} + \underset{\text{(solution)}}{\text{Sodium chloride}}$$

Suppose in an experiment to verify the law of conservation of mass, the data obtained is as follows:

(i) Mass of barium chloride taken = 41.6 g
(ii) Mass of sodium sulfate taken = 28.4 g
(iii) Mass of barium sulfate formed = 46.6 g
(iv) Mass of sodium chloride formed = 23.4 g

Here, barium chloride and sodium sulfate are reactants.

Total mass of reactants = 41.6 g + 28.4 g = 70.0 g   ... (1)

Here, barium sulfate and sodium chloride are products.

Total mass of products = 46.6 g + 23.4 g = 70.0 g   ... (2)

The total mass of the products (70 g) is equal to the total mass of the reactants (70 g). Hence, the given data clearly verifies the law of conservation of mass.

TEST YOUR KNOWLEDGE

1. Magnesium carbonate decomposes on heating to form magnesium oxide and carbon dioxide. When 8.4 g of magnesium carbonate is decomposed completely, 4 g of magnesium oxide is formed. Calculate the mass of the carbon dioxide formed. Which law of chemical combination will you use to solve this problem?

Solution: This problem can be solved by using the law of conservation of mass in chemical reactions. Here, magnesium carbonate is the reactant while magnesium oxide and carbon dioxide are the products.

$$\text{Magnesium carbonate} \xrightarrow{\Delta} \text{Magnesium oxide} + \text{Carbon dioxide}$$

Now, using the law of conservation of mass:

Mass of products = Mass of reactants

that is, Mass of magnesium oxide + Mass of carbon dioxide = Mass of magnesium carbonate
So,   4 + Mass of carbon dioxide = 8.4
And,   Mass of carbon dioxide = 8.4 – 4 = 4.4 g
Hence the mass of carbon dioxide formed is 4.4 g.

2. Sodium carbonate reacts with ethanoic acid to form sodium ethanoate, carbon dioxide and water. In an experiment, 2.65 g of sodium carbonate reacts with 3 g of ethanoic acid to form 4.1 g of sodium ethanoate, 1.1 g of carbon dioxide and 0.45 g of water. Show that this data verifies the law of conservation of mass.

Solution: This problem can be solved by using the law of conservation of mass in chemical reactions. In it, sodium carbonate and ethanoic acid are the reactants while sodium ethanoate, carbon dioxide and water are the products and the reaction is as follows:

$$\text{Sodium carbonate} + \text{Ethanoic acid} \longrightarrow \text{Sodium ethanoate} + \text{Carbon dioxide} + \text{Water}$$

(Reactants) (Products)

(i) Total mass of the reactants = Mass of sodium carbonate + Mass of ethanoic acid
= 2.65 + 3
= 5.65 g   ... (1)

(ii) Total mass of the products = Mass of sodium ethanoate + Mass of carbon dioxide + Mass of water
= 4.1 + 1.1 + 0.45
= 5.65 g   ... (2)

We can see that
Total mass of the reactants = Total mass of the products = 5.65 g.
It means the given data verifies the law of conservation of mass.

3.2.2 Law of Constant Proportions

Lavoisier, along with Joseph Proust and other scientists, observed that many compounds are composed of two or more elements and each such compound has the same elements in the same proportions, irrespective of where the compound comes from and how it is prepared.

In a compound like water, the ratio of the mass of hydrogen to the mass of oxygen is always 1 : 8, whatever may be source (river, well) of water. It means if 18 g of water is decomposed, 2 g of hydrogen and 16 g of oxygen are always formed. Similarly in the case of ammonia, nitrogen and hydrogen are always present in the ratio 14 : 3 by mass, whatever be the method or the source from which it is obtained.

This led to the law of constant proportions, which is also known as the **law of definite proportions**. According to this law, in a pure chemical substance or compound, the elements are always present in definite proportions by mass. This means that whatever be the source from which it is obtained or the method by which it is prepared, a pure chemical compound is always made up of the same elements in the same mass percentage.

For example, it is important to note that whatever sample of carbon dioxide is taken, that the ratio of carbon and oxygen is always the same. That is, 12 : 32 or 3 : 8 in the methods given here.

(i) By burning carbon in oxygen:
$$C(s) + O_2(g) \rightarrow CO_2(g)\uparrow$$

(ii) By heating sodium bicarbonate:
$$2NaHCO_3(s) \xrightarrow{\Delta} Na_2CO_3(s) + CO_2(g)\uparrow + H_2O(l)$$

(iii) By heating calcium carbonate:
$$CaCO_3(s) \xrightarrow{\Delta} CaO(s) + CO_2(g)\uparrow$$

(iv) By reacting calcium carbonate with hydrochloric acid:

$$CaCO_3(s) + 2HCl(aq) \rightarrow CaCl_2(aq) + CO_2(g)\uparrow + H_2O(l)$$

TEST YOUR KNOWLEDGE

1. When 6 g of carbon is burnt in 16 g of oxygen, 22 g of carbon dioxide is produced. What mass of carbon dioxide will be formed when 6 g of carbon is burnt in 60 g of oxygen? Which law of chemical combination will govern your answer?

Solution: This problem can be solved by using the law of constant proportions. As carbon and oxygen combine in the fixed proportion of 3 : 8 by mass to give 22 g of carbon dioxide, the same mass of carbon dioxide (22 g) will be obtained even if we burn 6 g of carbon in 60 g of oxygen. The extra amount of oxygen (60 – 16 = 44 g) will remain unreacted here.

2. Hydrogen and oxygen combine in the ratio of 1 : 8 by mass to form water. What mass of oxygen gas would be required to react completely with 4.5 g of hydrogen gas?

Solution: Hydrogen and oxygen always combine in the fixed ratio of 1 : 8 by mass. Therefore

1 g of hydrogen gas requires 8 g of oxygen gas

So, 4.5 g of hydrogen gas requires 8 × 4.5 g of oxygen gas
= 36 g of oxygen gas
Hence, 36 g of oxygen gas is needed to react completely with 4.5 g of hydrogen gas.

Competition Edge

Law of Combining Volumes: This was proposed by Gay-Lussac and it is applicable for gases. According to this law, when gases react with each other, they bear a simple whole number ratio with one another as well as with the product under the conditions of the same temperature and pressure.

For example, the volumes of hydrogen and chlorine which combine to form HCl are in the ratio 1 : 1 : 2.

$$H_2(g) \quad + \quad Cl_2(g) \quad \rightarrow \quad 2HCl(g)$$
1 unit volume   1 unit volume   2 unit volume

Ratio = 1 : 1 : 2

Competition Edge

Law of Multiple Proportions: This was proposed by Dalton in 1804 and verified by Berzelius. According to this law, different weights of an element that combine with a fixed weight of another element bear a simple numerical ratio.

For example, in CO, CO_2, the weight of oxygen which combines with 12 g of carbon is in the ratio of 1 : 2.

3.3 Dalton's Atomic Theory

Both the laws of chemical combination were experimental laws and the only logical explanation of these laws was that matter must be composed of minute 'unit particles', which can take part in chemical combination in fixed whole numbers. Such 'unit particles' of matter were known as atoms.

In order to provide appropriate explanations of these laws, the British chemist John Dalton worked on a basic theory about the nature of matter. Dalton suggested the idea of divisibility of matter and said that the smallest particles of matter are atoms.

3.3.1 Features of Dalton's Atomic Theory

- All matter is composed of very tiny particles known as atoms (Fig. 3.2).

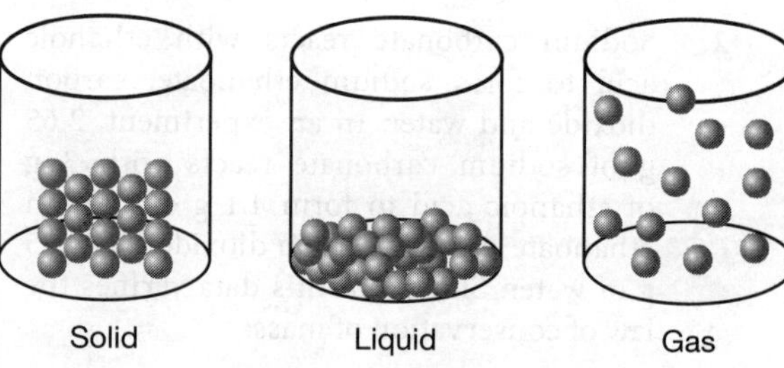

Fig. 3.2 According to Dalton's atomic theory, all matter (whether a solid, liquid or gas) is made up of very tiny particles called 'atoms'

- Atoms are indivisible particles, which can neither be created nor destroyed in a chemical reaction.
- Atoms are of various types. There are as many types of atoms as there are elements.
- The atoms of a given element are identical in mass, size and chemical properties.
- Atoms of different elements have different mass, size and chemical properties.

- Atoms combine in the ratio of small whole numbers to form compounds and the 'number' and 'type' of atoms in a given compound is also fixed. For example, in H_2O, the ratio of H : O is 2 : 1 and in ammonia (NH_3) the ratio of N : H is 1 : 3.
- Atoms of the same elements can combine in more than one ratio to form more than one compound. For example, sulfur can combine with oxygen in the ratios of 1 : 2 and 1 : 3 to give SO_2 and SO_3, respectively. Similarly, carbon can combine with oxygen in the ratios 1 : 1 and 1 : 2 to give CO and CO_2, respectively.

3.3.2 Significance of Dalton's Atomic Theory

Dalton's theory was the first modern attempt to describe the behaviour of matter or properties of matter in terms of atoms. It was also utilised to explain the laws of chemical combination in terms of atoms.

- The postulate of Dalton's atomic theory that "elements consist of atoms and that an atom can neither be created nor be destroyed" can be utilised to explain the law of conservation of mass. Let us understand it by taking the example of decomposition of magnesium carbonate.

  Magnesium carbonate ($MgCO_3$) is made up of 1 magnesium atom, 1 carbon atom and 3 oxygen atoms. The products obtained after its decomposition are magnesium oxide (MgO) and carbon dioxide (CO_2). The products on being taken together also comprise 1 magnesium atom, 1 carbon atom and 3 oxygen atoms. As the total number of the various types of atoms in the products (MgO and CO_2) and the reactant ($MgCO_3$) remains the same (5), the total mass of the products and the reactants is also the same (84).

- The postulate of Dalton's atomic theory that "elements consist of atoms having fixed mass, and that the number and type of atoms of each element in a given compound is fixed" can be utilised to explain the law of constant proportions. Let us understand it by taking the example of ammonia.

  According to Dalton's atomic theory, hydrogen consists of hydrogen atoms (H) and nitrogen consists of nitrogen atoms (N). It also states that three hydrogen atoms always combine with one nitrogen atom to form ammonia (NH_3). As an ammonia molecule always has the same number of hydrogen and nitrogen atoms, each atom having a fixed mass, the masses of hydrogen and nitrogen in ammonia are also in constant proportion.

3.3.3 Drawbacks of Dalton's Atomic Theory

In modern times, most of Dalton's statements have been found to be not fully correct or incorrect. Some of the drawbacks are given below:

- Atoms were thought to be indivisible. However, we can now say that under special circumstances, atoms can be further divided into smaller particles such as electrons, protons and neutrons. This means that atoms are themselves composed of these sub-particles.
- This theory states that all the atoms of an element have exactly the same mass, but it is now known that atoms of the same element can have different masses. For example, in the case of isotopes $_1^1H, _1^2H, _1^3H$.
- This theory states that atoms of different elements have different masses, but it is now known that even atoms of different elements can have the same mass. For example, in the case of isobars $_6^{14}C, _7^{14}N$.
- It failed to explain why atoms of different elements differ from one other. That is, it cannot tell us anything about the internal structure of the atom.
- It could not explain the nature of the forces that hold different atoms together in a molecule.

Competition Edge

We might think atoms are so insignificant in size that there is no need to bother about them. However, our entire world is made up of atoms. We may not be able to see them, but they are present and constantly affect whatever we do. They cannot be viewed using simple optical microscopes. However, using modern techniques such as scanning tunnelling microscopy, it is possible to produce magnified images of the surfaces of elements showing atoms.

3.4 ATOMS

Atoms are the building blocks of all matter and an atom is the smallest particle of an element which may or may not exist independently but which can take part in a chemical reaction. Atoms of most elements are very reactive and do not exist in the free state. They exist in combination with atoms of the same element or another element, as molecules. This means that the molecule is the basic unit of matter in any state because it is the smallest identity that can exist independently. For example, hydrogen and oxygen exist as diatomic molecules H_2 and O_2, respectively.

Atoms are very small (Table 3.1); they are smaller than anything that we can imagine or compare with. Millions of atoms when stacked would barely make a layer as thick as this sheet of paper. The size of an atom is indicated by its radius, which is called the **atomic radius**. The atomic radius is measured in nanometres (nm).

$$1/10^9\,m = 1\,nm$$

$$1\,m = 10^9\,nm$$

Table 3.1 Relative sizes of atoms

Radii (in m)	10^{-10}	10^{-9}	10^{-8}	10^{-4}	10^{-2}	10^{-1}
Example	Hydrogen atom	Water molecule	Hemoglobin molecule	Grain of sand	Ant	Watermelon

3.4.1 Symbols of Elements

A 'symbol' is something which represents or stands for something else. When we say 'symbols of elements', it means 'the symbols of the atoms of the elements'. In order to represent the elements, instead of using the entire name, scientists often use abbreviated names; these are known as symbols. Hence, a symbol may be defined as an abbreviation used for an element.

Dalton was the first scientist to use symbols for elements (Table 3.2). When he used a symbol for an element, he also meant a definite quantity of that element, that is, one atom of that element.

Table 3.2 Dalton's symbols for elements

Element	Dalton's symbol	Element	Dalton's symbol
Hydrogen	⊙	Copper	Ⓒ
Carbon	●	Silver	Ⓢ
Oxygen	○	Gold	Ⓖ
Phosphorus	⊗	Lead	Ⓛ
Sulfur	⊕	Mercury	☼
Iron	Ⓘ	Platinum	Ⓟ

Modern Method or IUPAC System of Symbols

As Dalton's symbols for elements were difficult to draw and inconvenient to use, they are now only of historical importance (they are not used at all). J.J. Berzelius of Sweden proposed that the first letter or the first letter and another letter of the name of the element be used as its symbol. His idea led to the modern symbols of elements.

IUPAC (International Union of Pure and Applied Chemistry) approves the names of elements. The symbols of elements are generally either the first letter or the first two letters or the first and the third letters of the name of an element. The first letter of a symbol is always written as a capital letter (uppercase) and the second letter as a small letter (lowercase). For example,

  (i) Hydrogen, H
 (ii) Aluminium, Al and not AL
(iii) Cobalt, Co and not CO

In the modern method, the symbols of elements can be classified as follows:

- Some symbols are derived from the first letter of the names of the elements (Table 3.3).

Table 3.3 Symbols derived from the first letter of the element

Element	Symbol	Element	Symbol
Hydrogen	H	Fluorine	F
Carbon	C	Phosphorus	P
Nitrogen	N	Sulfur	S
Oxygen	O	Iodine	I

- Some symbols are derived from the first two letters of the names of the elements (Table 3.4).

Table 3.4 Symbols derived from the first two letters of the name of the element

Element	Symbol	Element	Symbol
Aluminium	Al	Silicon	Si
Barium	Ba	Argon	Ar
Lithium	Li	Calcium	Ca
Beryllium	Be	Nickel	Ni
Neon	Ne	Cobalt	Co

- Some symbols are derived from the first and the third letters of the names of the elements (Table 3.5).

Table 3.5 Symbols derived from the first and the third letters of the name of the element

Element	Symbol	Element	Symbol
Arsenic	As	Manganese	Mn
Magnesium	Mg	Zinc	Zn
Chlorine	Cl	Rubidium	Rb
Chromium	Cr		

- Some symbols are derived from the Latin or Greek names of the elements (Table 3.6).

Table 3.6 Symbols derived from the Latin or Greek names of the element

Element	Latin name	Symbol
Copper	Cuprum	Cu
Potassium	Kalium	K
Iron	Ferrum	Fe
Gold	Aurum	Au
Sodium	Natrium	Na
Silver	Argentum	Ag
Mercury	Hydragyrum	Hg
Tin	Stannum	Sn
Lead	Plumbum	Pb
Antimony	Stibium	Sb

Significance of the symbol of an element: A symbol represents:

- The name of the element
- One atom of the element
- A definite mass of the element (equal to atomic mass expressed in grams)
- One mole of atoms of the element, that is, the symbol also represents 6.022×10^{23} atoms of the element

For example, 'O' represents oxygen, one atom of oxygen, 16 g of oxygen, one mole of oxygen atoms and 6.02×10^{23} atoms of oxygen.

3.4.2 Atomic Mass

One of the most remarkable concepts of Dalton's atomic theory was the introduction of atomic mass. According to this theory, each element has a characteristic atomic mass. This could explain the law of constant proportions so well that scientists were prompted to measure the atomic mass of an atom. As determining the mass of an individual atom was a relatively difficult task, relative atomic masses were determined by using the laws of chemical combinations and the compounds formed.

For example, consider a compound carbon monoxide (CO) formed from carbon and oxygen. It was observed experimentally that 12 g of carbon combines with 16 g of oxygen to form CO. That is, carbon combines with 4/3 times its mass of oxygen. Let us define the atomic mass unit [earlier abbreviated as amu, but now written as u (unified mass)] as equal to the mass of one carbon atom. Then we would assign carbon an atomic mass of 1.0 u and oxygen an atomic mass of 1.33 u. However, it is more convenient to have these numbers as whole numbers or as near to a whole number as possible. While searching for various atomic mass units, scientists initially took 1/16 of the mass of an atom of naturally occurring oxygen as the unit. This was considered relevant for two reasons:

- Oxygen reacts with a large number of elements and forms compounds.
- This atomic mass unit gave the masses of most of the elements as whole numbers.

In 1961, the International Union of Chemists universally accepted the carbon-12 isotope as the standard reference for measuring atomic mass in terms of atomic mass units (Table 3.7). One atomic mass unit is a mass unit equal to exactly one-twelfth (1/12th) the mass of one atom of carbon-12. The relative atomic masses of all elements have been found with respect to an atom of carbon-12. That is, the atomic mass of an element may be defined as the average relative mass of an atom of the element as compared with the mass of an atom of carbon (C-12 isotope) taken as 12 amu.

$$\text{Atomic mass} = \frac{\text{Mass of 1 atom of the element}}{\frac{1}{12} \text{ of the mass of an atom of carbon-12}}$$

For example, the atomic mass of magnesium is 24 u, which indicates that one atom of magnesium is 24 times heavier than $\frac{1}{12}$ of a carbon-12 atom. The atomic mass of aluminium is 27 u, which indicates that one atom of aluminium is 27 times heavier than $\frac{1}{12}$ of a carbon-12 atom.

Table 3.7 Atomic mass of some elements

Element	Symbol	Atomic mass
Hydrogen	H	1 u
Carbon	C	12 u
Nitrogen	N	14 u
Oxygen	O	16 u
Sodium	Na	23 u
Magnesium	Mg	24 u
Aluminium	Al	27 u
Phosphorus	P	31 u
Sulfur	S	32 u
Chlorine	Cl	35.5 u
Potassium	K	39 u
Calcium	Ca	40 u
Iron	Fe	56 u
Copper	Cu	63.5 u
Zinc	Zn	65.3 u

Competition Edge

The gram atomic mass of an element is defined as the quantity of the element whose mass expressed in grams is numerically equal to its atomic mass. To find the gram atomic mass, we keep the numerical value the same as the atomic mass, but simply change the units from u to g. For example, the atomic mass of aluminium is 27 u. Its gram atomic mass is 27 g.

The gram atomic mass of isotopes is expressed as the average gram atomic mass using the relation given below:

$$\text{(Gram atomic mass)}_{ave} = \frac{M_1 X_1 + M_2 X_2}{X_1 + X_2}$$

M_1 and M_2 are relative masses of isotopes, and X_1 and X_2 are the relative % content.

Due to the presence of isotopes, the average atomic mass can be fractional. For example, in the case of chlorine, it is 35.5 g. Chlorine contains two types of atoms having relative masses 35 and 37 and their relative abundance is 3 : 1. In such cases, the atomic mass of the element is the average of the relative masses of the different isotopes of the element.

$$\text{Atomic mass of chlorine} = \frac{35 \times 3 + 37 \times 1}{4} = 35.5$$

TEST YOUR KNOWLEDGE

1. Which of the following symbols of elements are correct?
   (i) Cobalt: CO   (ii) Carbon: c
   (iii) Aluminium: AL   (iv) Helium: He
   (v) Sodium: So   (vi) Iron: Fe
   (vii) Gold: Gd

Solution: Only helium and iron have the correct symbols. The correct symbols of these elements are as follows:
   (i) Cobalt: Co   (ii) Carbon: C
   (iii) Aluminium: Al   (iv) Helium: He
   (v) Sodium: Na   (vi) Iron: Fe
   (vii) Gold: Au

2. The mass of one steel screw is 4.11 g. Find the mass of one mole of these steel screws. Compare this value with the mass of the earth (5.98×10^{24} kg). Which is heavier and by how many times?

Solution:
Mass of 1 screw = 4.11 g
Mass of 1 mole of screws = $4.11 \times 6.022 \times 10^{23}$
= 2.475×10^{24} g = 2.475×10^{21} kg

$$\frac{\text{Mass of the earth}}{\text{Mass of 1 mole of screws}} = \frac{5.98 \times 10^{24}\,\text{kg}}{2.475 \times 10^{21}\,\text{kg}} = 2416$$

It shows that the mass of the earth is 2416 times more than the mass of one mole of screws.

3.4.3 How Do Atoms Exist?

The atoms of only a few elements like noble gases are chemically unreactive and exist in the free state or as single atoms; for example, helium (He), neon (Ne) and argon (Ar). The atoms of most elements are chemically very reactive and do not exist in the free state or as single atoms; they exist in the combined state.

Atoms usually exist: as molecules and as ions. These molecules and ions aggregate in large numbers to form the matter that we can see, feel or touch. For example, you cannot see the individual iodine molecules (I_2) with your eyes as these are quite small, but you can see iodine crystals as a violet-coloured solid as it is a collection of millions of iodine molecules held together.

Similarly, you can see potassium chloride compound as a white powder as it is formed by millions of potassium ions and chloride ions.

Molecules

In general, a molecule is a group of two or more atoms that are chemically bonded together or tightly held together by attractive forces. A molecule is defined as the smallest particle of an element or a compound that is capable of independent existence and shows all the properties of that substance. Thus, a molecule is the smallest identity that can exist individually. For example, hydrogen exists as H_2 (dihydrogen) and not as H. Atoms of the same element or of different elements can join together to form molecules. For example, H_2, N_2 and NH_3.

Types: There are two types of molecules:

- **Molecules of elements**: The molecules of an element are constituted of the same type of atoms (homoatomic), that is, one, two or more atoms of the same elements existing as one species in the free state. Depending upon the number of atoms present in the molecule, they are further classified as follows:

(a) **Monoatomic:** Molecules of many elements (noble gases) like argon (Ar) and helium (He) are made up of only one atom of that element (Fig. 3.3).

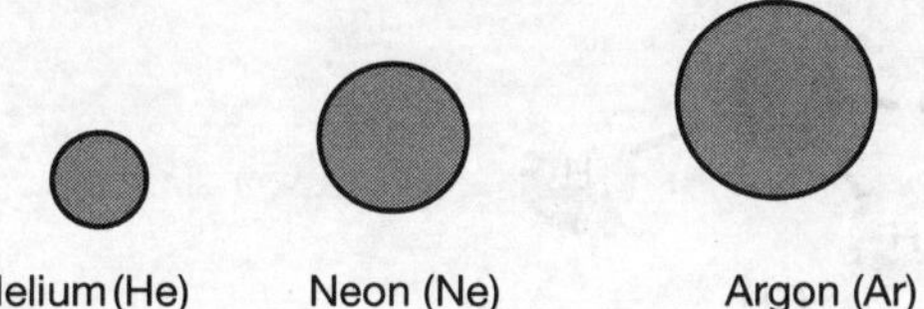

Fig. 3.3 Monoatomic elements

(b) **Polyatomic:** Molecules of most non-metals are made of two or more atoms and they are further classified as follows:
  - **Diatomic:** Such a molecule is formed by two atoms (Fig. 3.4); for example, O_2, N_2, H_2, Cl_2.
  - **Triatomic:** Such a molecule is formed by three atoms (Fig. 3.5); for example, O_3 (ozone).
  - **Tetratomic:** Such a molecule is formed by four atoms; for example, P_4.
  - **Octatomic:** Such a molecule is formed by eight atoms; for example, S_8.

    In the case of molecules of elements, mostly the atomicity is not indicated while writing the name. For example, H_2 is known as a hydrogen molecule and O_2 is known as an oxygen molecule. In the modern system, they can be named dihydrogen and dioxygen, respectively.

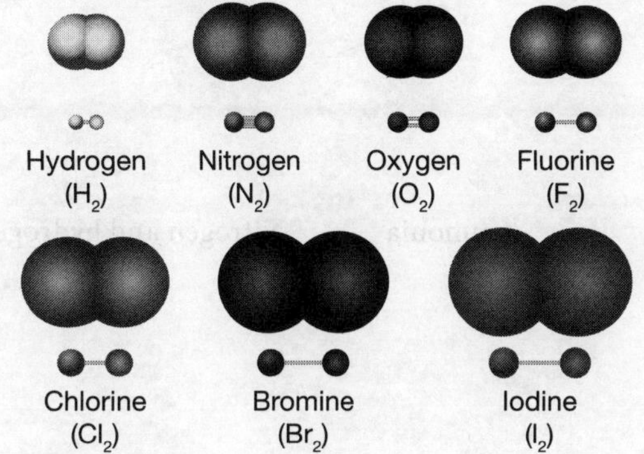

Fig. 3.4 Diatomic elements

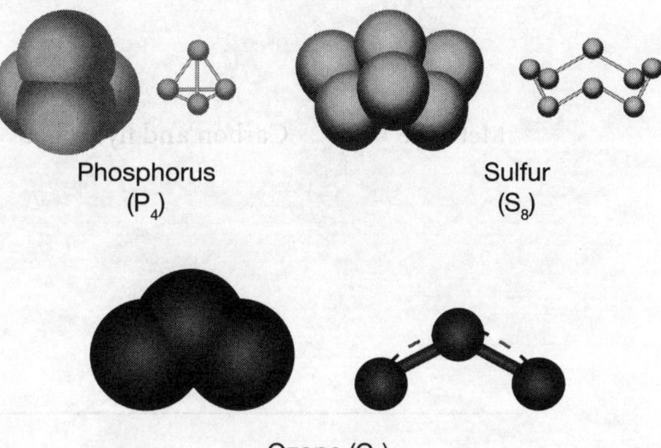

Fig. 3.5 Triatomic, tetratomic and octatomic elements

- **Molecules of compounds**: Atoms of different elements (heteroatomic) join together in definite proportions to form molecules of compounds (Table 3.8); here, a molecule of a compound means two or more atoms of different elements that combine together in a definite proportion by mass to form a species that can exist freely. For example, hydrogen chloride is a compound. A molecule of hydrogen chloride (HCl) contains two different types of atoms: hydrogen (H) and chlorine (Cl).

Table 3.8 Molecules of some compounds

Compound	Combining elements	Formula and Structure	Ratio by mass
Water	Hydrogen and oxygen	Water molecule H_2O	1 : 8
Ammonia	Nitrogen and hydrogen	NH_3	14 : 3
Carbon dioxide	Carbon and oxygen	CO_2	3 : 8
Methane	Carbon and hydrogen	CH_4	4 : 1

Competition Edge

Buckminsterfullerenes (C_{60}) are spherical carbon allotropes where 60 atoms are assembled in pentagons and hexagons, in a geometry similar to a soccer ball.

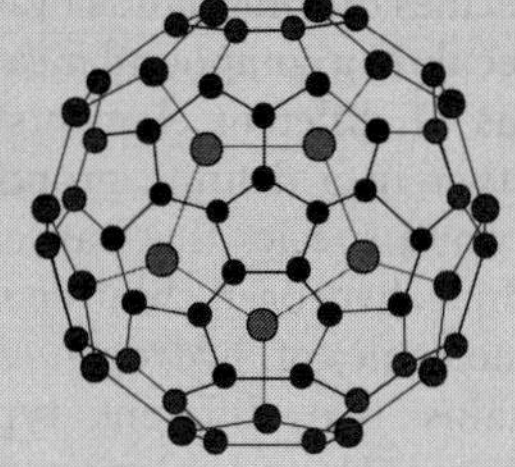

Atomicity: The number of atoms constituting a molecule is known as its atomicity. For example,

(i) when atomicity is one (monoatomic) – He, Ar, Ne, H, etc.

(ii) when atomicity is two (diatomic) – H_2, N_2, O_2, etc.

(iii) when atomicity is three (triatomic) – CO_2, SO_2, etc.

(iv) when atomicity is four (tetratomic) – SO_3, NH_3, etc.

The compounds composed of molecules are known as molecular compounds. In order to understand the naming of such compounds, let us first understand the term electronegativity. The atoms of a molecule are held together by forces of attraction; that is, a chemical bond. The bond is simply a shared pair of electrons between the atoms. The tendency of an atom in a molecule to attract the shared pair of electrons towards itself is known as **electronegativity**. Normally, a more electronegative atom attracts more electrons; for example, halogen or oxygen attracts more electrons than hydrogen.

While writing the name:

(i) The less electronegative element is named first, followed by the name of the more electronegative element. For example, HCl is known as hydrogen chloride.

(ii) The name of the more electronegative element generally ends with 'ide'; for example H_2S is hydrogen sulfide.

(iii) Sometimes, common names are preferred over scientific names. For example H_2O is known as water and not as hydrogen oxide. Similarly, NH_3 is known as ammonia and not as nitrogen trihydride.

(iv) The number of atoms of each element are generally indicated in the name by adding prefixes like 'di', 'tri', 'tetra', 'penta', etc. to show the number of atoms: 2, 3, 4, 5, etc. For example, CCl_4 is known as carbon tetrachloride and PCl_5 is known as phosphorus pentachloride.

(v) If the first element is hydrogen, such prefixes are not used for indicating the number of its atoms. For example, H_2S is simply called hydrogen sulfide and not dihydrogen sulfide.

Molecular Mass

Just as the atomic mass of an element is the average relative mass of its atoms as compared with that of an atom of C-12 isotope taken as 12, molecular mass is defined in the same manner as follows

The molecular mass of a substance (element or compound) is the average relative mass of its molecules as compared with that of an atom of C-12 isotope taken as 12. In other words, the molecular mass of a substance represents the number of times the molecule of that substance is heavier than 1/12ᵗʰ of the mass of an atom of C-12 isotope.

Molecular mass can be expressed in terms of atomic mass units (amu) or unified mass (u) on the atomic scale.

Calculation of molecular mass: As molecules are composed of two or more atoms of the same or different elements and each atom has a definite atomic mass, the molecular mass of a molecule can be found out by adding the atomic masses of all the atoms present in one molecule of the substance. For example,

(i) A molecule of oxygen has the formula O_2. Hence, Molecular mass of oxygen = 2 × Atomic mass of oxygen = 2 × 16 u = 32 u.

(ii) A molecule of water has the formula H_2O. Hence, Molecular mass of H_2O = 2 × Atomic mass of hydrogen + 1 × Atomic mass of oxygen = 2 × 1.0 u + 16 u = 18 u.

Significance of the Formula of a Substance

- It represents the name of the substance.
- It represents one molecule of the substance.
- It gives the number of atoms of each element present in one molecule.
- It gives the names of all the elements present in the molecule.
- It represents a definite mass of the substance which is equal to the molecular mass expressed in grams.
- It also represents one mole of molecules of the substance, that is, 6.022×10^{23} molecules of the substance.

For example, let us take the significance of the formula CO_2:

- CO_2 represents carbon dioxide.
- CO_2 represents one molecule of carbon dioxide.
- CO_2 tells us that carbon dioxide contains two elements, carbon and oxygen.
- CO_2 tells us that one molecule of carbon dioxide contains 1 atom of carbon and 2 atoms of oxygen.
- CO_2 represents 18 grams of carbon dioxide which is equal to the molecular mass of carbon dioxide expressed in grams.
- CO_2 also represents one mole of molecules of carbon dioxide, that is, 6.022×10^{23} molecules of carbon dioxide.

TEST YOUR KNOWLEDGE

1. Find the molecular masses of the following:
   (i) $C_6H_{12}O_6$ (ii) $Fe_2(SO_4)_3$
   Given atomic masses: C = 12 u, H = 1 u, O = 16 u, Fe = 56 u, S = 32 u, Cu = 63.5

Solution: (i) Molecular mass of $C_6H_{12}O_6$
   = 6 × Atomic mass of C + 12 × Atomic mass of H + 6 × Atomic mass of O
   = 6 × 12 + 12 × 1 + 6 × 16 = 72 + 12 + 96
   = 180 u

(ii) Molecular mass of $Fe_2(SO_4)_3$
     $= 2 \times$ Atomic mass of Fe $+ 3 \times$ Atomic mass
     of S $+ 12 \times$ Atomic mass of O
     $= 2 \times 56 + 3 \times 32 + 12 \times 16 = 112 + 96 + 192$
     $= 400$ u

2.   Find the molecular masses of the following:
     (i) $CuSO_4.5H_2O$ (ii) $C_{12}H_{22}O_{11}$
     Given atomic masses: C = 12 u, H = 1 u, O =
     16 u, S = 32 u, Cu = 63.5

Solution: (i) Molecular mass of $CuSO_4.5H_2O$
     $=$ Atomic mass of Cu $+$ Atomic mass of S $+ 4$
     $\times$ Atomic mass of O $+ 5 \times (2 \times$ Atomic mass
     of H $+ 1 \times$ Atomic mass of O)
     $= 63.5 + 32 + 4 \times 16 + 5 (2 \times 1 + 16)$
     $= 63.5 + 32 + 64 + 90 = 249.5$ u

(ii) Molecular mass of $C_{12}H_{22}O_{11}$
     $= 12 \times$ Atomic mass of C $+ 22 \times$ Atomic mass
     of H $+ 11 \times$ Atomic mass of O

     $= 12 \times 12 + 22 \times 1 + 11 \times 16$
     $= 144 + 22 + 176 = 342$ u

3.   Classify each of the following on the basis
     of their atomicity:
     (a) F_2 (b) NO_2 (c) N_2O (d) C_2H_6 (e) P_4
     (f) H_2O_2 (g) P_4O_{10} (h) O_3 (i) HCl (j) CH_4
     (k) He (l) Ag

Solution:
Monoatomic with atomicity (1): He, Ag
Diatomic with atomicity (2): F_2, HCl
Polyatomic with atomicity > 2
Atomicity (3): NO_2, N_2O, O_3
Atomicity (4): P_4, H_2O_2
Atomicity (5): CH_4
Atomicity (8): C_2H_6
Atomicity (14): P_4O_{10}

Quick Review

- **Law of conservation of mass:** Matter can neither be created nor destroyed in a chemical reaction.
- **Law of constant proportions:** In a pure chemical substance or compound, the elements are always present in definite proportions by mass.
- **Law of combining volumes:** When gases react with each other, they bear a simple whole number ratio with one another as well as with the product under conditions of the same temperature and pressure.
- ***Avogadro's law:** Under similar conditions of temperature and pressure, equal volumes of gases contain an equal number of molecules.
- The building blocks of all matter are atoms. An atom is the smallest particle of an element which may or may not exist independently, but which can take part in a chemical reaction.
- The atomic mass of an element may be defined as the average relative mass of an atom of the element as compared with the mass of an atom of carbon (C-12 isotope) taken as 12 amu.
- Gram atomic mass is defined as the quantity of the element whose mass expressed in grams is numerically equal to its atomic mass.
- The number of atoms constituting a molecule is known as its atomicity.
- As molecules are composed of two or more atoms of the same or different elements and each atom has a definite atomic mass, the molecular mass of a molecule can be found out by adding the atomic masses of all the atoms present in one molecule of the substance.
  For example, Molecular mass of $C_aH_bO_c = a \times 12 + b \times 1 + z \times 16$.
     Here a, b and c represent the number of C, H and O atoms, respectively, while 12, 1 and 16 represent the atomic masses of C, H and O, respectively.

Exercise 3.1

Questions marked * are practical based.

Section A: Multiple Choice Questions
(1 Mark)

1. Who discovered the law of constant proportions?
   (a) Ernest Rutherford   (b) John Dalton
   (c) Antoine L. Lavoisier   (d) Neils Bohr

2. Which postulate of Dalton's atomic theory can explain the law of definite proportions?
   (a) Atoms are indivisible particles, which can neither be created nor be destroyed in a chemical reaction
   (b) The relative numbers and kinds of atoms are constant in a given compound
   (c) Atoms combine in the ratio of small whole numbers to form compounds
   (d) Atoms of different elements have different masses and chemical properties

3. The chemical symbol for sodium is:
   (a) So   (b) Sd
   (c) NA   (d) Na

4. The chemical symbol for cobalt is:
   (a) co   (b) CO
   (c) Co   (d) Cb

5. Which of the following represents 1 amu?
   (a) Mass of hydrogen molecule
   (b) 1/12 of mass of C-12 atom
   (c) Mass of O-12 atom
   (d) Mass of C-12 atom

6. What is the atomic mass of magnesium?
   (a) 24   (b) 32
   (c) 16   (d) 35.5

7. The chemical symbol for nitrogen gas is:
   (a) Ni   (b) N_2
   (c) N^+ (d) N

8. How many atoms are present in a molecule of H_2O_2?
   (a) 3   (b) 1
   (c) 4   (d) 5

9. Which of the following pairs is of polyatomic molecules?
   (a) Oxygen, nitrogen
   (b) Helium, argon
   (c) Water, fluorine
   (d) Phosphorus, sulfur

10. What is the molecular mass of CO_2?
    (a) 44   (b) 42
    (c) 40   (d) 50

11. Which pair of molecules has the same molecular mass?
    (a) N_2, O_2 (b) N_2, CO
    (c) CO_2, SO_2 (d) NH_3, CH_4

12. Which of the following statements are true about an atom?
    (i) Atoms cannot exist independently
    (ii) Atoms are the basic units from which molecules and ions are formed
    (iii) Atoms are always neutral in nature
    (iv) Atoms aggregate in large numbers to form the matter that we can see, feel or touch
    (a) (i), (ii)   (b) (ii), (iii)
    (c) (i), (ii), (iii)   (d) (ii), (iii), (iv)

13. A molecule of salt found to contain sodium and chlorine combined together in the ratio of 23 : 35.5 by mass. What is the formula of salt?
    (a) Na_2Cl (b) $NaCl_2$
    (c) Na_3Cl_2 (d) $NaCl$

14. During an experiment, hydrogen (H_2) and oxygen (O_2) gases reacted in an electric arc to produce water as follows:

$$2H + O_2 \rightarrow 2H_2O$$

The experiment is repeated three times and data tabulated as shown below:

Experiment	Mass of H_2 reacted	Mass of O_2 reacted	Mass of H_2O produced
1	2 g	16 g	18 g
2	4g	32 g	36 g
3	–	–	9 g

During the third experiment, the researcher forgot to list the masses of H_2 and O_2 used. So, if the law of constant proportion is correct, find the mass of O_2 used during the third experiment.
(a) 4 g   (b) 8 g
(c) 16 g   (d) 32 g

15. Match the formula units given in Column I with their masses given in Column II.

Column I	Column II
1. Calcium oxide	(i) 102 u
2. Magnesium chloride	(ii) 100 u
3. Aluminium phosphide	(iii) 58 u
4. Calcium carbonate	(iv) 95 u
5. Aluminium oxide	(v) 56 u

	1	2	3	4	5
(a)	(v)	(iv)	(iii)	(i)	(ii)
(b)	(v)	(iv)	(ii)	(iii)	(i)
(c)	(v)	(iv)	(iii)	(ii)	(i)
(d)	(iv)	(v)	(iii)	(ii)	(i)

Assertion–Reason Questions

Direction: In the following question two statements (Assertion) A and Reason (R) are given Mark.
(a) if A and R both are correct and R is the correct explanation of A;
(b) if A and R both are correct but R is not the correct explanation of A;
(c) A is true but R is false;
(d) A is false but R is true

Assertion	Reason
1. Pure water obtained from different sources such as rivers, wells, springs and seas always contains hydrogen and oxygen in the ratio of 1 : 8 by mass.	1. A chemical compound always contains elements combined in a fixed proportion by mass.
2. Atomic mass has no unit but is expressed in amu.	2. It is the average mass of an atom taking into account the relative abundance of its isotopes.
3. Atomic mass of aluminium is 27.	3. An atom of aluminium is 27 times heavier than 1/12th of the mass of a carbon-12 atom.
4. Elements that are made up of only one kind of atom are classified as pure substances.	4. Hydrogen, oxygen and nitrogen are elements.
5. The percentage of copper in the different oxides formed by copper remains the same.	5. A chemical compound is always made up of the same elements combined together in a definite proportion by mass.
6. The number of carbons atoms in one mole of carbon is the same as the number of atoms in one mole of sodium carbonate (Na_2CO_3).	6. One mole of carbon atom contains Avogadro number of atoms and one mole of Na_2CO_3 contains Avogadro number of molecules.
7. Molecular weight of SO_2 is double that of O_2.	7. One mole of SO_2 contains double the number of molecules present in one mole of O_2.

Section B: Very Short Answer Questions
(2 Marks)

1. Name the scientist who laid the foundations of the chemical sciences. How did he do so?

2. Define the law of conservation of mass.

3. Define the law of constant proportion.

4. Which postulate of Dalton's atomic theory is the result of the law of conservation of mass?

5. Which organisation approves the names of elements all over the world? Write the symbol for gold.

6. Write the symbols for tungsten and iron.

7. Name the element which is used as the reference for atomic mass.

8. 'The atoms of most elements cannot exist independently'. Name two atoms which exist as independent atoms.

9. Define the term atomicity.

10. What is the atomicity of argon?

11. Write the atomicity of the following:
(i) Phosphorus, (ii) Sulfur

12. Define atomic mass unit.

13. The relative atomic mass of an oxygen atom is 16. Explain its meaning.

14. Give one reason why scientists choose 1/16 of the mass of an atom of naturally occurring oxygen as the atomic mass unit.

15. What is molar mass? What are its units?

16. Distinguish between molecular mass and molar mass.

17. What is meant by the term chemical formula?

18. How many atoms are present in (i) H_2S molecule and (ii) H_3PO_4?

Section C: Short Answer Questions
(3 Marks)

1. Name the scientists whose experimentation established the laws of chemical combination and name the laws.

2. Give any two drawbacks of Dalton's atomic theory.

3. Explain why/why not the law of constant compositions is true for all types of compounds.

4. Write the chemical symbols of two elements:
   (i) Which are formed from the first letter of the elements' name
   (ii) Whose names have been taken from the names of the elements in Latin
   (iii) Which are formed from the first two letters of the elements' name

5. Write the correct symbols of the following elements which are written incorrectly:
(i) FE (Iron), (ii) AL (Aluminium), (iii) AR (Argon), (iv) AG (silver), (v) NA (sodium), (vi) CO (Cobalt)

6. What do you mean by a molecule? Give some examples.

7. How would you differentiate between a molecule of an element and a molecule of a compound? Write one example of each type.

8. Classify the following compounds as diatomic, triatomic and polyatomic:

HCl, H_2, H_2O, NH_3, CH_3OH, PCl_5, O_3

9. Hydrogen and oxygen combine in the ratio of 1 : 8 by mass to form water. What mass of oxygen gas would be required to react completely with 4 g of hydrogen gas?

10. When 3 g of carbon is burnt in 8 g of oxygen, 11 g of carbon dioxide is produced. What mass of carbon dioxide will be formed when 6 g of carbon is burnt in 40 g of oxygen? Which law of chemical combination will govern your answer?

11. *$BaCl_2$ solution is taken in a small conical flask. A small ignition tube having Na_2SO_4 solution is suspended in it by a thread and the flask is corked such that the open mouth of the tube remains above the level of the solution in the flask. The system is weighed. Now the tube is loosened so that its contents mix in the solution of the flask. Weigh the flask again. What observations will you make and what result will you draw?

12. *2.925 g of NaCl was dissolved in water. How much minimum $AgNO_3$ by mass should be dissolved in water and mixed into the NaCl solution so that all the chloride ions are completely precipitated as AgCl. If AgCl after filtering and drying is found to be 7.175 g, what is present in the filtrate and how much by mass? (Atomic mass: Ag = 108, N = 14, O = 16, Na = 23, Cl = 35.5).

Section D: Long Answer Questions
(5 Marks)

1. Give the postulates of Dalton's atomic theory.

2. (i) Why does the atomic mass of an element not represent the actual mass of its atom?
   (ii) "The atomic mass of an element is in fractions." What does this mean?

3. Answer the following:
   (i) Symbol ⊖ was used by Dalton for which element?
   (ii) Symbol ☉ was used by Dalton for which element?
   (iii) What is the order of the size of atoms?
   (iv) What is the molecule C-60 known as and what is its shape?
   (v) Why is copper represented by the symbol Cu, though it does not have 'u' in its name?

4. State any three examples in each case and write their chemical formulae:
   (i) Molecules having the same kind of atoms only.
   (ii) Molecules having two different kinds of atoms.
   (iii) Molecules having three different kinds of atoms.

5. (i)   Write the difference between molecular mass and molar mass.
    (ii) Find the molecular mass of the following compounds:
    (1) Phenol (C_6H_5OH)
    (2) Phosphoric acid (H_3PO_4)
    (3) Ammonium chloride (NH_4Cl)
    (4) Washing soda ($Na_2CO_3.10H_2O$)
    [Atomic weight: H = 1 u, O = 16 u, N = 14 u, C = 12 u, P = 31 u, Na = 23 u, Cl = 35.5 u)

Section E: Case Study or Passage-Based Questions   (4 Marks)

1. In chemistry, symbols represent elements. It is simple to use the symbol of an element rather writing the whole name. Dalton introduced symbols for representing elements for the first time, but they were difficult to draw and inconvenient to use, so modern symbols were introduced. These symbols are defined as "a shorthand representation of the name of an element." In the beginning, the names of elements were derived from the name of the place where they were first found. Nowadays, IUPAC approves the names and symbols of the elements.

Element	P	Q	R	S	T
Latin name of the element	Cuprum	Kalium	Ferrum	Natrium	Argentum

(i)   Who invented the modern symbols of elements?

(ii) What is the name of element P. Also write its correct symbol.

(iii) What is the name of element S. Also write its correct symbol.

(iv) Name the element R and write its correct symbol.

(v)   An element is used for making jewellery and is a good conductor of heat and electricity. State the Latin name of that element.

2. The molecular mass of a substance is the sum of the atomic masses of all the atoms in a molecule of the substance. It is, therefore, the relative mass of a molecule expressed in atomic mass units (u). Depending upon the number of atoms of the same or different elements present in the molecule, it can be monoatomic, diatomic, triatomic, tetratomic or polyatomic. The formula unit mass is calculated in the same manner as the molecular mass. It is the sum of the atomic masses of all the atoms in a formula unit of a compound.

The atomic masses of a few elements are given below.

Symbol of element	H	C	O	S	Na	Ca
Atomic mass (u)	1	12	16	32	23	40

(i)   Which of the following is an example of a polyatomic molecule?
    (a) H_2
    (b) O_3
    (c) S_8
    (d) I_2

(ii) The relative molecular mass of H_2O is:
    (a) 34 u
    (b) 18 u
    (c) 10 u
    (d) 40 u

(iii) How many kinds of atoms are present in a molecule of calcium carbonate ($CaCO_3$)?
    (a) 3
    (b) 4
    (c) 5
    (d) 6

(iv) What is the ratio by mass of the combining elements C : H : O in the compound methanol (CH_3OH)?
    (a) 4 : 1 : 3
    (b) 4 : 3 : 1
    (c) 1 : 3 : 4
    (d) 3 : 1 : 4

(v)   What is the formula unit mass of $CaCO_3$?
    (a) 80 u
    (b) 100 u
    (c) 50 u
    (d) 148 u

High Order Thinking Skills (HOTS) Questions

1. Why was oxygen-16 replaced by carbon-12 as the reference for measuring atomic masses?

2. (i) Carbon dioxide produced by heating sodium hydrogen carbonate is dry whereas that produced by the action of dilute hydrochloric acid on sodium hydrogen carbonate is moist. What do you think about the difference in the composition of carbon dioxide in the two cases?
    (ii) Which law is proved by this example? Define the law.

3. When 3 g of magnesium is burnt in 2 g of oxygen, 5 g of magnesium oxide is produced. What mass of magnesium oxide will be formed when 6 g of

magnesium is burnt in 15 g of oxygen? Also find the amount of oxygen that remains unreacted here. Which law of chemical combination will govern your answer? State the law.

4. State the law of conservation of mass. Is this law applicable to chemical reactions? Elaborate your answer with the help of an example.

5. (i) 20 g of silver nitrate solution is added to 20 g of sodium chloride solution. What change in mass do you expect after the reaction and why?

   (ii) In one experiment, 10 g of magnesium ribbon on burning in oxygen gave 16.6 g of magnesium oxide. In another experiment, magnesium oxide obtained on heating magnesium carbonate was found to contain 60% magnesium. Show that these data illustrate the law of constant composition.

6. *Shashwat was performing an experiment related to the laws of chemical combination in the science laboratory under the guidance of his chemistry teacher Dr. AK Singhal. He found that when he burned 0.1 gram of hydrogen gas in 0.8 grams of oxygen gas in a closed container, he obtained 0.9 grams of water. He repeated this experiment many times but obtained the same results every time.

   (i) Write a balanced chemical equation for the reaction between hydrogen and oxygen to form water.

   (ii) What are the reactants and products in the reaction here?

   (iii) Which law of chemical combination is illustrated by the fact that when Shashwat burned 0.1 g of hydrogen in 0.8 g of oxygen, he obtained 0.9 g of water?

   (iv) What mass of water will be obtained if 2 g of hydrogen is burned in 20 g of oxygen? Which law of chemical combination will govern your answer?

   (v) What values are displayed by Shashwat in this episode?

Answers

Section A: Multiple Choice Questions

1. (c)	2. (b)	3. (d)	4. (c)
5. (b)	6. (a)	7. (b)	8. (c)
9. (d)	10. (a)	11. (b)	12. (c)
13. (d)	14. (b)	15. (c)	

Assertion–Reason Questions

1. (a)	2. (b)	3. (a)	4. (c)
5. (d)	6. (b)	7. (c)	

Section B: Very Short Answer Questions

1. Antoine Laurent Lavoisier, by establishing two important laws of chemical combination.

2. According to it "Mass can neither be created nor be destroyed in a chemical reaction." In other words, the mass of the reactants must be equal to the mass of the products.

3. According to it "In a pure chemical substance, the elements are always present in definite proportions by mass."

4. Atoms are indivisible particles, which cannot be created or destroyed in a chemical reaction.

5. International Union of Pure and Applied Chemistry (IUPAC). Au.

6. (i) Tungsten (W) (ii) Iron (Fe).

7. Carbon.

8. Noble gases like helium (He) and argon (Ar) exist as independent atoms.

9. The number of atoms present in one molecule of an element or a compound is known as its atomicity.

10. Monoatomic.

11. (i) Tetratomic, (ii) Polyatomic (octatomic).

12. One atomic mass unit is a mass unit equal to exactly one twelfth (1/12th) the mass of one atom of carbon-12.

13. The relative atomic mass of an atom is the average mass of the atom, as compared to 1/12th of the mass of one carbon-12 atom.

14. Initially 1/16th of the mass of naturally occurring oxygen was taken as the atomic mass unit because this unit gave the masses of most elements as whole numbers.

15. The mass of one mole of a substance is called its molar mass. Its unit is gram per mole ($g\ mol^{-1}$).

16. The molecular mass of a substance is the sum of the atomic masses of all the atoms in a molecule, whereas the mass of 1 mole of any substance is called its molar mass.

17. The chemical formula of a compound is a symbolic representation of its composition and the actual number of atoms in one molecule of a pure substance, which may be an atom or a compound.

18. (i) In H_2S, 3 atoms are present
    (ii) In H_3PO_4, 8 atoms are present

Section C: Short Answer Questions

1. Experiments performed by Antoine Laurent Lavoisier and Joseph L. Proust established two laws of chemical combination:
    (i) Law of conservation of mass
    (ii) Law of constant proportions

2. The two drawbacks of Dalton's atomic theory were:
    (i) According to modern theory, an atom is not the ultimate indivisible particle of matter. Today, we know that atoms are divisible, that is, they are themselves made up of particles (protons, electrons, neutrons, etc.).
    (ii) In the case of isotopes of an element, the assumption that atoms of the same element have the same mass does not hold good.

3. No, the law of constant composition is not true or valid for all types of compounds as it is valid only when the compounds are obtained from a single isotope. For example, carbon exists in two common isotopes, ^{12}C and ^{14}C. When it forms CO_2 from ^{12}C, the ratio of masses is 12 : 32 = 3 : 8, but from ^{14}C, the ratio will be 14 : 32 = 7 : 16 which is not same as in the first case.

4. (i) N (Nitrogen), F(Fluorine)
    (ii) Fe (Ferrum), Cu (Cuprum)
    (iii) Ca (Calcium), He (Helium)

5. (i) Fe, (ii) Al, (iii) Ar, (iv) Ag, (v) Na, (vi) Co

6. A molecule is the smallest particle of an element or a compound capable of independent existence under ordinary conditions. It shows all the properties of the substance. For example, the molecule of oxygen is O_2, ozone is O_3, phosphorus is P_4.

7. The molecule of an element is made up of only one kind of atom, such as O_2, N_2, F_2, P_4, S_8. The molecule of a compound is made up of two or more different kinds of atoms in a fixed ratio, such as H_2O, CO_2, H_2S, NH_3, CH_4.

8. Diatomic: HCl, H_2
    Triatomic: H_2O, O_3
    Polyatomic: NH_3, CH_3OH, PCl_5

9. 1 g of hydrogen reacts with oxygen = 8 g
    4 g of hydrogen reacts with oxygen
    = 8 × 4 g = 32 g

10. As 3 g of carbon produces carbon dioxide
    = 11 g
    6 g of carbon produces carbon dioxide
    = 11 × 6/3 = 22 g
    As 3 g of carbon needs 8 g of oxygen
    6 g carbon needs 8 × 6/3 = 16 g
    Hence, the remaining oxygen 40 g – 16 g = 24 g
    (does not take part in the reaction).
    The law of definite proportion is governed by the above data.

11. (i) On mixing these, a white precipitate appears. It is due to the formation of barium sulfate and the following reaction takes place here.

$$BaCl_2(aq) + Na_2SO_4(aq) \rightarrow \underset{\substack{\text{Barium sulfate} \\ \text{(White precipitate)}}}{BaSO_4(ppt.)} + 2NaCl(aq)$$

    (ii) As the mass of the flask with contents before and after the reaction will be found to be the same, this proves the law of conservation of mass.

12. The reaction is $AgNO_3 + NaCl \rightarrow AgCl + NaNO_3$
    Molar masses   170   58.5   (ppt.)
    58.5 g NaCl will react completely and exactly with 170 g of $AgNO_3$
    So 2.925 g NaCl will react with $AgNO_3$ = 8.5 g
    Hence, total mass of reactants = 2.925 + 8.5 = 11.425 g

After filtering off AgCl, $NaNO_3$ will be present in the filtrate.

By using the law of conservation of mass

Total mass of reactants = Total mass of products

11.425 g = Mass of AgCl + Mass of $NaNO_3$

Hence, mass of $NaNO_3$ = 11.425 – 7.175 = 4.25 g

Section D: Long Answer Questions

1. Refer to Section 3.3.1, Features of Dalton's Atomic Theory.

2. (i) Atoms of different elements are very small and their actual masses are extremely small. To solve this problem, we consider the relative atomic mass of the element. The relative atomic mass of hydrogen is 1 u and its corresponding gram atomic mass is 1 g.

   (ii) If the atomic mass of an element is in fractions this means that it exists in the form of isotopes. The atomic mass of such an element is the average of the atomic masses of its isotopes and is generally in fractions.

3. (i) Phosphorus

   (ii) Mercury

   (iii) Radii of atoms are of the order 10^{-10} m or 10^{-8} cm

   (iv) C-60 is known as Buckminsterfullerene and its shape is like a football.

   (v) For copper, 'Cu' is used, which is taken from the Latin word cuprum.

4. (i) Hydrogen (H_2), Oxygen (O_2), Nitrogen (N_2)

   (ii) Hydrogen chloride (HCl), Water (H_2O), Ammonia (NH_3)

   (iii) Calcium carbonate ($CaCO_3$), Calcium hydroxide ($Ca(OH)_2$), Sodium hydroxide (NaOH)

5. (i) Refer to Molecular Mass, under Section 3.4.3, How Do Atoms Exist?

   (ii) (1) Molecular mass of C_6H_5OH
       = 6 × Atomic mass of C + 6 × Atomic mass of H + 1 × Atomic mass of O
       = 6 × 12 + 6 × 1 + 1 × 16 = 72 + 6 + 16
       = 94 u

   (2) Molecular mass of H_3PO_4
       = 3 × Atomic mass of H + 1 × Atomic mass of P + 4 × Atomic mass of O
       = 3 × 1 + 1 × 31 + 4 × 16 = 3 + 31 + 64
       = 98 u

   (3) Molecular mass of NH_4Cl
       = 1 × Atomic mass of N + 4 × Atomic mass of H + 1 × Atomic mass of Cl
       = 1 × 14 + 4 × 1 + 1 × 35.5 = 14 + 4 + 35.5 = 53.5 u

   (4) Molecular mass of $Na_2CO_3.10H_2O$
       = 2 × Atomic mass of Na + 1 × Atomic mass of C + 3 × Atomic mass of O + 10 × (2 × Atomic mass of H + 1 × atomic mass of O)
       = 2 × 23 + 1 × 12 + 3 × 16 + 10 (2 × 1 + 1 × 16)
       = 46 + 12 + 48 + 180 = 286 u

Section E: Case Study or Passage-Based Questions

1. (i) J J Berzelius   (ii) Copper (Cu)
   (iii) Sodium (Na)   (iv) Iron (Fe)
   (v) Argentum (Ag)

2. (i) (c) (ii) (b) (iii) (c) (iv) (d) (v) (b)

High Order Thinking Skills (HOTS) Questions

1. Naturally occurring oxygen is a mixture of atoms of masses 16, 17 and 18 (isotopes of oxygen). Two atomic scales have emerged. Physicists assigned an exact value of 16 to oxygen-16 isotopes as the reference on the physical scale. Chemists assigned an exact value of 16 to the average atomic mass of oxygen on the chemical scale. Hence, a unified scale was made taking 12 with the mass of the C-12 isotope as the reference.

2. (i) There will be no difference in the composition of CO_2 in the two cases, that is, in both cases, C and O will be present in the same ratio by mass (12 : 32 or 3 : 8).

   (ii) This proves the law of definite proportion or constant composition. According to this law, a compound is always made up of the same elements combined together in a fixed ratio by mass.

3. When 3 g of magnesium is burnt in 2 g of oxygen, 5 g of magnesium oxide is produced. It means magnesium and oxygen are combined in the ratio of 3 : 2 to form magnesium oxide.
   As 3 g of magnesium gives 5 g of magnesium oxide

6 g magnesium gives $5 \times 6/3 = 10$ g of magnesium oxide

As 3 g of magnesium needs 2 g of oxygen

6 g of magnesium needs $6/3 \times 2 = 4$ g of oxygen

The oxygen left or unreacted $= 15 - 4 = 11$ g

It is governed by the law of definite proportion, which states that in a chemical substance, the elements are always present in definite proportions by mass.

4. According to the law of conservation of mass "mass can neither be created nor destroyed in a chemical reaction".

    Yes, this law is applicable to chemical reactions. In all chemical reactions, exchange of reactants takes place when products are formed. There is no loss or gain of mass.

    For example, in the following reaction, the total mass of the reactants is equal to the total mass of the products formed.

$$\text{Calcium carbonate} \xrightarrow[\text{(chemical reaction)}]{\text{Heat}} \text{Calcium} + \text{Carbon}$$
$$\text{100 g} \qquad \qquad \text{oxide} \quad \text{dioxide}$$
$$\text{56 g} \qquad \text{44 g}$$
$$56 + 44 = 100 \text{ g}$$

5. (i) No change in mass will take place because the law of conservation of mass holds good.

    (ii) In the first experiment,

    Mass of Mg = 10 g

    Mass of MgO formed = 16.6 g

    So % of Mg in MgO $= \dfrac{10 \text{ g}}{16.6 \text{ g}} \times 100 = 60\%$

    And % of oxygen in MgO $= 100 - 60 = 40\%$

    In the second experiment also, MgO contains 60% Mg and 40% O. It means the given data illustrates the law of constant composition.

6. (i) The balanced chemical equation for the reaction between hydrogen and oxygen to form water is as follows:

$$2H_2 + O_2 \rightarrow 2H_2O$$

    (ii) The reactants in this reaction are hydrogen and oxygen and the product is water.

    (iii) Here, Total mass of reactants = Mass of hydrogen + Mass of oxygen

    $= 0.1 \text{ g} + 0.8 \text{ g} = 0.9$ g

    So Total mass of product = Total mass of water = 0.9 g

    The mass of the product (0.9 g) is equal to the total mass of the reactants (0.9 g). So the given experimental data illustrates the 'law of conservation of mass' in chemical reactions.

    (iv) This answer will be governed by the law of constant proportions. It has been given in this question that when 2 g of hydrogen is burned in 20 g of oxygen, 18 g of water is always obtained. Now, since hydrogen and oxygen combine in the fixed proportion of 1 : 8 by mass to produce 18 g of water, the same mass of water (18 g water) will be obtained even if we burn 2 g of hydrogen in 20 g of oxygen. The extra oxygen $(20 - 16 = 4$ g) will be unreacted here.

    (v) The various values displayed by Shashwat in this episode are knowledge of the law of conservation of mass and law of constant proportions in chemical reactions, and application of knowledge in solving problems.

Hints

Section A: Multiple Choice Questions

1. Antione L Lavoisier established the law of constant proportions.

2. 'The relative number and kinds of atoms are constant in a given compound'. This postulate explains the law of definite proportions.

3. The chemical symbol for sodium is derived from its Latin name 'natrium'. In a 'two-letter' symbol, the first letter is the 'capital letter' but the second letter is the 'small letter'. Therefore, its symbol is 'Na'.

5. 1 amu = 1/12th of the mass of the C-12 atom

6. The atomic mass of magnesium is 24 u.

7. Nitrogen is a diatomic molecule; therefore, it exists as N_2 molecule.

8. In the H_2O_2 molecule, four atoms (2 atoms of H + 2 atoms of O) are present.

10. The molecular mass of CO_2 is = [(Atomic mass of C) + (2 × Atomic mass of O)] = 12 + 2 × 16 = 44 u.

11. Both N_2 and CO have a molecular mass of 28.

12. The correct statement is: The molecules and ions aggregate together in large numbers to form the matter. We cannot see the individual molecules/ions with our eyes; we can only see the substances that are a big collection of molecules/ions.

13.

Element	Mass ratio (x)	Atomic mass (y)	Mole Ratio (x/y)	Simplest ratio
Na	23	23	23/23 = 1	1
Cl	35.5	35.5	35.5/35.5 = 1	1

Thus, the formula of salt is NaCl.

14. The ratio of the mass of hydrogen to the mass of oxygen is always 1 : 8. Thus, if 9 g of water(H_2O) is produced, then 1 g of hydrogen(H_2) and 8 g of oxygen (O_2) are used in the third experiment.

Assertion-Reason Questions

5. The two oxides of copper are CuO and Cu_2O, so the percentages of copper differ.

3.5 IONS

As we know, an atom has sub-atomic particles such as electrons (negatively charged), protons (positively charged) and neutrons (neutral), and in an atom, the number of protons and electrons is equal, so it is neutral. If electrons are removed or added, the neutrality of the atom is lost as the number of protons and electrons will not be equal. The new substance formed that has a charge is known as an ion. An ion is formed by the loss or gain of electrons. We can say that compounds composed of metals and non-metals can contain charged species which are known as ions.

There are two types of ions:

- **Cation:** A positively charged ion is known as a cation. It is formed by the loss of one or more electrons by an atom. The ions of all the metal elements are cations. As we know, in a neutral atom, the number of electrons and protons is always equal. However, in a cation, the number of electrons become less than those of protons. For example (Fig. 3.6),

Li	$\xrightarrow{-1\ electron}$	Li^+
Lithium atom		Lithium ion
Protons = 3 (+ charge)		Protons = 3 (+charge)
Electrons = 3 (– charge)		Electrons = 2 (–charge)
Overall charge = 0		Overall charge = 1+

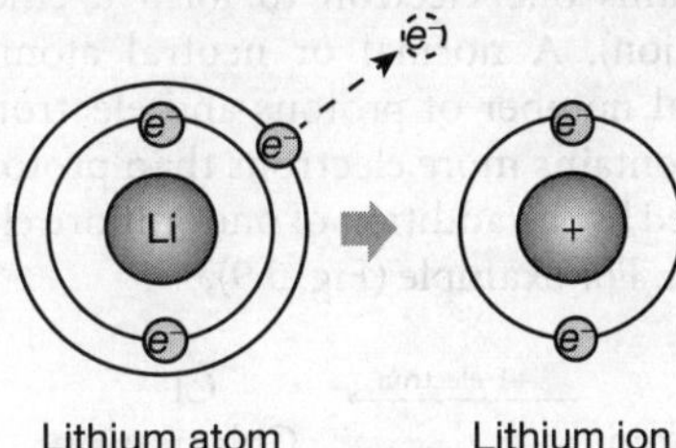

Fig. 3.6 Lithium atom and lithium ion

Both Li and Li^+ have 3 protons. However, the number of electrons in Li and Li^+ are 3 and 2, respectively.

A sodium atom loses one electron to form a sodium ion (Fig. 3.7), Na^+ (cation).

Na	$\xrightarrow{-1\ electron}$	Na^+
Sodium atom		Sodium ion
Protons = 11 (+ charge)		Protons = 11 (+charge)
Electrons = 11 (– charge)		Electrons = 10 (–charge)
Overall charge = 0		Overall charge = 1+

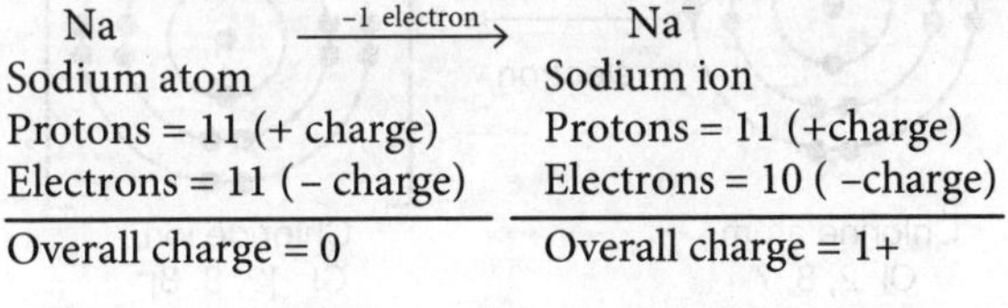
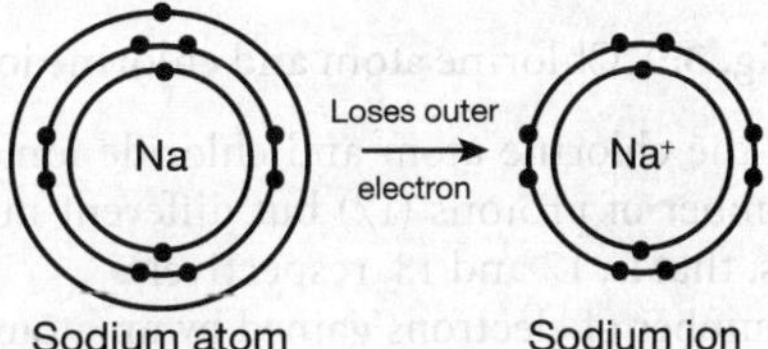

Fig. 3.7 Sodium atom and sodium ion

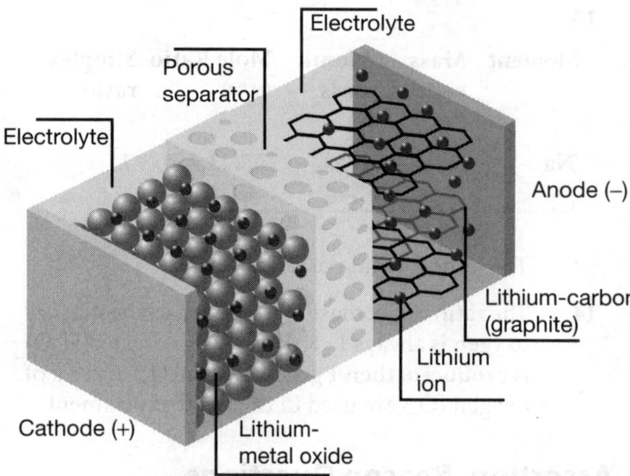

Fig. 3.8 Lithium battery used in mobiles

- **Anion**: A negatively charged ion is known as an anion. An anion is formed by the gain of one or more electrons by an atom. The ions of all the non-metal elements are anions. For example, a chlorine atom gains one electron to form a chloride ion, Cl^- (anion). A normal or neutral atom contains an equal number of protons and electrons, but an anion contains more electrons than protons since it is formed by the addition of one or more electrons to an atom. For example (Fig. 3.9),

$$Cl \xrightarrow{+1\ electron} Cl^-$$

Cl	Cl⁻
Chlorine atom	Chlorine ion
Protons = 17 (+ charge)	Protons = 1 7 (+ charge)
Electrons = 17 (– charge)	Electrons = 1 8 (– charge)
Overall charge = 0	Overall charge = 1 –

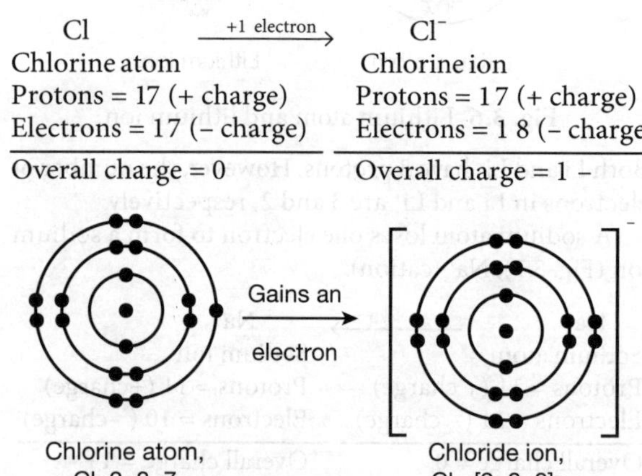

Fig. 3.9 Chlorine atom and chlorine ion

Here, the chlorine atom and chloride ion have the same number of protons (17) but different number of electrons, that is, 17 and 18, respectively.

The number of electrons gained by an atom is equal to the charge assigned on the ion formed. For example, oxygen (O) gains two electron to form oxide ion (O^{2-}) and nitrogen gains three electrons to form nitride ion (N^{3-}).

3.5.1 Simple Ions and Compound Ions

Simple ions: These are ions formed from single atoms. For example, the sodium ion Na^+ is a simple ion because it is formed from a single sodium atom, Na. The chloride ion (Cl^-) is a simple ion as it is formed from a single atom of chlorine (Cl). Some other examples are Mg^{2+}, Ca^{2+}, Al^{3+}, Fe^{2+}, O^{2-}, N^{3-}.

Compound ions (polyatomic ions): These are ions formed from groups of joined atoms. For example, ammonium ion NH_4^+ is a compound ion which is made up of two types of atoms: nitrogen and hydrogen. The carbonate ion CO_3^{2-} is a compound ion which is made up of two types of atoms: carbon and oxygen.

Table 3.9 lists some common simple and compound ions.

Table 3.9   Some common ions

Simple Ions		Compound Ions	
Name of ion	**Formula**	**Name of ion**	**Formula**
Sodium ion	Na^+	Ammonium ion	NH_4^+
Potassium ion	K^+	Hydroxide ion	OH^-
Silver ion	Ag^+	Nitrite ion	NO_2^-
Calcium ion	Ca^{2+}	Nitrate ion	NO_3^-
Magnesium ion	Mg^{2+}	Sulphite ion	SO_3^{2-}
Copper (II) ion	Cu^{2+}	Sulphate ion	SO_4^{2-}
Nickel (II) ion	Ni^{2+}	Phosphate ion	PO_4^{3-}
Zinc ion	Zn^{2+}	Dichromate ion	$Cr_2O_7^{2-}$
Iron (II) ion	Fe^{2+}	Carbonate ion	CO_3^{2-}
Iron (III) ion	Fe^{3+}	Bicarbonate ion	HCO_3^-
Aluminium ion	Al^{3+}	Cyanide ion	CN^-
Halide ion	X^-	Acetate ion	CH_3COO^-
Oxide ion	O^{2-}	Oxalate ion	$C_2O_4^{2-}$
Sulfide ion	S^{2-}	Perchlorate ion	ClO_4^-
Nitride ion	N^{3-}	Chlorate ion	ClO_3^-

Ionic Compounds

Compounds which are composed of ions are known as ionic compounds (Table 3.10). In an ionic compound, the positively charged ions (cations) and negatively charged ions (anions) are held together by strong electrostatic forces of attraction (Coulombic forces). These forces are called **ionic or electrovalent bonds**. As an ionic compound has an equal number of positive and negative ions, the overall or net charge is always zero.

Table 3.10 Some common ionic compounds

Name/Formula/Ions	Crystal structure
Sodium chloride NaCl Na^+ and Cl^-	 *Source:* Looking for all the world..., NASA Expedition 6 crew, 2010, https://commons.wikimedia.org/wiki/File:Close_Up_View_Of_Sodium_Chloride_Crystals.jpg. Licensed under: PD NASA
Potassium chloride KCl K^+ and Cl^-	*Source:* Sharp crystals, totally gemmy, of this stable potassium chloride..., Robert M Lavinsky, https://commons.wikimedia.org/wiki/File:Sylvite-mrz218a.jpg. Licensed under: CC-BY-SA-3.0
Calcium chloride $CaCl_2$ Ca^{2+} and Cl^-	*Source:* Calcium chloride, Firetwister, 2005, https://commons.wikimedia.org/wiki/File:Calcium_chloride_CaCl2.jpg. Licensed under: CC-BY-SA-3.0-migrated

(Continued)

Table 3.10 (Continued)

Name/Formula/Ions	Crystal structure
Magnesium chloride $MgCl_2$ Mg^{2+} and Cl^-	 *Source:* A picture of crystals..., User: Albireo8, 2006, https://commons.wikimedia.org/wiki/File:Iron(II)_sulfate_crystals.jpg. Licensed under: PD-self
Magnesium oxide MgO Mg^{2+} and O^{2-}	*Source:* Tlenek magnezu, Adam Redzikowski, 2015, https://commons.wikimedia.org/wiki/File:Magnesium_oxide_sample.jpg. Licensed under: CC-BY-SA-4.0
Aluminium oxide Al_2O_3 Al^{3+} and O^{2-}	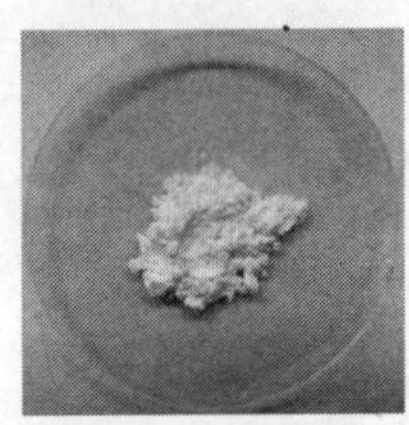 *Source:* Aluminium oxide over white background, Aariuser I, 2021, https://commons.wikimedia.org/wiki/File:Aluminium_oxide_A.jpg. Licensed under: CC-BY-SA-2.0
Sodium hydroxide NaOH Na^+ and OH^-	 *Source:* A base used in...., Hari vinayak santosh, 2016, https://commons.wikimedia.org/wiki/File:Sodium_hydroxide_image_.jpg. Licensed under: CC-BY-SA-4.0

(Continued)

Table 3.10 (Continued)

Name/Formula/Ions	Crystal structure
Copper sulfate $CuSO_4$ Cu^{2+} and SO_4^{2-}	*Source:* Homegrown Copper (II) Sulfate…, Crystal Titan, 2015, https://commons.wikimedia.org/wiki/File:Copper_Sulfate_Crystals.jpg. Licensed under: CC-BY-SA-4.0
Sodium nitrate $NaNO_3$ Na^+ and NO_3^-	*Source:* Sodium nitrate, Rasbak, 2007, https://commons.wikimedia.org/wiki/File:Chilisalpeter_(Sodium_nitrate).jpg. Licensed under: CC-BY-SA-3.0-migrated
Sodium sulfate Na_2SO_4 Na^+ and SO_4^{2-}	*Source:* A sample of anhydrous sodium sulfate…, Walkerma, 2005, https://commons.wikimedia.org/wiki/File:Sodium_sulfate.jpg. Licensed under: PD-self
Iron sulfate $FeSO_4$ Fe^{2+} and SO_4^{2-}	*Source:* A picture of crystals…, User: Albireo8, 2006, https://commons.wikimedia.org/wiki/File:Iron(II)_sulfate_crystals.jpg. Licensed under: PD-self

(Continuod)

Table 3.10 (Continued)

Name/Formula/Ions	Crystal structure
Magnesium sulfate $MgSO_4$ Mg^{2+} and SO_4^{2-}	*Source:* A sample of anhydrous…, Walkerma, 2005, https://commons.wikimedia.org/wiki/File:Magnesium_sulfate_anhydrous.jpg. Licensed under: PD-self
Potassium nitrate KNO_3 K^+ and NO_3^-	*Source:* Sample of Potassium nitrate, Walkerma, 2005, https://commons.wikimedia.org/wiki/File:Potassium_nitrate.jpg. Licensed under: PD-self
Ferric chloride $FeCl_3$ Fe^{3+} and Cl^-	*Source:* Iron(III) chloride hexahydrate, Benjah-bmm27, 2007, https://commons.wikimedia.org/wiki/File:Iron(III)-chloride-hexahydrate-sample.jpg. Licensed under: PD-user

For example, potassium chloride (KCl) is an ionic compound which is made up of an equal number of positively charged potassium ions (K^+) and negatively charged chloride ions (Cl^-). Sodium sulfate (Na_2SO_4) is made up of an equal number of positively charged sodium ions (Na^+) and negatively charged sulfate ions (SO_4^{2-}).

Formula Unit of Ionic Compounds

The simplest combination of ions that gives an electrically neutral unit is known as the formula unit of the ionic compound. *The 'formula unit' of an ionic compound* can be considered as the smallest unit of that compound; it is the equivalent of a 'molecule' of the compound. For example, the formula unit of sodium chloride compound is NaCl (Fig. 3.10) which has one Na^+ ion and one Cl^- ion. The formula unit of sodium carbonate compound is Na_2CO_3 which has two Na^+ ions and one CO_3^{2-} ion.

An ionic compound is made up of an extremely large number of positively charged ions and negatively charged ions joined together by electrostatic forces. Sodium chloride is an ionic compound having a large equal number of sodium ions Na^+ and chloride ions, Cl^-. Hence the actual formula of the sodium chloride compound must be $(Na^+)_n (Cl^-)_n$ or $(Na^+Cl^-)_n$; here n is a very large number (minimum 6). Now we can say that NaCl is the simplest formula of sodium chloride and not its actual formula.

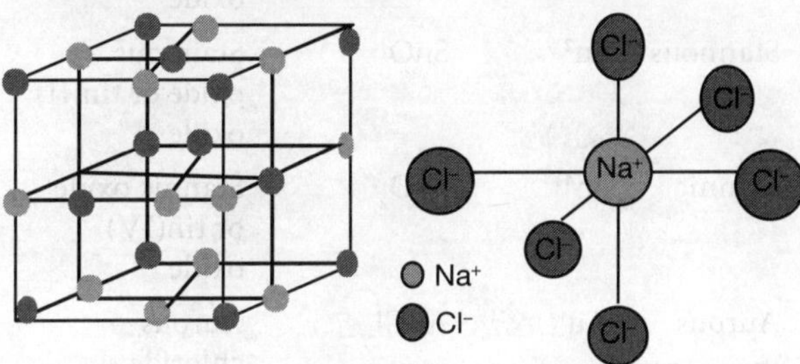

Fig. 3.10 Crystal lattice of NaCl

TEST YOUR KNOWLEDGE

1. Write the cations and anions present (if any) in the following compounds:
    (a) CH_3COONa (b) KCl
    (c) Na_2CO_3 (d) NH_4NO_3

Solution:
(a) CH_3COONa: CH_3COO^-, Na^+
(b) KCl: K^+, Cl^-
(c) Na_2CO_3: $2Na^+$, CO_3^{2-}
(d) NH_4NO_3: NH_4^+, NO_3^-

3.6 VALENCY

The word valency means 'the power to combine'. That is, the combining capacity of an element is known as its valency. It can be measured in terms of the number of H-atoms or double the number of O-atoms that can combine with one atom of an element. As hydrogen has less combining capacity than any other element, its valency is taken as 1, the standard value. For example, in CH_4 the valency of carbon is 4, in NH_3 the valency of nitrogen is 3 and in water (H_2O) the valency of oxygen is 2.

In the modern view, valency is equal to the number of electrons that an atom can share or lose or gain during a chemical reaction. The number of electrons present in the valence orbit of an atom are called **valence electrons**. In order to attain the octet state (8 valence electrons), every atom either loses or gains electrons or share electrons.

Generally, elements having 1, 2, 3 valence electrons are metals and have a tendency to lose electrons and their valency is equal to the number of electrons lost.

For example,

$$\underset{2,\,8,\,1}{Na} - e^- \rightarrow \underset{2,\,8}{Na^+} \quad \text{(monovalent)}$$

$$\underset{2,\,8,\,2}{Mg} - 2e^- \rightarrow \underset{2,\,8}{Mg^{2+}} \quad \text{(divalent)}$$

$$\underset{2,\,8,\,3}{Al} - 3e^- \rightarrow \underset{2,\,8}{Al^{3+}} \quad \text{(trivalent)}$$

Elements having 4 valence electrons are non-metals like C and Si, and they undergo sharing of electrons. Their common valency is 4. For example, in CH_4 and CCl_4, carbon has a valency of 4. Elements with 5, 6 and 7 electrons in their valence orbit are non-metals and have a tendency to gain electrons to complete their octet.

For example,

$$\underset{(2,\,5)}{N} + 3e^- \rightarrow \underset{2,\,8}{N^{3-}} \quad \text{(trivalent)}$$

$$\underset{(2,\,6)}{O} + 2e^- \rightarrow \underset{2,\,8}{O^{2-}} \quad \text{(divalent)}$$

$$\underset{(2,\,8,\,7)}{Cl} + e^- \rightarrow \underset{2,\,8,\,8}{Cl^-} \quad \text{(trivalent)}$$

In general, Valency = Number of valence electrons (up to 4). If there are more than 4 valence electrons, the valency is given as, Valence electrons – 8.

Valency is not negative or positive; it is just a number.

Number of electrons in outermost shell	1	2	3	4	5	6	7	8
Valency	1	2	3	4	3	2	1	0

3.6.1 Valency of Ions

The valency of any ion is equal to the charge present on that ion and it may be 1, 2, 3, 4 and the ions are known as monovalent, bivalent, trivalent and tetravalent, respectively.

For example,

Monovalent ion: H^+, Na^+, K^+, Ag^+, F^-, OH^-

Di or Bivalent: Mg^{2+}, Ca^{2+}, Cu^{2+}, Fe^{2+}, Zn^{2+}, O^{2-}, S^{2-}, SO_4^{2-}

Trivalent: Al^{3+}, Fe^{3+}, PO_4^{3-}

3.6.2 Variable Valency

Some elements exhibit more than one valency in their ions (or compounds). When an atom of an element loses not only its valence electron but also electrons from its penultimate orbit, variable valency is observed (Table 3.11).

If an element has two different positive valences, we use the suffix 'ous' for the lower valency state and the suffix 'ic' for the higher valency state. For example, iron can exist as Fe^{2+} (ferrous) or Fe^{3+} (ferric) in its compounds.

Nowadays, in place of 'ous' or 'ic', we write the valency in roman numbers within brackets. For example, $SnCl_2$ (stannous chloride) can be written as tin(II) chloride and $SnCl_4$ (stannic chloride) can be written as tin(IV) chloride.

Table 3.11 Some basic ions exhibiting variable valency and their compounds

Name of ions	Formula	Formula of Compound	Name of Compound
Cuprous	Cu^+	Cu_2O	Cuprous oxide or copper(I) oxide
Cupric	Cu^{2+}	CuO	Cupric oxide or copper(II) oxide
Mercurous	Hg_2^{2+}	Hg_2Cl_2	Mercurous chloride or mercury(I) chloride

(Continued)

Table 3.11   (Continued)

Name of ions	Formula	Formula of Compound	Name of Compound
Mercuric	Hg^{2+}	$HgCl_2$	Mercuric chloride or mercury(II) chloride
Ferrous	Fe^{2+}	$FeCl_2$	Ferrous chloride or iron(II) chloride
Ferric	Fe^{3+}	$FeCl_3$	Ferric chloride or iron(III) chloride
Plumbous	Pb^{2+}	PbO	Plumbous oxide or lead(II) oxide
Plumbic	Pb^{4+}	PbO_2	Plumbic oxide or lead(IV) oxide
Stannous	Sn^{2+}	SnO	Stannous oxide or tin(II) oxide
Stannic	Sn^{4+}	SnO_2	Stannic oxide or tin(IV) oxide
Aurous	Au^+	$AuCl$	Aurous chloride or gold(I) chloride
Auric	Au^{3+}	$AuCl_3$	Auric chloride or gold(III) chloride

Acidic and Basic Radicals: An acidic radical may be an atom or a group of atoms having a negative charge. That is, anions or electronegative radicals are acidic radicals. A basic radical may be an atom or group of atoms having a positive charge. That is, cations or electropositive radicals are basic radicals. For example, in ammonium sulfate $(NH_4)_2SO_4$, the acidic radical is SO_4^{2-} with a combining power of 2 and the basic radical is NH_4^+ with a combining power of 1.

3.7 Chemical Formulae

A compound is represented in abbreviated form using a chemical formula. The chemical formula of a compound represents the composition of a molecule of the compound in terms of the symbols of the elements present in it. In other words, we can say that the formula of a compound tells us about the type of atoms as well as the number of atoms of various elements present in one molecule of the compound.

In the chemical formula of a compound, the elements are represented by their symbols and the number of atoms of every element is indicated by writing the digits 2, 3, 4, 5, etc., as subscripts or lower figures on the right-hand side of the symbol.

For example, water is a compound having 2 atoms of hydrogen and 1 atom of oxygen. So the formula of water can be given as H_2O. Here, the subscript 2 indicates 2 atoms of hydrogen while oxygen has no subscript, indicating 1 atom of oxygen. Ammonia is a compound having 3 atoms of hydrogen and 1 atom of nitrogen. So the formula of ammonia can be given as NH_3. Here, the subscript 3 indicates 3 atoms of hydrogen while nitrogen has no subscript, indicating 1 atom of nitrogen.

3.7.1 Formulae of Elements

The chemical formula of an element is a statement of the composition of its molecule in which the symbol tells us about the element and the subscript tells us how many atoms are present in one molecule. For example, one molecule of hydrogen element contains two atoms of hydrogen; hence the formula of hydrogen is H_2.

3.7.2 Formulae of Ionic Compounds

Ionic compounds can be formed by a combination of metals and non-metals. While writing the formula of an ionic compound, the metal element is written on the left hand side and the non-metal element on the right-hand side. In naming an ionic compound, the metal element is named as it is; however, the name of the non-metal element is changed to end with 'ide'. For example:

  (i) CaO $(Ca^{2+}O^{2-})$ is named calcium oxide
  (ii) KCl (K^+Cl^-) is named potassium chloride
  (iii) CuS $(Cu^{2+}S^{2-})$ is named copper sulfide
  (iv) AlN $(Al^{3+}N^{3-})$ is named aluminium nitride

3.7.3 Formulae of Molecular Compounds

Molecular compounds can be formed by a combination of two different non-metal elements. While writing the formula of a molecular compound, the less electronegative non-metal element is written on the left-hand side while the more electronegative non-metal element is written on the right-hand side. In naming a molecular compound, the name of the less electronegative non-metal is written as it is; however, the name of the more electronegative non-metal is changed to end with 'ide'. For example (Fig. 3.11):

  (i) HCl (H^+Cl^-) is named hydrogen chloride
  (ii) HI (H^+I^-) is named hydrogen iodide
  (iii) H_2S $(2H^+S^{2-})$ is named hydrogen sulfide

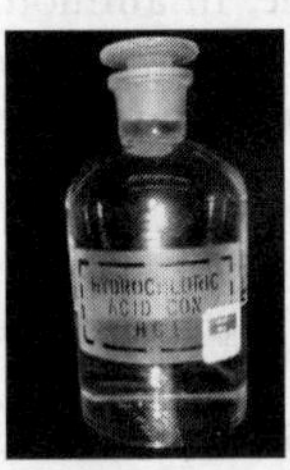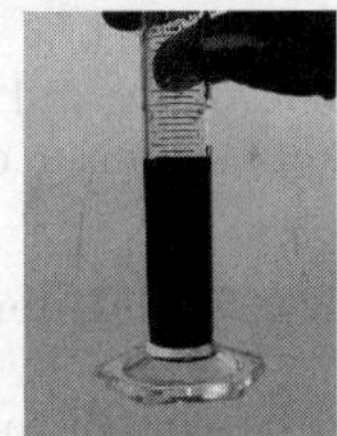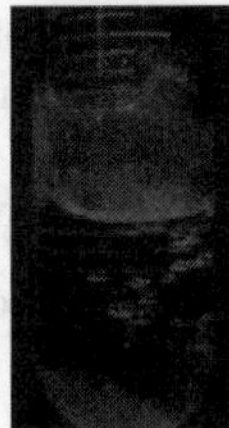

Hydrogen chloride   Hydrogen iodide   Hydrogen sulfide

Fig. 3.11 Some important molecular compounds

In case more than one atom of an element is present in a molecular compound, the number of atoms is indicated by using appropriate prefixes in the molecular formula. For 1, 2, 3, 4, 5 and 6 atoms, the prefixes used are mono, di, tri, tetra, penta and hexa, respectively. For example,

(i) CO is named carbon monoxide
(ii) CO_2 is named carbon dioxide
(iii) SO_2 is named sulfur dioxide
(vi) SO_3 is named sulfur trioxide
(v) PCl_3 is named phosphorus trichloride
(vi) PCl_5 is named phosphorus pentachloride
(vii) CCl_4 is named carbon tetrachloride
(viii) SF_6 is named sulfur hexafluoride

3.7.4 Writing the Formula of a Compound

The chemical formulae of different compounds can be written easily, but for this we need to learn the symbols and combining capacity or valency of the elements. Valency can be used to find out how the atoms of an element will combine with the atom(s) of another element to form a chemical compound. By knowing the valences of elements, we can write the formula of any compound by balancing the valences of the different atoms which are present in that compound. For example, if a compound is made up of two elements hydrogen and nitrogen, we must adjust the number of hydrogen atoms and the number of nitrogen atoms in such a way that the total valences of the hydrogen and nitrogen atoms become equal, to obtain the correct formula of the compound. For example, in ammonia the valency of nitrogen atoms is three and that of hydrogen atoms is one. So we will take one nitrogen atom and three hydrogen atoms to obtain the formula NH_3 for ammonia.

To write the chemical formula for a binary atomic compound:

Step 1: First, write the symbols of all the elements which form the compound.

Step 2: Below the symbol of every element, write down its valency. The valences or charges on the ions must be balanced.

Step 3: Finally, cross-over the valences of the combining atoms. That is, with the first atom, write the valency of the second atom as a subscript and with the second atom, write the valency of the first atom as a subscript.

For example,

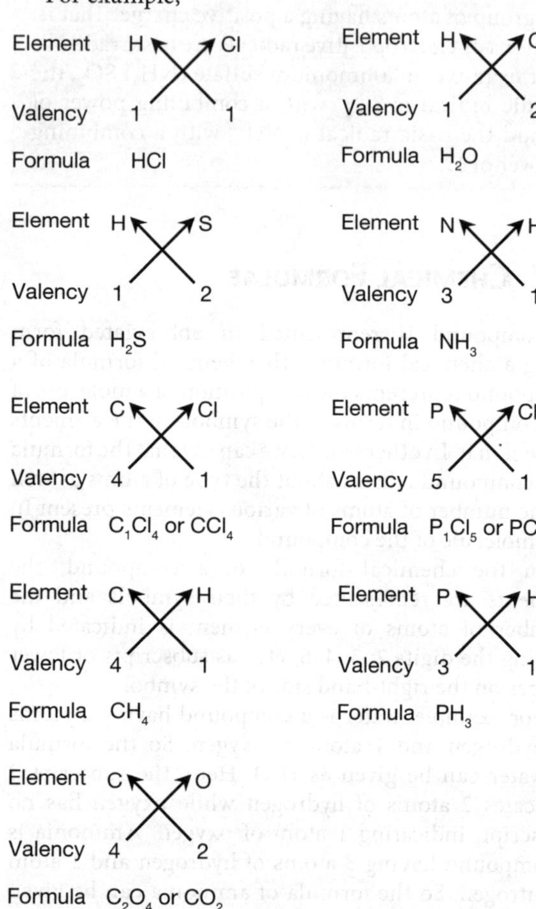

To write the formula of a simple ionic compound: Apply the rules discussed above. For example,

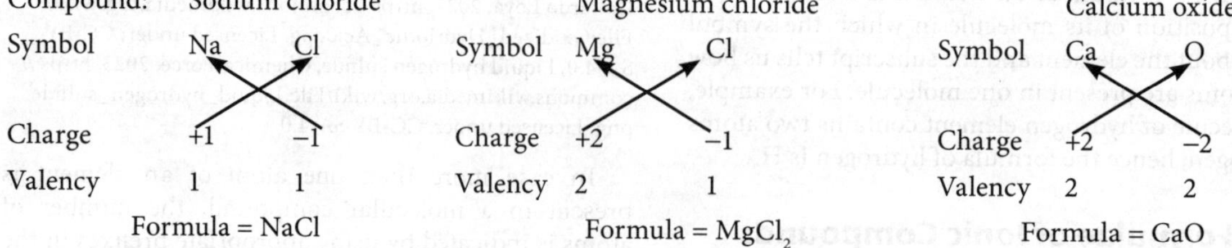

To write the formula of a polyatomic ionic compound: In the case of ionic compounds, follow these steps to write the formula:

Step 1: Write the symbols of the formulae of the ions of the compound side by side, with the positive ion on the left-hand side and the negative ion on the right-hand side.

Step 2: Enclose the polyatomic ion within brackets.

Step 3: Write the valency of each ion below its symbol.

Step 4: Reduce the valency numerals to a simple ratio by dividing with a common factor, if any.

Step 5: Cross the valences but do not write the charges positive or negative of the ions.

Formula of calcium nitrate.
Step 1: Writing the formula of the ions:

$$Ca^{2+} \qquad NO_3^{-}$$

Step 2: Ca^{2+} $(NO_3)^{-}$

Step 3: Ca^{2+} $(NO_3)^{-}$
        2   1

Step 4: Not applicable, because the ratio is already simple.

Step 5:

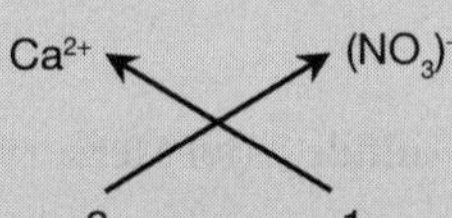

Thus, the formula of calcium nitrate is $Ca(NO_3)_2$.

Formula of calcium phosphate.
Step 1: Writing the formula of the ions:

$$Ca^{2+} \qquad PO_4^{\,3-}$$

Step 2: Ca^{2+} $(PO_4)^{3-}$

Step 3: Ca^{2+} $(PO_4)^{3-}$
        2   3

Step 4: Not applicable, because the ratio is already simple.

Step 5:

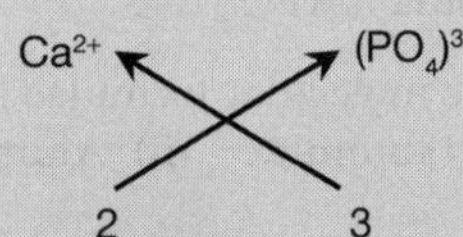

Thus, the formula of calcium phosphate is $Ca_3(PO_4)_2$.

Some more examples,

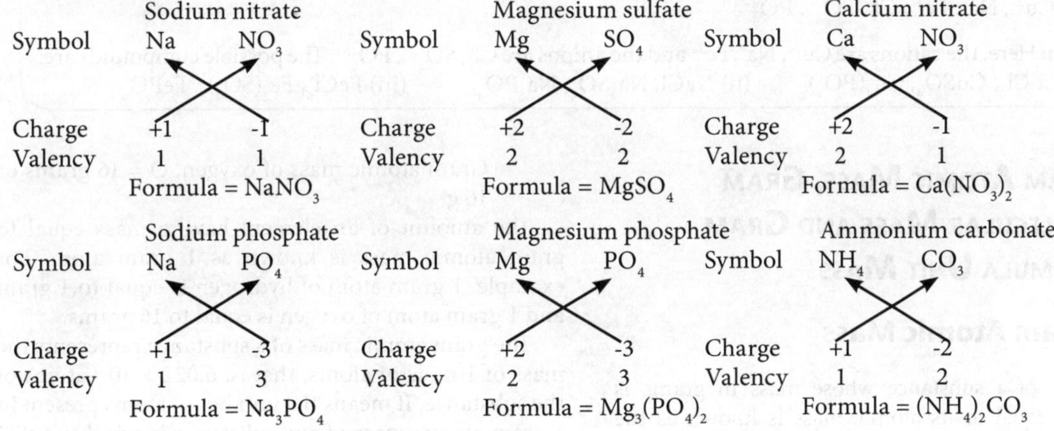

Significance of the formula of a substance:
- It represents the name of the substance.
- It represents one molecule of the substance.
- It gives the names of all the elements present in the molecule.
- It gives the number of atoms of each element present in one molecule.
- It represents a definite mass of the substance.
- It also represents one mole of molecules of the substance. That is, the formula also represents 6.022×10^{23} molecules of the substance.

TEST YOUR KNOWLEDGE

1.   An element A is trivalent. Write the formula of its:
(i) chloride, (ii) oxide, (iii) sulfate.

Solution:

Symbol	A	Cl		Symbol	A	O		Symbol	A	SO_4
Charge	+3	−1		Charge	+3	−2		Charge	+3	−2
Valency	3	1		Valency	3	2		Valency	3	2

Formula = ACl_3 Formula = A_2O_3 Formula = $A_2(SO_4)_3$

2.   A element A forms the oxide M_2O_3. Write the formula of its phosphate.

Solution: In A_2O_3 the total charge on the metal atom must be equal to +6 as $3O^{2-}$ ions are present or the total negative charge is −6.

Now the charge on one metal ion = +6/2 = +3

Symbol	A	PO_4
Charge	+3	−3
Valency	3	3

Formula = APO_4

3.   Write the molecular formulae for the following compounds:
(i) Copper(II) bromide   (ii) Aluminium(III) nitrate   (iii) Iron(III) sulfide   (iv) Mercury(II) chloride

Solution:
(i) Copper(II) bromide: $CuBr_2$ (ii) Aluminium(III) nitrate: $Al(NO_3)_3$
(iii) Iron(III) sulfide: Fe_2S_3 (iv) Mercury(II) chloride: $HgCl_2$

4.   Write the molecular formulae for all the compounds that can be formed by the combination of the following ions:
Cu^{2+}, Na^+, Fe^{3+}, Cl^-, SO_4^{2-}, PO_4^{3-}

Solution: Here, the cations are Cu^{2+}, Na^+, Fe^{3+} and the anions are Cl^-, SO_4^{2-}, PO_4^{3-}. The possible compounds are:
(i) $CuCl_2$, $CuSO_4$, $Cu_3(PO_4)_2$ (ii) $NaCl$, Na_2SO_4, Na_3PO_4 (iii) $FeCl_3$, $Fe_2(SO_4)_3$, $FePO_4$

3.8 Gram Atomic Mass, Gram Molecular Mass and Gram Formula Unit Mass

3.8.1 Gram Atomic Mass

The amount of a substance whose mass in grams is numerically equal to its atomic mass is known as the gram atomic mass of that substance. The atomic mass of a substance expressed in grams is known as its gram atomic mass. In order to write the gram atomic mass of a substance, we write its atomic mass and then replace the atomic mass unit 'u' with the word 'gram' or its symbol 'g'.

For example:
(i) Atomic mass of carbon, C = 12 u
Gram atomic mass of carbon, C = 12 grams or 12 g
(ii) Atomic mass of oxygen, O = 16 u
Gram atomic mass of oxygen, O = 16 grams or 16 g

The amount of an element having mass equal to gram atomic mass is known as 1 gram atom. For example, 1 gram atom of hydrogen is equal to 1 gram and 1 gram atom of oxygen is equal to 16 grams.

The gram atomic mass of a substance represents the mass of 1 mole of atoms, that is, 6.022×10^{23} atoms of the substance. It means the number of atoms present in 1 gram atomic mass of any substance is equal to 6.022×10^{23} atoms. For example, 12 g or u of carbon has 6.022×10^{23} carbon atoms.

3.8.2 Gram Molecular Mass

As molecules are made up of two or more atoms of the same/different elements, the molecular mass may be calculated as the sum of the atomic masses of all the atoms in a molecule of that substance. It is therefore the relative mass of a molecule expressed in atomic mass units (u). The molecular mass of a substance expressed in grams is known as its gram molecular mass.

For example, Molecular mass of oxygen, O_2 = 32 u
Gram molecular mass of oxygen, O_2 = 32 grams
Molecular mass of nitrogen, N_2 = 28 u
Gram molecular mass of nitrogen, N_2 = 28 grams

In the case of polyatomic molecules, the molecular mass can be found out by adding the atomic masses of all the elements. For example, ammonia has the formula NH_3. It consists of one atom of N and three atoms of H. The atomic mass of N and H are 14.0 and 1, respectively. Therefore,

Molecular mass of NH_3 = Atomic mass of N + 3 × Atomic mass of H
= 14 + 3 × 1 = 17 u

Sulfuric acid has the formula H_2SO_4. It consists of two H, one S and four O atoms. The atomic masses of H, S and O are 1,32 and 16, respectively. Therefore,

Molecular mass of H_2SO_4 = (2 × Atomic mass of H) + (1 × Atomic mass of S) + (4 × Atomic mass of O)
= (2 × 1) + (1 × 32) + (4 × 16) = 98 u

The amount of the substance having mass equal to its gram molecular mass is known as 1 gram molecule. For example, 2 g hydrogen (H_2) is equal to 1 g molecule of H_2 and 32 g oxygen (O_2) is equal to 1 g molecule of O_2.

The gram molecular mass of a substance represents the mass of 1 mole of the substance, that is, 6.022×10^{23} molecules of the substance. It means the number of molecules present in 1 gram molecular mass of any substance is equal to 6.022×10^{23} molecules. For example, 44 g or u of carbon dioxide has 6.022×10^{23} carbon dioxide molecules.

Competition Edge

Relative molecular mass (RMM) expresses how many times a molecule of a substance is heavier than $\dfrac{1}{12}$ th of the mass of an atom of carbon (carbon-12).

Thus, Molecular mass = $\dfrac{\text{Mass of a molecule}}{\dfrac{1}{12}\text{th mass of a carbon atom (carbon-12)}}$

3.8.3 Formula Unit Mass

The formula unit mass of a substance is the sum of the atomic masses of all the atoms in a formula unit of a compound. Formula unit mass is calculated in the same manner as molecular mass. The only difference is that we use the word formula unit for those substances whose constituent particles are ions. For example, sodium chloride, as discussed above, has the formula unit NaCl. Its formula unit mass can be calculated as 1 × 23 + 1 × 35.5 = 58.5 u.

Formula unit mass expressed in grams is known as gram formula unit mass and this amount is known as one gram formula unit. For example, the gram formula unit mass of NaCl is equal to 58.5 u.

TEST YOUR KNOWLEDGE

1. Calculate the formula mass of sodium carbonate (Na_2CO_3) and sodium sulfate (Na_2SO_4). (Given, atomic masses: Na = 23 u; C = 12 u; O = 16 u; S = 32 u)

Solution: We know that the formula mass of a compound is the sum of the atomic masses of all the element present in it. So,

Formula mass of Na_2CO_3 = Mass of 2 Na atoms + Mass of one C atom + Mass of 3 O atoms
= 2 × 23 + 12 + 3 × 16
= 46 + 12 + 48 = 106 u

Hence, the formula mass of sodium carbonate is 106 u.

Formula mass of Na_2SO_4 = Mass of 2 Na atoms + Mass of one S atom + Mass of 4 O atoms
= 2 × 23 + 32 + 4 × 16
= 46 + 32 + 64 = 142 u

Hence, the formula mass of sodium sulfate is 142 u.

2. Calculate the molar masses of the following substances:
(i) C_2H_4, (ii) H_2SO_4, (iii) H_3PO_4, (iv) $HClO_4$
(Atomic masses: C = 12 u; H = 1 u; S = 32 u; P = 31 u; Cl = 35.5 u; O = 16 u)

Solution: The molar mass of a molecule is equal to the sum of the atomic masses of all the elements present in the molecule.
So
(i) Molar mass of C_2H_4 = Atomic mass of C × 2 + Atomic mass of H × 4
= 12 × 2 + 1 × 4 = 24 + 4 = 28 g/mol
(ii) Molar mass of H_2SO_4 = Atomic mass of H × 2 + Atomic mass of S × 1 + Atomic mass of O × 4
= 1 × 2 + 32 × 1 + 16 × 4 = 2 + 32 + 64 = 98 g/mol

(iii) Molar mass of H_3PO_4 = Atomic mass of H $\times$ 3 + Atomic mass of P $\times$ 1 + Atomic mass of O $\times$ 4
$$= 1 \times 3 + 31 \times 1 + 16 \times 4 = 3 + 31 + 64$$
$$= 98 \text{ g/mol}$$

(iv) Molar mass of $HClO_4$ = Atomic mass of H $\times$ 1 + Atomic mass of Cl $\times$ 1 + Atomic mass of O $\times$ 4
$$= 1 \times 1 + 35.5 \times 1 + 16 \times 4 = 1 + 35.5 + 64$$
$$= 100.5 \text{ g/mol}$$

3.9 MOLE

In order to understand the concept of mole, let us consider the reaction of hydrogen and oxygen to form water.

$$2H_2 + O_2 \rightarrow 2H_2O$$

This reaction predicts that

(i) Two molecules of hydrogen combine with one molecule of oxygen to form two molecules of water, or

(ii) 4 u of hydrogen molecules combine with 32 u of oxygen molecules to form 36 u of water molecules.

Fig. 3.12 This figure shows 12 grams of carbon element having 6.022×10^{23} carbon atoms, or 1 mole of carbon atoms

It means that the quantity of a substance can be characterised by its mass or the number of molecules. However, a chemical reaction equation indicates directly the number of atoms or molecules taking part in that reaction. Hence, it is more convenient to refer to the quantity of a substance in terms of the number of its molecules or atoms, rather than their masses. That is why a new unit called mole was introduced around 1896 by Wilhelm Ostwald who derived the term from the Latin word *'moles'* meaning a 'heap' or a 'pile'. A substance may be considered as a heap of atoms or molecules. One mole of any species (atoms, molecules,

ions or particles) is that quantity in number having a mass equal to its atomic or molecular mass in grams. For example, 12 g of carbon (Fig. 3.12) and 44 g of CO_2 both represent 1 mole.

The number of particles (atoms, molecules or ions) present in 1 mole of any substance is fixed, with a value of 6.022×10^{23}. This is an experimentally obtained value, given by Amedeo Avogadro and this number is known as the **Avogadro constant or Avogadro number**. It is represented by N_0 or N_A.

1 mole of any species = 6.022×10^{23} in number

1 mole of atoms = 6.022×10^{23} atoms

1 mole of molecules = 6.022×10^{23} molecules

1 mole of electrons = 6.022×10^{23} electrons

Mole of Atoms:

1 Mole of atoms of an element = Gram atomic mass of the element

For example, the atomic mass of nitrogen (N) is 14 u. So the gram atomic mass of nitrogen is 14 grams. Hence, 1 Mole of nitrogen atoms = Gram atomic mass of nitrogen = 14 grams

It is important to note that the symbol of an element represents 1 mole of atoms of that element. For example, N represents 1 mole of nitrogen atoms and 2 N represents 2 moles of nitrogen atoms.

Mole of Molecules:

1 Mole of molecules of a substance = Gram molecular mass of the substance

For example, the molecular mass of nitrogen (N_2) is 28 u, so the gram molecular mass of oxygen is 28 grams.

Hence, 1 Mole of nitrogen molecules = Gram molecular mass of nitrogen = 28 grams

It is important to note that the molecular formula of a substance represents 1 mole of molecules of that substance. For example, N_2 represents 1 mole of nitrogen molecules and $2N_2$ represents 2 moles of nitrogen molecules.

Mole in Terms of Volume or Molar Volume: In gaseous substances, it is observed that one mole of any such substance at the standard conditions of temperature and pressure, that is, STP condition 0°C or 273 K and 1 atm pressure, occupies the same volume which is known as molar volume and its value is equal to 22.4 L or 22400 mL or 22.4 dm^3.

It means that at STP, one mole of a substance has a volume of 22.4 L (molar volume). It is also known as gram molecular volume as it is the volume possessed by 6.023×10^{23} molecules.

Summary of Mole

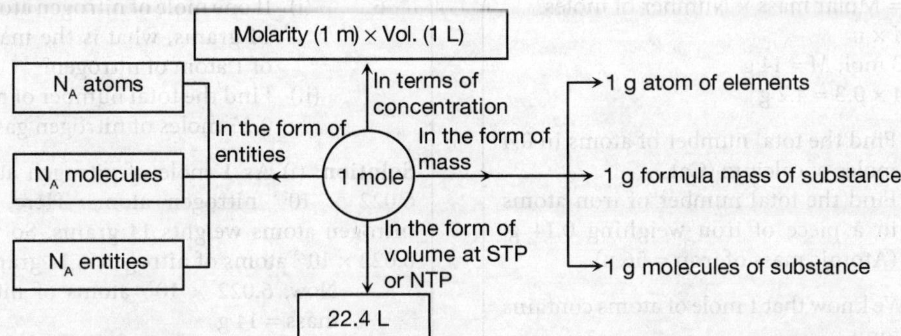

Quick Methods to Find Mole

(i) Number of moles $(n) = \dfrac{\text{Mass of element (w)}}{\text{Atomic mass (M)}}$

(ii) Number of moles $(n) = \dfrac{\text{Mass of compound (w)}}{\text{Molecular mass (M)}}$

(iii) Number of moles $(n) =$
$$\dfrac{\text{Given number of atoms or entities (N)}}{\text{Avogadro number } (N_A)}$$

These relations can also be interchanged as follows:

Mass of element $(w) = n \times M$

Mass of molecule $(w) = n \times M$

Number of particles of element, $N = n \times N_A$

Number of moles $(n) = \dfrac{\text{Given volume (V)}}{\text{Volume at NTP (22.4 L)}}$

Total number of molecules = Moles × Avogadro number

Total number of atoms = Moles × Avogadro number × Number of atoms in 1 molecule

---

TEST YOUR KNOWLEDGE

1. (i) How many moles are 10 grams of calcium? (Atomic mass of calcium = 40 u).

(ii) What is the mass of 0.4 moles of aluminium atoms? (Atomic mass of Al = 27 u)

Solution: (i) As 40 g of calcium = 1 mole of calcium

$$10 \text{ g of calcium} = \dfrac{1}{40} \times 10 \text{ mole}$$
$$= \dfrac{1}{4} \text{ mole} = 0.25 \text{ mole}$$

Hence, 0.25 moles are present in 10 grams of calcium.

(ii) As the atomic mass of aluminium is given to be 27 u, 1 mole of aluminium atoms has a mass of 27 grams.

Now, 1 mole of aluminium atoms = 27 g

4 moles of aluminium atoms = 27 × 0.4 g = 10.8 g

Hence, the mass of 4 moles of aluminium atoms is 10.8 grams.

2. (i) An ornament of silver contains 40 g of silver. Calculate the moles of silver present. (Atomic mass of silver = 108 u)

(ii) How many moles of CO_2 are present in 5.62 g?

Solution: (i) Moles of silver, $n = \dfrac{m}{M}$

Mass of silver, $m = 40$ g

Molar mass of silver, $M = 108$ g

$$\therefore n = \dfrac{40}{108} = 0.370 \text{ mol}$$

(ii) Molecular mass of $CO_2 = 12 + 2 + 16 = 44$ u

Molar mass of CO_2 (M) = 44 g

Mass of CO_2 (m) = 5.62 g

$$\text{Moles of } CO_2 \; n = \dfrac{m}{M} = \dfrac{5.62}{44} = 0.127 \text{ mol}$$

Note: You can also use the unitary method to solve it.

3. Calculate the mass of
(i) 0.3 moles of N_2 gas
(ii) 0.3 moles of N atoms

Solution: (i) 0.3 moles of N_2 gas

Mass = Molar mass × Number of moles

$m = M \times n$

$M = 28$ g, $n = 0.3$

$\therefore m = 28 \times 0.3 = 8.4$ g

(ii) Mass = Molar mass × Number of moles
$m = M \times n$
$n = 0.3$ mol, $M = 14$ g
$m = 14 \times 0.3 = 4.2$ g

4.   (i) Find the total number of atoms in 0.4 moles of calcium (Ca).
(ii) Find the total number of iron atoms in a piece of iron weighing 0.14 g. (Atomic mass of iron = 56 u)

Solution: (i) We know that 1 mole of atoms contains 6.022×10^{23} atoms.

Now, 1 mole of calcium atoms = 6.022×10^{23} atoms
0.4 mole of calcium atoms = $6.022 \times 10^{23} \times 0.4$ atoms
= 2.4088×10^{23} atoms
Hence, 0.4 moles of calcium element contain 2.4088×10^{23} atoms.

(ii) 56 g of iron contains = 6.022×10^{23} atoms
0.14 g of iron contains
= $\dfrac{6.022 \times 10^{23}}{56} \times 0.14 = 1.5 \times 10^{21}$

Hence, a piece of iron metal having a mass of 0.14 grams contains 1.5×10^{21} atoms of iron.

5.   (i) If one mole of nitrogen atoms weighs 14 grams, what is the mass in grams of 1 atom of nitrogen?
(ii) Find the total number of molecules in 0.25 moles of nitrogen gas (N_2).

Solution: (i) As 1 mole of nitrogen atoms means 6.022×10^{23} nitrogen atoms. Here, 1 mole of nitrogen atoms weights 14 grams. So the mass of 6.022×10^{23} atoms of nitrogen is 14 grams.

Now, 6.022×10^{23} atoms of nitrogen have mass = 14 g

So 1 atom of nitrogen has mass = $\dfrac{14}{6.022 \times 10^{23}}$ g = 2.32×10^{-23} g

Hence, the absolute mass of 1 atom of nitrogen is 2.32×10^{-23} gram.

(ii) 1 mole of nitrogen contains = 6.022×10^{23} molecules
0.25 moles of nitrogen contain = $6.022 \times 10^{23} \times 0.25 = 1.505 \times 10^{23}$ molecules
Hence, 0.25 moles of nitrogen contain 1.505×10^{32} molecules.

Quick Review

- In the modern view, valency is equal to the number of electrons that an atom can share or lose or gain during a chemical reaction. The number of electrons present in the valence orbit of an atom is called valence electrons.
- A radical is an atom or group of atoms having the same or different elements that behave like a single unit with a positive or negative charge.
- An acidic radical is an atom or group of atoms having a negative charge. Anions or electronegative radicals are acidic radicals. A basic radical is an atom or group of atoms having a positive charge. Cations or electropositive radicals are basic radicals.
- The chemical formula of a compound represents the composition of a molecule of the compound in terms of the symbols of the elements present in it.
- The molecular mass of a substance is the sum of the atomic masses of all the atoms in a molecule of that substance. It is therefore the relative mass of a molecule expressed in atomic mass units (u).
- One mole of any species (atoms, molecules, ions or particles) is that quantity in number having a mass equal to its atomic or molecular mass in grams. For example, 12 g carbon and 44 g of CO_2 both represent 1 mole.
- Moles of atoms: We know that 1 mole of atoms of an element = Gram atomic mass of the element = 6.022×10^{23} atoms
1 Mole of atoms = Gram atomic mass
1 Mole of atoms = 6.022×10^{23} atoms
Gram atomic mass = 6.022×10^{23} atoms
- Moles of molecules:
1 Mole of molecules = Gram molecular mass
1 Mole of molecules = 6.022×10^{23} molecules
Gram molecular mass = 6.022×10^{23} molecules

Exercise 3.2

Questions marked * are practical based.

Section A: Multiple Choice Questions
(1 Mark)

1. What is the valency of CO_3^{2-}?
   (a) 1 (b) 3
   (c) 2 (d) 4

2. The relative atomic mass of chlorine is 35.5. What is the mass of 2 moles of chlorine atoms?
   (a) 71.0 g (b) 142 g
   (c) 17.8 g (d) 35.5 g

3. Which one of the following is the correct formula of chromium(III) hydroxide?
   (a) Cr_3OH (b) $Cr_3(OH)$
   (c) $CrOH_3$ (d) $Cr(OH)_3$

4. What mass of dinitrogen oxide N_2O occupies a volume of 18 dm³ measured at STP?
   (a) 24g (b) 22g
   (c) 44g (d) 33g

5. Which one of the following is the correct formula of cerium(IV) sulfate? (The symbol for cerium is Ce.)
   (a) $Ce(SO)_4$ (b) $Ce(SO_4)_2$
   (c) $Ce(SO_4)_4$ (d) Ce_4SO_4

6. When magnesium hydroxide reacts with hydrochloric acid, magnesium chloride and water are formed. Which one of the following gives the correct balancing numbers in the equation?

	Magnesium hydroxide	Hydrochloric acid	Magnesium chloride	Water
A	1	1	1	1
B	1	2	1	1
C	1	2	1	2
D	1	2	2	2

7. If the formula of nitrous acid is HNO_2, which one of the following is the formula of calcium nitrite?
   (a) $CaNO_2$ (b) Ca_2NO_2
   (c) $Ca(NO_2)_2$ (d) $Ca(NO)_2$

8. If the formula of iron(III) arsenate is $FeAsO_4$, the formula of iron(II) arsenate will be:

   (a) $Fe(AsO_4)_2$ (b) $Fe_2(AsO_4)_3$
   (c) $Fe(AsO_4)_3$ (d) $Fe_3(AsO_4)_2$

9. In athletics, banned drugs such as nandrolone have been taken illegally to improve performance. Nandrolone has the molecular formula $C_{18}H_{26}O_2$. What is the relative molecular mass of nandrolone?
   (a) 150 (b) 46
   (c) 306 (d) 274

10. 18 g of water contains the same number of molecules as
    (a) 16 g of oxygen gas
    (b) 18 g of ammonia
    (c) 2 g of hydrogen gas
    (d) 14 g of nitrogen gas

11. One mole of oxygen (O_2) and one mole of nitrogen (N_2) at STP:
    (a) Contain the same elements
    (b) Have the same mass
    (c) Weigh the same
    (d) Occupy 24 dm³ volume

12. What mass of methane (CH_4) occupies the same volume, measured at STP as 11 g of carbon dioxide?
    (a) 176 g (b) 264 g
    (c) 4 g (d) 16 g

13. Calculate the number of aluminium ions which are present in 0.0027 g of aluminium oxide:
    (a) 6.022×10^{15} Al^{3+} ions
    (b) 6.022×10^{19} Al^{3+} ions
    (c) 6.022×10^{14} Al^{3+} ions
    (d) 6.022×10^{23} Al^{3+} ions

14. The number of molecules of sulfur (S8) present in 16 g of solid sulfur will be:
    (a) 3.76×10^{16} molecules
    (b) 3.76×10^{22} molecules
    (c) 2.76×10^{14} molecules
    (d) 2.76×10^{22} molecules

15. Which of the following correctly represents 360 g of water?
    (i) 2 moles of H_2O
    (ii) 20 moles of water
    (iii) 6.022×10^{23} molecules of water
    (iv) 1.2044×10^{25} molecules of water
    (a) (i) (b) (i) and (iv)
    (c) (ii) and (iii) (d) (ii) and (iv)

Assertion–Reason Questions

Instruction: In the following question two statements (Assertion) A and Reason (R) are given Mark.
(a) If A and R both are correct and R is the correct explanation of A;
(b) If A and R both are correct but R is not the correct explanation of A;
(c) A is true but R is false;
(d) A is false but R is true

Assertion	Reason
1. The valences of cations of magnesium and aluminium are +2 and +3, respectively.	1. The valency of any ion is equal to the charge present on the ion.
2. 4 g methane and 8 g oxygen gas have the same number of moles.	2. The number of molecules in them are different.
3. The formula of calcium phosphate is $CaPO_4$.	3. In a neutral molecule, total positive charge and total negative charge must be equal.
4. In mercurous and mercuric ions, the positive charge present is +1 and +2, respectively.	4. The correct formulae of mercurous chloride and mercuric chloride are Hg_2Cl_2 and $HgCl_2$, respectively.
5. The volume occupied by one mole of any gas at STP is known as molar volume.	5. At STP, molar volume of 8 g of methane is 22.4 L.
6. The formula unit mass of calcium chloride is 120 u.	6. Formula unit mass is used mainly for ionic compounds like NaCl.

Section B: Very Short Answer Questions   (2 Marks)

1. What are ions? Write the formulae of two divalent cations and anions.

2. Why do atoms form ions?

3. What are polyatomic ions? Give two examples.

4. What is the number of electrons in an Mg atom and an Mg^{2+} ion?

5. Name the compound $Al_2(SO_4)_3$ and mention the ions present in it.

6. Write the valency of sulfur in H_2S, SO_2 and SO_3.

7. What is the difference between 2H and H_2?

8. What is the valency of calcium in $CaCO_3$?

9. The valency of an element A is 4. Write the formula of its oxide.

10. Write the names of the following compounds:
     (i) K_3PO_4,   (ii) Ca_3N_2

11. Avogadro's number represents how many particles?

12. The formula of the carbonate of a metal M is M_2CO_3. Write the formula of its chloride.

13. MNO_3 is the formula of the nitrate of metal M. Write the formula of its oxide.

14. Where do we use the words mole and mol?

15. State the number of hydrogen atoms in 1 g of hydrogen.

Section C: Short Answer Questions   (3 Marks)

1. What are ionic and molecular compounds? Give examples.

2. What are polyatomic ions? Give examples.

3. Give one example each of
    (i) Monovalent cation
    (ii) Bivalent cation
    (iii) Monovalent anion
    (iv) Bivalent anion

4. Write the cations and anions present (if any) in the following compounds:
    (i) CH_3COONa
    (ii) $NaCl$
    (iii) H_2
    (iv) NH_4NO_3

5. What do you mean by a chemical formula? Give some examples.

6. Define formula unit mass. Calculate the formula unit mass of NaCI (atomic mass of Na = 23 u, Cl = 35.5 u)

7. Write the molecular formulae of all the compounds that can be formed by the combination of following ions:
    Cu^{2+}, Na^+, Fe^{3+}, Cl, SO_4^{2-}, PO_4^{3-}

8. Write down the formula of:
   (i) Sodium oxide
   (ii) Aluminium chloride
   (iii) Sodium sulfide
   (iv) Magnesium hydroxide

9. Write the names of the following compounds:
   (i) NiS,
   (ii) $Mg(NO_3)_2$
   (iii) Na_2SO_4
   (iv) $Al(NO_3)_3$

10. Write the chemical names of the following compounds:
    (i) K_2SO_4
    (ii) $Mg_3(PO_4)_2$
    (iii) ZnS (iv) Na_3N

11. Define one mole and illustrate its relationship with Avogadro constant.

12. Sample X contains 1 gram molecule of oxygen molecules and Sample Y contains 1 mole of oxygen molecules. What is the ratio of the number of molecules in both the samples?

13. What is 'molar volume'? What is its value?

14. What is the mass of:
    (i) 0.2 moles of oxygen atoms
    (ii) 0.5 moles of water molecules

Section D: Long Answer Questions
(5 Marks)

1. (i) Give one point of difference between an atom and an ion.
   (ii) Give one example each of a polyatomic cation and an anion.
   (iii) Identify the correct chemical name of $FeSO_3$: Ferrous sulfate, Ferrous sulfide, Ferrous sulfite.
   (iv) Write the chemical formula for the chloride of magnesium.
   (v) Write the chemical formula of calcium phosphate.

2. (i) Give two examples in each of the following cases:
   (1) a divalent anion, (2) a trivalent cation, (3) a monovalent anion
   (ii) Calculate the mass of the following:
   (1) 2 moles of carbon dioxide (2) 6.022×10^{23} molecules of carbon dioxide

3. (i) Write the chemical formulae of the following:
   (1) Calcium oxide, (2) Magnesium bromide, (3) Aluminium hydroxide, (4) Ferric hydroxide
   (ii) Calculate the number of molecules of sulfur (S_8) present in 25.6 g of solid sulfur.   (Atomic mass S = 32 u)

4. (i) Define molecular mass and Avogadro's constant.
   (ii) Calculate the number of molecules in 50 g of $CaCO_3$. (Atomic mass of Ca = 40 u, C = 12 u and O = 16 u)
   (iii) If one mole of sodium atom weighs 23 g, what is the mass (in g) of one atom of sodium?

5. Find the mass of:
   (i) 1 mole of nitrogen atoms
   (ii) 4 moles of aluminium atoms (Atomic mass of aluminium = 27)
   (iii) 10 moles of sodium sulfite (Na_2SO_3)

Section E: Case Case Study or Passage-Based Questions   (4 Marks)

1. Formula unit mass is the sum of the atomic masses of all the atoms present in a formula unit of a compound and it is calculated in the same manner as molecular mass. The mole is the base unit of amount of substance in the international system of units (SI). 1 mole is equal to the atomic weight of an atom or molecular weight of a molecule or number of entities equal to Avogadro number or volume at STP equal to 22.4 L.

   (i) Formula unit is used for the substance whose constituent particles are ________

   (ii) What is the formula unit mass of HCl?

   (iii) The number of particles present in 1 mole of any substance is equal to ________

   (iv) What is the mass of 1 mole of oxygen atoms?

   (v) Calculate the number of moles for 84 g of nitrogen atoms.

   (vi) Write the formula of baking powder and find its molar mass.

2. The chemical formula of an ionic compound simply shows the ratio of cations and anions present in the structure of the compound. The valency of an element is equal to the number of hydrogen atoms or chlorine atoms or double the number of oxygen atoms which combines with 1 atom of the element.

While writing the formula of binary compounds, we cross over the valences or the charges present on the given ions. For example,

Symbol   K (NO_3)

          Formula = KNO_3

Charge   +1   1–

(i) If a potassium ion has +1 unit charge and dichromate (Cr_2O_7) has –2 unit charge, what is the correct formula of potassium dichromate?
(a) KCr_2O_7 (b) $K_2Cr_2O_7$
(c) $(K_2Cr_2O_7)_2$ (d) $K_4(Cr_2)O_7$

(ii) The correct formula of sodium sulfate and sodium phosphate are, respectively:
(a) $NaSO_4$, $NaPO_4$
(b) Na_2SO_4, Na_2PO_4
(c) Na_2SO_4, Na_3PO_4
(d) Na_2SO_4, $NaPO_4$

(iii) The mass of 1 mole of Na_2CO_3 is equal to:
(a) 53 g   (b) 106 g
(c) 212 g   (d) 152 g

(iv) The formula of mercurous chloride and mercuric chloride are, respectively:
(a) $HgCl_2$, $HgCl_4$ (b) $HgCl_2$, Hg_2Cl_2
(c) Hg_2Cl_2, $HgCl_2$ (d) Hg_2Cl_2, Hg_2Cl_4

(v) The molar mass of ferrous sulfate and ferric sulfate are, respectively:
(a) 76 g, 200 g   (b) 152 g, 400 g
(c) 400 g, 152 g   (d) 152 g, 200 g

High Order Thinking Skills (HOTS) Questions

1. The difference in the mass of 100 moles each of sodium atoms and sodium ions is 5.48002 g. Compute the mass of an electron.

2. Explain why the number of atoms in one mole of hydrogen gas is double the number of atoms in one mole of helium gas.

3. Shanvi took 5 moles of carbon atoms in a container and Shashwat also took 5 moles of sodium atoms in another container of the same weight
(i) Whose container is heavier?
(ii) Whose container contains more atoms?

4. How many molecules of water are present in a drop of water which has a mass of 50 mg?

5. (i) Why is the statement 'one mole of oxygen' not correct? What should be the correct statement?
(ii) Find the number of carbon atoms present in a black dot marked on white paper with graphite pencil as a full stop at the end of a sentence. Remember that the dot has a mass of 1 atto-gram or 10^{-18} g.

6. Answer the following questions about one mole of sulfuric acid $[H_2SO_4]$?
(i) Find the number of gram atoms of hydrogen in it.
(ii) How many atoms of hydrogen does it have?
(iii) How many atoms (in grams) of hydrogen are present for every gram atom of oxygen in it?
(iv) Calculate the total number of atoms in H_2SO_4.
[Given: 1 Mole of H_2SO_4 = Gram molecular mass = 6.023×10^{23} molecules]

7. (i) Convert 12.044×10^{21} molecules of sulfur dioxide into moles.
(ii) Find the number of water molecules present in a drop of water weighing 0.09 g.

8. *The students of class IX made placards showing the symbols of some elements and the valences of these elements. These placards got mixed up, as shown in the figure below:

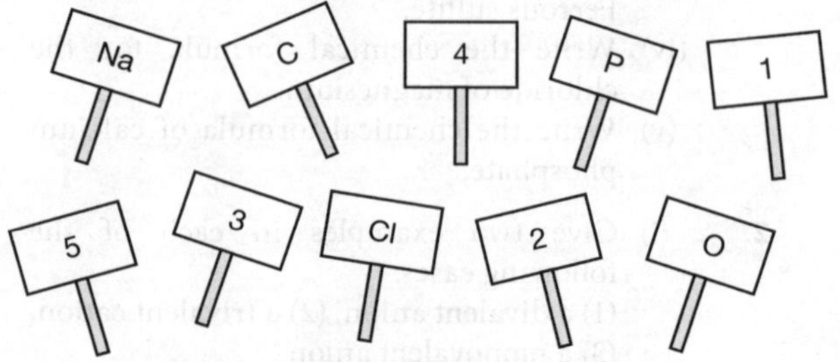

On the basis of these placards the chemistry teacher, Mr. Gupta, asked a student Shanvi to answer the following questions:

   (i) Choose the symbol of an element from among the placards which shows valences of 3 and 5. What is the name of this element? Also state if it is a metal or a non-metal.

  (ii) Choose the symbol of an element from the given placards which shows a valency of 2. What is the name of this element? Also state if it is a metal or a non-metal.

 (iii) Predict the formula and name of the compound formed between the element of valency 3 and element of valency 2.

 (iv) Predict the formula and name the compound that is formed between the element (metal) showing a valency of 1 and non-metal showing a valency of 2.

  (v) What values are displayed by Shanvi in this episode?

Answers

Section A: Multiple Choice Questions

1. (c)	**2.** (a)	**3.** (d)	**4.** (d)
5. (b)	**6.** (c)	**7.** (c)	**8.** (d)
9. (d)	**10.** (c)	**11.** (d)	**12.** (c)
13. (b)	**14.** (b)	**15.** (d)	

Assertion–Reason Questions

1. (b)	**2.** (c)	**3.** (d)	**4.** (d)
5. (c)	**6.** (d)		

Section B: Very Short Answer Questions

1. Ions are charged particles of atoms or groups of atoms.
Cations: Ba^{2+}, Mg^{2+}, Ca^{2+} Anions: O^{2-}, S^{2-}

2. Atoms form ions in order to become stable by acquiring the stable electronic configuration of the nearest noble gas.

3. A group of atoms having a charge is known as a polyatomic ion. Examples: NH_4^+, SO_4^{2-}.

4. Mg: 12e–: Mg^{2+}: 10e–

5. The compound is called aluminium sulfate; cation: Al^{3+}; anion: $(SO_4)^{2-}$

6. The valency of sulfur in H_2S, SO_2 and SO_3 are 2, 4, and 6, respectively.

7. 2H means two atoms of hydrogen, H_2 means a molecule of hydrogen which contains two atoms.

8. The valency of Ca in $CaCO_3$ is 2+ $[Ca^{2+}]$.

9. The formula of its oxide is A_2O_2 or AO_2.

10. (i) Potassium phosphate, (ii) Calcium nitride

11. Avogadro's number (N_0) represents 6.022×10^{23} particles.

12. The valency of the metal (M) in M_2CO_3 is (1+); the metal exists as an M^+ ion. Therefore, the formula of metal chloride is MCl.

13. M_2O

14. We use the word mole to define the number of atoms, molecules, ions or particles having a mass equal to its atomic or molecular mass in grams, while as a unit, we call it mol.

15. One gram of hydrogen = One mole = 6.022×10^{23} atoms

Section C: Short Answer Questions

1. (i) Molecular compound: Atoms of different elements join together in definite proportions to form molecules of compounds. For example, water (H_2O), ammonia (NH_3), carbon dioxide (CO_2).

  (ii) Ionic compound: Compounds composed of metals and non-metals contain charged species. The charged species are known as ions. An ion is a charged particle and can be negatively or positively charged. For examples, sodium chloride (NaCl), magnesium oxide (MgO).

2. Polyatomic ions: Two or more different atoms unite to form a charged particle, known as polyatomic ion.
Examples: $(PO_4)^{3-}$ Phosphate, $(NO_3)^-$ Nitrate, Sulphate (SO_4^{2-}).

3. (i) Na^+, (ii) Mg^{2+}, (iii) F^- and (iv) O^{2-}

4.

	(i)	(ii)	(iv)
Cations	Na^+	Na^+	NH_4^+
Anions	CH_3COO^-	Cl^-	NO_3^-

(iii) H_2 is a covalent compound having no ions.

5. The chemical formula of a compound shows its constituent elements and the number of atoms of each combining element.

For example, the chemical formula of ammonia is NH_3, water is H_2O and carbon dioxide is CO_2.

6. The formula unit mass is the same as molecular mass which is equal to the sum of the masses of atoms present in a formula unit.

Formula unit mass of NaCl = (23 + 35.5) = 58.5 u.

7. (i) $CuCl_2$, $CuSO_4$, $Cu_3(PO_4)_2$
   (ii) $NaCl$, Na_2SO_4, Na_3PO_4
   (iii) $FeCl_3$, $Fe_2(SO_4)_3$, $FePO_4$

8. (i) Na_2O, (ii) $AlCl_3$, (iii) Na_2S and (iv) $Mg(OH)_2$

9. (i) Nickel sulfide,
   (ii) Magnesium nitrate,
   (iii) Sodium sulfate,
   (iv) Aluminium nitrate

10. (i) Potassium sulfate,
    (ii) Magnesium phosphate,
    (iii) Zinc sulfide,
    (iv) Sodium nitride

11. One mole of any species (atoms, molecules, ions or particles) is that quantity in number having a mass equal to its atomic or molecular mass in grams.

The number of particles (atoms, molecules or ions) present in 1 mole of any substance is fixed, with a value of 6.022×10^{23}. This number is known as Avogadro's constant or number.

12. One-gram molecule is the same as one gram mole of a substance. Therefore, both the samples X and Y contain the same number of molecules $(6.022 \times 10)^{23}$ and the ratio is 1: 1.

13. The volume occupied by one mole of a gas under standard conditions of temperature and pressure, that is, STP conditions (0°C and 1 atmosphere) is known as the molar volume of the gas. Its value is 22.4 L at STP.

14. (i) 1 mole of oxygen atoms = 1×16 = 16 g

0.2 mole of oxygen atoms = $16 \text{ g} \times 0.2$ = 3.2 g

(ii) 1 mole of water (H_2O) molecules = 2×1 g + 1×16 g = 18 g

0.5 mole of water (H_2O) molecules = 18 g $\times 0.5$ = 9.0 g

Section D: Long Answer Questions

1. (i) An atom is neutral while an ion is charged. For example, Na (atom) and Na^+ (ion).
   (ii) NH_4^+, SO_4^{2-} are examples of polyatomic cation and polyatomic anion, respectively.
   (iii) $FeSO_3$ is ferrous sulfite having Fe^{2+} (ferrous) and SO_3^{2-} (sulfite).
   (iv) $MgCl_2$
   (v) $Ca_3(PO_4)_2$

2. (i) (1) O^{2-}, S^{2-}, (2) Fe^{3+}, Al^{3+}, (3) F^-Cl^-
   (ii) (1) As 1 mole of carbon dioxide = molecular weight of CO_2 = 44 g of CO_2
   2 mole of carbon dioxide = $44 \times 2 = 88$ g of CO_2
   (2) 1 mole of carbon dioxide = Avogadro number of molecules = 6.02×10^{23}
   As 1 mole of CO_2 = 44 g of CO_2
   6.02×10^{23} molecules have a weight equal to 44 g.

3. (i) (1) CaO, (2) $MgBr_2$, (3) $Al(OH)_3$, (4) $Fe(OH)_3$
   (ii) The molar mass of S_8 = 8×32 = 256
   1 mole of S_8 = 256 g
   As 256 g of S_8 has the number of molecules = 6.02×10^{23}
   25.6 g of S_8 has the number of molecules = $\dfrac{6.023 \times 10^{23} \times 25.6}{256} = 6.023 \times 10^{22}$

4. (i) Refer to Section 3.9, Mole Concept.
   (ii) As 100 g of calcium carbonate = 1 mol

50 g of calcium carbonate
$= \dfrac{50 \times 1}{100} = 0.5$ mol

   (iii) 1 mol of sodium = 23 g = 6.02×10^{23} atoms of sodium.
   This means weight of 6.023×10^{23} atoms = 23 g.
   Now weight of 1 atom of sodium
   $= \dfrac{23}{6.02 \times 10^{23}} = 3.81 \times 10^{-23}$ g

5. (i) 1 mole of nitrogen atoms = 1 × Gram atomic mass of nitrogen atom
= 1 × 14 g = 14 g

(ii) 4 moles of aluminium atoms = 4 × Gram atomic mass of aluminium atoms
= 4 × 27 g = 108 g

(iii) 10 moles of sodium sulfite (Na_2SO_3) = 10 (2 × Gram atomic mass of Na + 1 × Gram atomic mass of sulfur + 3 × Gram atomic mass of oxygen)
= 10(2 × 23 g + 1 × 32 g + 3 × 16 g) = 10 (46 g + 32 g + 48 g)
= 10 × 126 g = 1260 g

Section E: Case Study or Passage-Based Questions

1.
(i) Atoms
(ii) 36.5
(iii) 6.022×10^{23}
(iv) 32 g
(v) As 14 g of nitrogen atoms = 1 mole
84 g of nitrogen atoms = 84/14 = 6
(vi) $NaHCO_3$
Molar mass = 23 + 1 + 12 + 3 × 16 = 84.

2. (i) (b)   (ii) (c)   (iii) (b)   (iv) (c)   (v) (b)

High Order Thinking Skills (HOTS) Questions

1. A sodium atom and ion differ by one electron. For 100 moles each of sodium atoms and ions, there would be a difference of 100 moles of electrons.
Mass of 100 moles of electrons = 5.48002 g

2. The number of atoms in one mole of hydrogen gas is double the number of atoms in one mole of helium gas because the hydrogen molecule is diatomic, that is, a molecule of hydrogen consists of two atoms of hydrogen, whereas helium is monoatomic.

3. (i) Mass of carbon atom carried by Shanvi = (5 × 12) g = 60 g
Mass of sodium atoms carried by Shashwat = (5 × 23) g = 115 g
Hence, Shashwat's container is heavier.

(ii) Both the bags have the same number of atoms as they have the same number of moles of atoms. (Avogadro's law)

4. We know that:
1 mole of a compound = 6.023×10^{23} atoms = Gram molecular mass
Gram molecular mass of H_2O = 18 g
18 g = 6.023×10^{23} atoms

$$1\,g = \frac{6.023 \times 10^{23}}{18}\,atoms$$

Now 50 mg of $H_2O = \dfrac{6.023 \times 10^{23} \times 50 \times 10^{-3}}{18}$

$= 1.673 \times 10^{21}$ molecules

5. (i) 1 mole of oxygen does not tell us whether we have 1 mole of oxygen atoms or 1 mole of oxygen molecules, so it is incorrect and the correct statement should be either 1 mole of oxygen atoms or 1 mole of oxygen molecules.

(ii) 1 mole of carbon atoms = Gram atomic mass of carbon = 12 g
Also, 1 mole of carbon = 6.022×10^{23} atoms
Hence, 12 g of carbon have C-atoms = 6.022×10^{23}
10^{-18} g of carbon will have C-atoms
$= \dfrac{6.022 \times 10^{23}}{12} \times 10^{-18} = 5.02 \times 10^{4}$

6. (i) In H_2SO_4 → 2 gram atoms of hydrogen are present

(ii) One mole of H_2SO_4 contains = 6.023×10^{23} molecules
= 2 × 6.023×10^{23} atoms of H + 6.023×10^{23} atoms of S + 4 × 6.023×10^{23} atoms of O
∴ 2H = 2 × 6.023×10^{23} = 12.046×10^{23}

(iii) In H_2SO_4
For every 2 hydrogen there are 4 oxygen atoms
For 1 hydrogen = 2 oxygen atoms are present
For 1 oxygen = 2/4 hydrogen atoms are present
= 0.5 hydrogen atoms are present.

(iv) Total atoms in 1 mole of H_2SO_4 = $6.023 \times 10^{23} \times 5 = 3.0115 \times 10^{24}$ atoms

7. (i) We know that
As 6.022×10^{23} molecules of sulfur dioxide = 1 mole
12.044×10^{21} molecules of sulfur dioxide
$= \dfrac{1}{6.022 \times 10^{23}} \times 12.044 \times 10^{21} = 0.02$ mole

Hence, 12.044×10^{21} molecules of sulphur dioxide is equal to 0.02 mole.

(ii) As, Molecular mass of water, $H_2O = 2 + 16 = 18$ u

Gram molecular mass of water = 18 grams = 1 mole of water

Hence, 1 mole of water has a mass of 18 grams and it contains 6.022×10^{23} molecules of water.

As, 18 g water contains = 6.022×10^{23} molecules

$$0.09 \text{ g water contains} = \frac{6.022 \times 10^{23}}{18} \times 0.09$$

$= 3.011 \times 10^{21}$ molecules

Hence, a drop of water weighing 0.09 gram has 3.011×10^{21} molecules of water in it.

8. (i) The symbol of the element which exhibits the valences of 3 and 5 is P. This element is named as phosphorus and it is a non-metal.

(ii) The symbol of the element which exhibits a valency of 2 is O. This element is named as oxygen and it is also a non-metal.

(iii) The formula of the compound formed between phosphorus exhibiting a valency of 3 and oxygen with valency 2 can be worked out as follows:

Symbols: P   O P_2O_3

Valences: 3   2

Hence, the formula of this compound is P_2O_3 and the name of this compound is phosphorus trioxide.

(iv) The formula of the compound is Na_2O and it is named as sodium oxide.

Symbols: Na   O Na_2O

Valences: 1   2

(v) The values displayed by Shanvi in this episode are her knowledge of the symbols of elements and their valences, and her skill in applying knowledge to solve problems.

Hints

Section A: Multiple Choice Questions

1. The valency of CO_3^{2-} is 2 because valency of an ion is equal to charge on the ion and CO_3^{2-} contains a charge of -2.

5.
Ce SO_4 $= Ce_2(SO_4)_4 = Ce(SO_4)_2$

$+4$ -2

6. As the balanced equation is
$$Mg(OH)_2 + 2HCl \rightarrow MgCl_2 + 2H_2O$$

7.
Ca NO_2 $= Ca(NO_2)_2$

$+2$ -1

8.
Fe AsO_4 $= Fe_3(AsO_4)_2$

$+2$ -3

9. Molecular mass of $C_{18}H_{26}O_2 = 18 \times 12 + 26 \times 1 + 2 \times 16 = 216 + 26 + 32 = 274$

13. As number of aluminium ions in 1 mol or 27 g aluminium = 6.022×10^{23}

The number of aluminium ions in 0.0027 g

$$= \frac{6.022 \times 10^{23} \times 0.0027}{27} = 6.022 \times 10^{19}$$

14. As the number of molecules in 1 mol or 256 g S_8 = 6.022×10^{23}

The number of molecules in 16 g S_8

$$= \frac{6.022 \times 10^{23} \times 16}{256} = 3.76 \times 10^{22}$$

15. (i) 1 mole of water = 18 g
2 moles of water = 2×18 g = 36 g

(ii) 20 moles of water = 18×20 = 360 g

(iii) 6.022×10^{23} molecules of water = 1 mole
= 18 g

(iv) 1 mole of water = 6.022×10^{23} molecules
6.22×10^{23} molecules = 1 mole

$$1.2044 \times 10^{25} \text{molecules}$$
$$= \frac{1}{6.022 \times 10^{23}} \times 1.2044 \times 10^{25}$$
$$= 20 \text{ moles} = 20 \times 18 = 360 \text{ g}$$

Assertion–Reason Questions

2. Both 4 g methane and 8 oxygen gas have the same number of moles (1/4) and the same number of molecules.

3. As the correct formula of calcium phosphate is $Ca_3(PO_4)_2$.

5. As at STP, 8 g of methane will have a volume of 11.2 L.

6. As the correct formula mass of calcium chloride is 111 u.

Section E: Case Study or Passage-Based Questions

2. **(ii)**

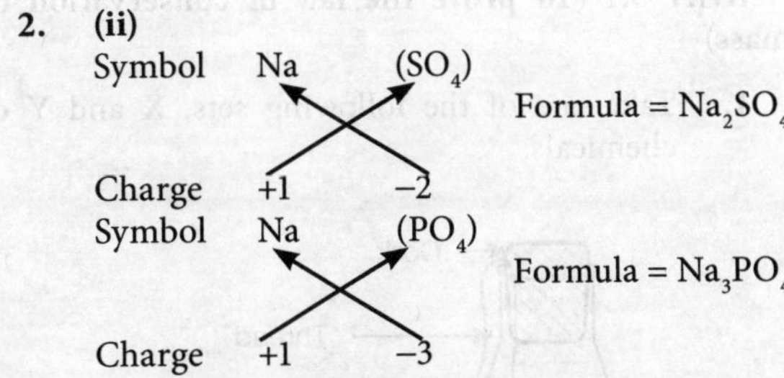

(iii) 1 mole of Na_2CO_3 = Molar mass of Na_2CO_3
$2 \times 23 + 1 \times 12 + 3 \times 16 = 106$ g

(v) Ferrous sulfate ($FeSO_4$) = $1 \times 56 + 1 \times 32 + 4 \times 16 = 56 + 32 + 64 = 152$ g

Ferric sulfate ($Fe_2(SO_4)_3$) = $2 \times 56 + 3(1 \times 32 + 4 \times 16) = 112 + 3(96) = 400$ g

Activities with Discussion and Conclusion

Activity 3.1 (To prove the law of conservation of mass)

1. Take one of the following sets, X and Y of chemicals:

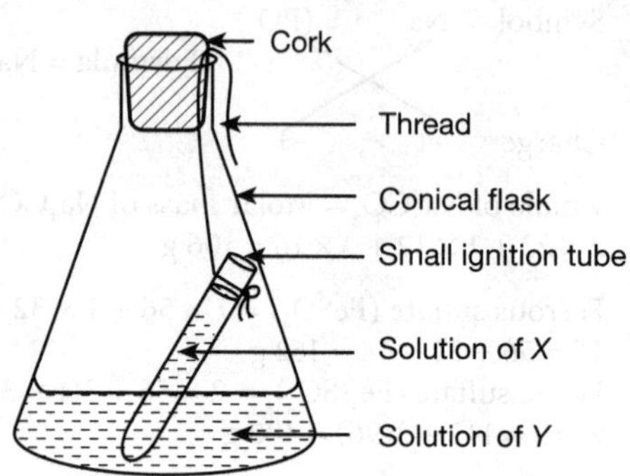

	X	Y
(i)	Copper sulfate	Sodium carbonate
(ii)	Barium chloride	Sodium sulfate
(iii)	Lead nitrate	Sodium chloride

2. Prepare separately a 5% solution of any one pair of substances listed under X and Y in water.

3. Take a little amount of solution of Y in a conical flask and some solution of X in an ignition tube.

4. Hang the ignition tube in the flask carefully; see that the solutions do not mix. Put a cork in the flask.

5. Weigh the flask with its contents carefully.

6. Now tilt and swirl the flask so that the solutions X and Y are mixed.

7. Weigh again.

Now Answer
- What happens in the reaction flask?
- Do you think that a chemical reaction has taken place?
- Why should we put a cork in the mouth of the flask?
- Does the mass of the flask and its contents change?

Discussion: (i) Mass of the reaction flask is found to remain unchanged, that is, the same as before tilting and swirling the flask.

(ii) The following chemical reaction takes place:

(a) Copper sulfate + Sodium carbonate → Copper carbonate + Sodium sulfate

$$CuSO_4(aq) + Na_2CO_3(aq) \rightarrow \underset{Ppt}{CuCO_3(s)\downarrow} + Na_2SO_4(aq)$$

(b) Barium chloride + Sodium sulfate → Barium sulfate + Sodium chloride

$$BaCl_2(aq) + Na_2SO_4(aq) \rightarrow \underset{Ppt}{BaSO_4(s)\downarrow} 2NaCl(aq)$$

(c) Lead nitrate + Sodium chloride → Lead chloride + Sodium nitrate

$$Pb(NO_3)_2(aq) + 2NaCl(aq) \rightarrow \underset{Ppt}{PbCl_2(s)\downarrow} + 2NaNO_3(aq)$$

(iii) The cork is put in the mouth of the flask so that on titling and swirling the flask, the solution does not splash out or if some gas is formed as one of the products, it does not escape out.

(iv) The mass of the flask and its contents does not change.

Conclusion: The above activity shows that in any chemical reaction, the total mass of the reactants is equal to the total mass of the products. In other words, mass can neither be created nor destroyed in a chemical reaction. This proves the law of conservation of mass.

Activity 3.2 (To calculate the ratio of atoms of elements in a compound)

1. Refer to the data given in the table below:

Compound	Combining elements	Ratio by mass
Water	Hydrogen, Oxygen	1 : 8
Ammonia	Nitrogen, Hydrogen	14 : 3
Carbon dioxide	Carbon, Oxygen	3 : 8

Atomic masses are : H = 1.0 u, O = 16.0 u, N = 14.0 u, C = 12.0 u

2. Find the ratio by number of the atoms of elements in the molecules of compounds given in the above table.

Discussion: (i) To calculate ratio by number of atoms of elements in water. The calculations are carried out in tabular form as follows:

Element	Ratio by mass	Atomic mass (u)	Mass ratio/ Atomic mass	Simplest whole number ratio
H	1	1	1/1 = 1	2
O	8	16	8/16 = 1/2	1

Thus, ratio by number of atoms for water is H : O = 2 : 1.

(ii) To calculate ratio by number of atoms of elements in ammonia:

Element	Ratio by mass	Atomic mass (u)	Mass ratio/ Atomic mass	Simplest ratio
N	14	14	14/14 = 1	1
H	3	1	3/1 = 3	3

Thus, ratio by number of atoms for ammonia is N : H = 1 : 3.

(iii) To calculate ratio by number of atoms of elements in carbon dioxide:

Element	Ratio by mass	Atomic mass (u)	Mass ratio/ Atomic mass	Simplest whole number ratio
C	3	12	3/12 = 1/4	1
O	8	16	8/16 = 1/2	2

Thus, ratio by number of atoms for carbon dioxide is C : O = 1 : 2.

Conclusion: In a compound, the elements are combined together in a simple whole number atomic ratio. This is one of the postulates of Dalton's atomic theory.

Exemplar Problems

Short Answer Questions

1. Which of the following represents a correct chemical formula? Name it.
   (a) $CaCl$
   (b) $BiPO_4$
   (c) $NaSO_4$
   (d) NaS

2. Write the molecular formulae for the following compounds:
   (a) Copper(II) bromide
   (b) Aluminium(III) nitrate
   (c) Calcium(II) phosphate
   (d) Iron(III) sulfide
   (e) Mercury(II) chloride
   (f) Magnesium(II) acetate

3. Write the molecular formulae of all the compounds that can be formed by the combination of the following ions:
   Cu^{2+}, Na^+, Fe^{3+}, Cl^-, SO_4^{2-}, PO_4^{3-}

4. Write the cations and anions present (if any) in the following compounds:
   (a) CH_3COONa
   (b) $NaCl$
   (c) H_2
   (d) NH_4NO_3

5. Give the formulae of the compounds formed from the following sets of elements:
   (a) Calcium and fluorine
   (b) Hydrogen and sulfur
   (c) Nitrogen and hydrogen
   (d) Carbon and chlorine
   (e) Sodium and oxygen
   (f) Carbon and oxygen

6. Which of the following symbols of elements are incorrect? Give their correct symbols.
   (a) Cobalt – CO
   (b) Carbon – c
   (c) Aluminium – AL
   (d) Helium – He
   (e) Sodium – So

7. Give the chemical formulae for the following compounds and compute the ratio by mass of the combining elements in each one of them:
   (a) Ammonia
   (b) Carbon monoxide
   (c) Hydrogen chloride
   (d) Aluminium fluoride
   (e) Magnesium sulfide

8. State the number of atoms present in each of the following chemical species:
   (a) CO_3^{2-}
   (b) PO_4^{3-}
   (c) P_2O_5
   (d) CO

9. What is the fraction of the mass of water due to neutrons?

10. Does the solubility of a substance change with temperature? Explain with the help of an example.

11. Classify each of the following on the basis of their atomicity:
    (a) F_2
    (b) NO_2
    (c) N_2O
    (d) C_2H_6
    (e) P_4
    (f) H_2O_2
    (g) P_4O_{10}
    (h) O_3
    (i) HCl
    (j) CH_4
    (k) He
    (l) Ag

12. You are provided with a fine white-coloured powder which is either sugar or salt. How would you identify it without tasting?

13. Calculate the number of moles of magnesium present in a magnesium ribbon weighing 12 g. Molar atomic mass of magnesium is 24 g mol^{-1}.

Long Answer Questions

1. Verify by calculating that
   (a) 5 moles of CO_2 and 5 moles of H_2O do not have the same mass.
   (b) 240 g of calcium and 240 g magnesium elements have a mole ratio of 3 : 5.

2. Find the ratio by mass of the combining elements in the following compounds:
   (a) $CaCO_3$,
   (d) C_2H_5OH,
   (b) $MgCl_2$,
   (e) NH_3,
   (c) H_2SO_4,
   (f) $Ca(OH)_2$

3. Calcium chloride, when dissolved in water, dissociates into its ions according to the following equation:

$$CaCl_2(aq) \rightarrow Ca^{2+}(aq) + 2Cl^-(aq)$$

   Calculate the number of ions obtained from $CaCl_2$ when 222 g of it is dissolved in water.

4. The difference in the mass of 100 moles each of sodium atoms and sodium ions is 0.0548002 g. Compute the mass of an electron.

5. Cinnabar (HgS) is a prominent ore of mercury. How many grams of mercury are present in 225 g of pure HgS? Molar mass of Hg and S are 200.6 g mol^{-1} and 32 g mol^{-1}, respectively.

6. The mass of one steel screw is 4.11 g. Find the mass of one mole of these steel screws. Compare this value with the mass of the earth $(5.98 \times 10^{24}$ kg). Which one of the two is heavier and by how many times?

7. A sample of vitamin C is known to contain 2.58×10^{24} oxygen atoms. How many moles of oxygen atoms are present in the sample?

8. Raunak took 5 moles of carbon atoms in a container and Krish also took 5 moles of sodium atoms in another container of the same weight.
   (a) Whose container is heavier?
   (b) Whose container has more atoms?

9. Fill in the missing data in the table.

Species property	H_2O	CO_2	Na atom	$MgCl_2$
No. of moles	2		–	0.5
No. of particles	–	3.01×10^{23}		–
Mass	36 g	–	115 g	–

10. The visible universe is estimated to contain 1022 stars. How many moles of stars are present in the visible universe?

11. What is the SI prefix for each of the following multiples and submultiples of a unit?
    (a) 10^3, (b) 10^{-1}, (c) 10^{-2}, (c) 10^{-2}, (d) 10^{-6}, (e) 10^{-9}, (f) 10^{-12}

12. Express each of the following in kilograms:
    (a) 5.84×10^{-3} mg
    (b) 58.34 g
    (c) 0.584 g
    (d) 5.873×10^{-21} g

13. Compute the difference in masses of 10^3 moles each of magnesium atoms and magnesium ions. (Mass of an electron = 9.1×10^{-31} kg)

14. Which has more atoms, 100 kg of N_2 or 100 g of NH_3?

15. Compute the number of ions present in 5.85 g of sodium chloride.

16. A gold sample contains 90% of gold and the rest is copper. How many atoms of gold are present in one gram of this sample?

17. What are ionic and molecular compounds? Give examples.

18. Compute the difference in masses of one mole each of aluminium atoms and one mole of its ions. (Mass of an electron is 9.1×10^{-28} g). Which one is heavier?

19. A silver ornament of mass 'm' gram is polished with gold equivalent to 1% of the mass of silver. Compute the ratio of the number of atoms of gold and silver in the ornament.

20. A sample of ethane (C_2H_6) gas has the same mass as 1.5×10^{20} molecules of methane (CH_4). How many C_2H_6 molecules does the sample of gas contain?

21. (a) In a chemical reaction, the sum of the masses of the reactants and products remains unchanged. This is called ______.

(b) A group of atoms carrying a fixed charge on them is called ______.

(c) The formula unit mass of $Ca_3(PO_4)_2$ is ______.

(d) Formula of sodium carbonate is ______ and that of ammonium sulfate is ______.

22. Complete the following crossword puzzle by using the name of the chemical elements. Use the data given in the table:

Across	Down
2. The element used by Rutherford during his α-scattering experiment.	**1.** A white lustrous metal used for making ornaments and which tends to get tarnished in the presence of moist air.
3. An element which forms rust on exposure to moist air.	**4.** Both brass and bronze are alloys of the element.
5. A very reactive non-metal stored under water.	**6.** The metal which exists in the liquid state at room temperature.
7. Zinc metal when treated with dilute hydrochloric acid produces a gas of this element which when tested with burning splinter produces a pop sound.	**8.** An element with symbol Pb.

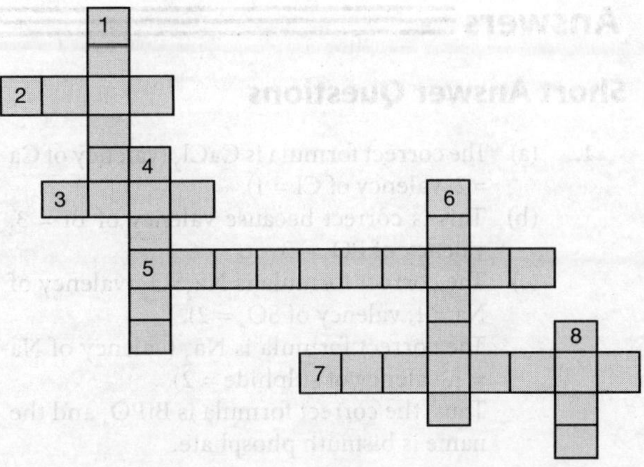

23. (a) In this crossword puzzle, the names of 11 elements are hidden. Symbols of these are given below. Complete the puzzle.

1. Cl   2. H   3. Ar   4. O   5. Xe
6. N   7. He   8. F   9. Kr   10. Rn
11. Ne

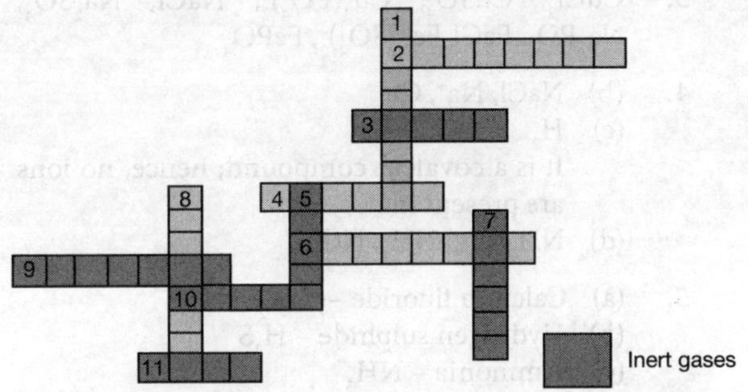

(b) Identify the total number of inert gases, their names and symbols from this crossword puzzle.

24. Write the formulae for the following and calculate the molecular mass for each one of them:

(a) Caustic potash   (b) Baking powder
(c) Limestone   (d) Caustic soda
(e) Ethanol   (f) Common salt

25. In photosynthesis, 6 molecules of carbon dioxide combine with an equal number of water molecules through a complex series of reactions to give a molecule of glucose having a molecular formula $C_6H_{12}O_6$. How many grams of water would be required to produce 18 g of glucose? Compute the volume of water so consumed assuming the density of water to be 1 g cm^{-3}.

Answers

Short Answer Questions

1. (a) The correct formula is $CaCl_2$ (valency of Ca = 2, valency of Cl = 1).
    (b) This is correct because valency of Bi = 3, valency of PO_4 = 3
    (c) The correct formula is Na_2SO_4 (valency of Na = 1, valency of SO_4 = 2).
        The correct formula is Na_2 (valency of Na = 1, valency of sulphide = 2).
        Thus, the correct formula is $BiPO_4$ and the name is bismuth phosphate.

2. (a) Copper(II) Bromide – $CuBr_2$
    (b) Aluminium(III) nitrate – $Al(NO_3)_3$
    (c) Calciuih(II) phosphate – $Ca_3(PO_4)_2$
    (d) Iron(III) sulfide – Fe_2S_3
    (e) Mercury(II) chloride – $HgCl_2$
    (f) Magnesium(II) acetate – $Mg(CH_3COO)_2$

3. $CuCl_2$, $CuSO_4$, $Cu_3(PO_4)_2$, NaCl, Na_2SO_4, Na_3PO_4, $FeCl_3$ $Fe_2(SO_4)_3$, $FePO_4$

4. (b) NaCl, Na^+, Cl^-
    (c) H_2
        It is a covalent compound; hence, no ions are present in it.
    (d) NH_4NO_3, NH_4^+, NO_3^-

5. (a) Calcium fluoride – CaF_2
    (b) Hydrogen sulphide – H_2S
    (c) Ammonia – NH_3
    (d) Carbon tetrachloride – CCl_4
    (e) Sodium oxide – Na_2O
    (f) Carbon monoxide – CO
    (g) Carbon dioxide – CO_2

6. (a) Cobalt – Co
    (b) Carbon – C
    (c) Aluminium – Al
    (d) Helium – He (correct)
    (e) Sodium – Na

7.

Chemical formula	Ratio by mass
(a) Ammonia (NH_3)	N(l) : H(3) 14 : 3
(b) Carbon monoxide (CO)	C(1) : O(1)12 : 16 or 3 : 4
(c) Hydrogen chloride (HCl)	H(l) : Cl(l)1 : 35.5 or 2 : 71
(d) Aluminium fluoride (AlF_3)	Al(l) : F(3)27 : 19 × 3 or 9 : 19
(e) Magnesium sulfide (MgS)	Mg(l) : S(l)24 : 32 or 3 : 4

8. (a) CO_3^{2-} = 1C + 3 (O) = 4
    (b) PO_4^{3-} = 1 P + 4 (O) = 5
    (c) P_2O_5 = 2 P + 5 (O) = 7
    (d) CO = 1 C + 1 (O) = 2

9. Solution: Mass of one mole (N_A) of neutrons = 1 g

    Mass of 1 neutron = $\dfrac{1}{N_A}$ g

    Mass of 1mole of H_2O = 18

    Mass of one molecule of water = $\dfrac{18}{N_A}$ g

    Number of neutrons in one atom of H = 0
    Number of neutrons in one atom of O = 8

    Mass of 8 neutrons (in H_2O) = $\dfrac{8}{N_A}$

    Fraction of mass of water due to neutrons

    $= \dfrac{8}{N_A} / \dfrac{18}{N_A}$ or $\dfrac{8}{18} = \dfrac{4}{9}$

10. Yes, solubility of a substance changes with temperature. It generally increases with temperature. More sugar can be dissolved in hot water as compared to cold water.

11. Monoatomic with atomicity (1) = He, Ag
    Diatomic with atomicity (2) = F_2, HCl
    Polyatomic with atomicity > 2
    Atomicity (3) = NO_2, N_2O, O_3
    Atomicity (4) = P_4, H_2O_2
    Atomicity (5) = CH_4
    Atomicity (8) = C_2H_6
    Atomicity (14) = P_4O_{10}

12. On heating, sugar powder is charred and becomes black while salt does not char. When dissolved in water, salt solution will conduct electricity because it is ionic while sugar solution will not conduct electricity because it is covalent.

13. Atomic mass of Mg = 24 g mol^{-1}
24 g of Mg = 1 mol
12 g of Mg = 12/24 = 0.5 mol

Long Answer Questions

1. (a) Mass of one mole CO_2 = 44 g

Mass of 5 moles of CO_2 = 44 × 5 = 200 g

Mass of 5 moles of H_2O = 18 × 5 = 90 g

(b) Molar mass of Ca = 40 g = 1 mole

Number of moles in 240 g of Ca = $\dfrac{240}{40}$ = 6

Molar mass of Mg = 24 g = 1 mole

Number of moles in 240 g of Mg = $\dfrac{240}{24}$ = 10

Mole ratio of Ca and Mg = 6 : 10 or 3 : 5

2. (a) $CaCO_3$ = Ca(1) : C(1) : O(3) = 40 : 12 : 16 × 3 = 40 : 12 : 48 = 10 : 3 : 12

(b) $MgCl_2$ = Mg(1) : Cl(2) = 24 : 35.5 × 2 = 24 : 71

(c) H_2SO_4 = H(2) : S(1) : O(4) = 2 : 32 : 64 = 1 : 16 : 32

(d) C_2H_5OH = C(2) : H(5) : O(1) : H(1) = C(2) : H(6) : O(1) = 24 : 6 : 16 = 12 : 3 : 8

(e) NH_3 = N(1) : H(3) = 14 : 3

(f) $Ca(OH)_2$ = Ca(1) : O(2) : H(2) = 40 : 32 : 2 = 20 : 16 : 1.

3. Solution: $CaCl_2 \rightarrow Ca^{2+} + 2Cl^-$
111 g (1 mole) → 1 mole + 2 moles
222 g (2 moles) → 2 moles + 4 moles
Total number of ions given by 2 moles of $CaCl_2$
= 6
Number of ions = Number of moles of ions $\times N_A$
= 6 × 6.022×10^{23}
= 3.6132 × 10^{24} ions

4. Number of electrons in Na atom = 11
Number of electrons in Na$^+$ = 10
For 1 mole of Na atom and Na$^+$ the difference in electrons = 1 mole
For 100 moles of Na atoms and Na ions the difference = 100 moles of electrons
Mass of 100 moles of electrons = 0.0548002 g

Mass of 1 mole of electron = $\dfrac{0.0548002}{100}$ g

Mass of 1 electron

$= \dfrac{0.0548002}{100 \times 6.022 \times 10^{23}} = 9.1 \times 10^{-28} = 9.1 \times 10^{-31}$ kg

5. Molar mass of HgS = 200.6 + 32 = 232.6 g
Mass of Hg in 232.6 g of HgS = 200.6 g
Mass of Hg in 225 g of HgS
$= \dfrac{200.6}{232.6} \times 225 = 194.04$ g

6. Mass of 21 screws = 4.11 g
Mass of 1 mole of screws = 4.11×6.022×10^{23}
= 2.475×10^{24}g
= 2.475×10^{21}kg

$\dfrac{\text{Mass of earth}}{\text{Mass of 1 mole screws}} = \dfrac{5.98 \times 10^{24} \text{kg}}{2.475 \times 10^{21} \text{kg}}$

= 2416

It shows that the mass of the earth is 2416 times more than the mass of one mole of screws.

7. Number of oxygen atoms in the sample = 2.58 × 10^{24} = 2.58 × 10^{24}
6.022 × 10^{23} atoms of oxygen = 1 mole
2.58 × 10^{24} atoms of oxygen

$= \dfrac{\text{Total atoms}}{\text{Atoms present in 1 mole}} = \dfrac{2.58 \times 10^{24}}{6.022 \times 10^{23}}$ moles

= 4.28 moles

8. (a) Mass of container containing 5 moles of C atoms = 5 × 12 = 60 g
Mass of container containing 5 moles of Na atoms = 5 × 23 = 115 g
Hence, Krish's container is heavier.

(b) Both containers have the same number of atoms since they contain the same number of moles.

9.

Species property	H$_2$O	CO$_2$	Na atom	MgCl$_2$
No. of moles	2	0.5	5	0.5
No. of particles	1.2044 × 10^{24}	3.011 × 10^{23}	5 × 6.022 × 10^{23} = 3.011 × 10^{24}	0.5 × 6.022 × 10^{23} × 3 = 9.033 × 10^{23}
Mass	36 g	22 g	115 g	47.5 g

10. 1 mole of stars = 6.022×10^{23} stars

$$\text{Number of moles of stars} = \frac{10^{22}}{6.023 \times 10^{22}}$$
$$= 0.0166 \text{ moles}$$

11. (a) 10^3 = kilo   (b) 10^{-1} = deci
(c) 10^{-2} = centi   (d) 10^{-6} = nano
(f) 10^{-12} = pico

12. (a) 5.84×10^{-3} mg $= \dfrac{5.84 \times 10^{-3}}{10^3 \times 10^3}$ kg $= 5.84 \times 10^{-9}$ kg

(b) $58.34 = \dfrac{58.34}{10^3} = 5.834 \times 10^{-2}$ kg

(c) 0.584 g $= \dfrac{0.584}{10^3}$ kg $= 5.84 \times 10^{-4}$ kg

(d) 5.873×10^{-21} g $= \dfrac{5.873 \times 10^{-21}}{10^3}$
$$= 5.873 \times 10^{-24} \text{ kg}$$

13. Mg^{+2} ion = 10 electron
Mg atom = 12 electrons
Difference between Mg^{2+} and Mg = 2 moles of electrons.
The difference between 10^3 moles of Mg and 10^3 moles of $Mg^{2+} = 10^3 \times 2$ moles of electrons.
Mass of $10^3 \times 2$ moles of electrons
$$= 2 \times 10^3 \times 6.022 \times 10^{23} \times 9.1 \times 10^{-31} \text{ kg}$$
$$= 1.096 \times 10^{-3} \text{ kg}$$

14. (i) 100 g of N_2 = 100/28 moles
Total number of molecules
$$= \frac{100}{28} \times 6.022 \times 10^{23} \text{ molecules}$$

Total number of atoms
$$= \frac{2 \times 100}{28} \times 6.022 \times 10^{23} \text{ atoms}$$
$$= 43.01 \times 10^{23} \text{ atoms}$$

(ii) 100 g of $NH_3 = \dfrac{100}{17}$ moles
Total number of molecules
$$= \frac{100}{17} \times 6.022 \times 10^{23} \text{ molecules}$$

Total number of atoms
$$= \frac{100}{17} \times 6.022 \times 10^{23} \times 4 \text{ atoms}$$
$$= 141.69 \times 10^{23} \text{ atoms}$$
Hence here, NH_3 will have more atoms.

15. 1 mole of NaCl = 23 + 35.5 = 58.5 g of NaCl
Number of moles in 5085 g of NaCl = 5085/58.5
= 0.1 moles
Each NaCl formula unit = $Na^+ + Cl^- = 2$ ions
Number of ions = $0.2 \times 6.022 \times 10^{23}$
= 1.2044×10^{23} ions

16. 1 g of gold sample contains $= \dfrac{90}{100} = 0.9$ g gold

Number of moles of gold
$$= \frac{\text{Mass of gold}}{\text{Atomic mass of gold}} = \frac{0.9}{197} = 0.0046$$
1 mole of gold contains = 6.022×10^{23} atoms
0.0046 mole of gold
$$= 6.022 \times 10^{23} \times 0.0046 = 2.77 \times 10^{21} \text{ atoms.}$$

17. Main ionic compounds are made up of ions. An ionic compound contains a cation which is a positive ion and an anion which is a negative ion. For example, sodium chloride is an ionic compound made up of Na^+ and Cl^- ions.
A molecular compound is made up of molecules; for example, ammonia (NH_3), carbon dioxide (CO_2).

18. Mass of 1 mole of aluminium atom = 27 g mol^{-1}

$Al - 3e^- \rightarrow Al^{3+}$ (3 moles of electrons)
Mass of 3 moles of electrons
$$= 3 \times (9.1 \times 10^{-28}) \times 6.022 \times 10^{23} \text{ g} = 0.00164 \text{ g}$$

Molar mass of
$Al^{3+} = 27 - 0.00164$ g $mol^{-1} = 26.9984$ g mol^{-1}

Difference = 27 − 26.9984 = 0.0016 g
1 mole of Al atoms is heavier than 1 mole of Al^{3+} ions.

19. Mass of silver = 'm' g

Mass of gold $= \dfrac{m}{100}$ g

Number of atoms of silver
$$= \frac{\text{Mass}}{\text{Atomic mass}} \times N_A = \frac{m}{108} \times N_A$$

Number of atoms of gold $= \dfrac{m}{100 \times 197} \times N_A$

Ratio of number of atoms of gold to silver = Au : Ag

$= \dfrac{m}{100 \times 197} \times N_A : \dfrac{m}{108} \times N_A$

$= 108 : 100 \times 197 = 108 : 19700 = 1 : 182.41$

20. Molar mass of $CH_4 = 12 + 4 = 16$ g mol^{-1}

1 mole of $CH_4 = 16$ g $= 6.022 \times 10^{23}$ molecules

Thus, 6.022×10^{23} molecules of CH_4 have mass $= 16$ g

1.5×10^{20} molecules of CH_4 will have mass

$= \dfrac{16}{6.022 \times 10^{23}} \times 1.5 \times 10^{20} = 4 \times 10^{-3}$ g

Molar mass of $C_2H_6 = 2 \times 12 + 6 = 30$ g mol^{-1}

1 mole of $C_2H_6 = 30$ g $= 6.022 \times 10^{23}$ molecules

Thus, 30 g of C_2H_6 have molecules $= 6.022 \times 10^{23}$

4×10^{-3} g of C_2H_6 will have molecules

$= \dfrac{6.022 \times 10^{23}}{30} \times 4 \times 10^{-3} = 0.803 \times 10^{20}$

$= 8.03 \times 10^{19}$

21. (a) Law of conservation of mass
 (b) Polyatomic ion
 (c) 310
  $(3 \times Ca) + (2 \times P) + (8 \times O)$
  $= (3 \times 40) + (2 \times 31) + (8 \times 16)$
  $= (120 + 62 + 128) = 310$ u
 (d) Na_2CO_3, $(NH_4)_2SO_4$

22.

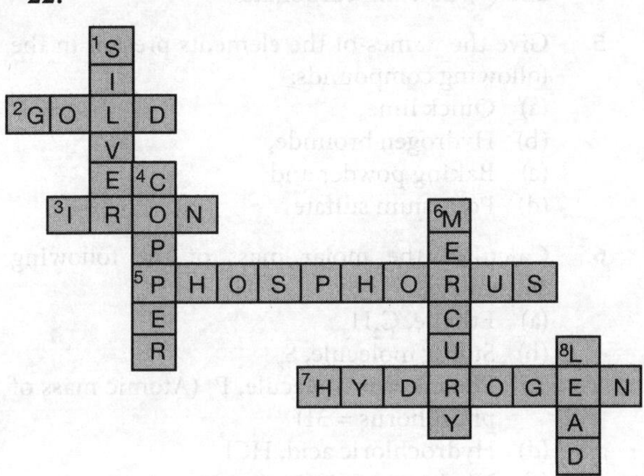

23.

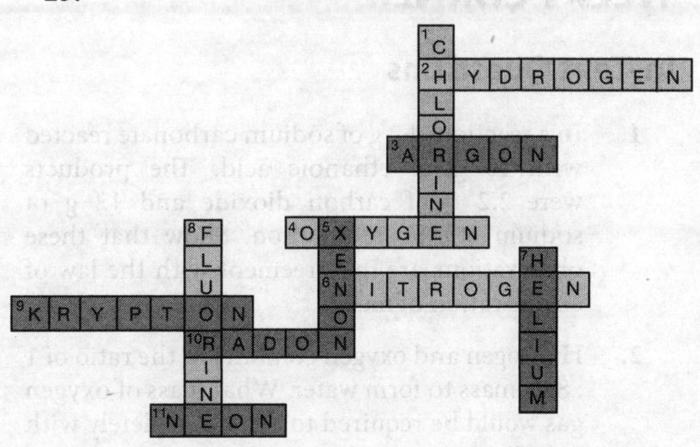

24. (a) Caustic potash, KOH $= (39 + 16 + 1) =$ 56 g mol^{-1}

$= (39 + 16 + 1) = 56$ g mol^{-1}

 (b) Baking powder, $NaHCO_3$
  $= (23 + 1 + 12 + 48) = 84$ g mol^{-1}

 (c) Limestone, $CaCO_3$
  $= (40 + 12 + 48) = 100$ g mol^{-1}

 (d) Caustic soda, NaOH
  $= (23 + 16 + 1) = 40$ g mol^{-1}

 (e) Ethanol, C_2H_5OH
  $= (2 \times 12 + 6 + 16) = 46$ g mol^{-1}

 (f) Common salt, NaCl

25. $6CO_2 + 6H_2O \xrightarrow[\text{Sunlight}]{\text{Chlorophyll}} C_6H_{12}O_6 + 6O_2$

1 mole (180 g) of glucose needs 6 moles (6×8 g) of H_2O

1 g of glucose needs $= \dfrac{108}{180}$ g of H_2O

18 g glucose needs $= \dfrac{108}{180} \times 18 = 10.8$ g

Volume of water used = Mass/Density

$= \dfrac{10.8}{1} = 10.8$ cm^3

NCERT CORNER

In-Text Questions

1. In a reaction, 4.2 g of sodium carbonate reacted with 10 g of ethanoic acid. The products were 2.2 g of carbon dioxide and 12 g of sodium ethanoate solution. Show that these observations are in agreement with the law of conservation of mass.

2. Hydrogen and oxygen combine in the ratio of 1 : 8 by mass to form water. What mass of oxygen gas would be required to react completely with 3 g of hydrogen gas?

3. Which postulate of Dalton's atomic theory is the result of the law of conservation of mass?

4. Which postulate of Dalton's atomic theory can explain the law of definite proportions?

5. Define atomic mass unit.

6. Why is it not possible to see an atom with the naked eye?

7. Write down the formulae of
    (i) Aluminium oxide
    (ii) Aluminium chloride
    (iii) Hydrogen sulfide
    (iv) Calcium hydroxide

8. Write down the names of compounds represented by the following formulae:
   (i) $Al_2(SO_4)_3$, (ii) $MgCl_2$, (iii) K_2SO_4, (iv) KNO_3, (v) $CaCO_3$

9. Calculate the relative molecular mass of (a) water (H_2O), (b) HNO_3 and (c) H_3PO_4.

10. Calculate the formula unit mass of $CaCl_2$.

11. Calculate the formula unit masses of ZnO, Na_2O, K_2CO_3.
    Given, atomic masses of Zn = 65 u, Na = 23 u, K = 39 u, C = 12 u and O = 16 u.

12. Calculate the number of moles for the following:
    (i) 52 g of He
    (ii) 12.044×10^{23} atoms of He

13. Calculate the mass of the following:
    (i) 0.5 moles of N_2 gas
    (ii) 0.5 moles of N atoms
    (iii) 3.011×10^{23} N atoms
    (iv) 6.022×10^{23} N_2 molecules

14. Calculate the number of particles in each of the following:
    (i) 46 g of Na atoms (number from mass) molecules (number of molecules from mass)
    (ii) 8 g O_2
    (iii) 0.1 mole of carbon atoms (number from given moles)

15. If one mole of carbon atoms weighs 12 grams, what is the mass (in grams) of 1 atom of carbon?

16. Which has more atoms, 100 grams of sodium or 100 grams of iron (given, atomic mass of Na = 23 u, Fe = 56 u)?

Exercises

1. A 0.24-g sample of compound of oxygen and boron was found by analysis to contain 0.096 g of boron and 0.144 g of oxygen. Calculate the percentage composition of the compound by weight.

2. When 3.0 g of carbon is burnt in 8.00 g oxygen, 11.00 g of carbon dioxide is produced. What mass of carbon dioxide will be formed when 3.00 g of carbon is burnt in 50.00 g of oxygen? Which law of chemical combination will govern your answer?

3. What are polyatomic ions? Give examples.

4. Write the chemical formulae of the following compounds:
   (i) Magnesium chloride, (ii) Calcium oxide, (iii) Copper nitrate, (iv) Aluminium chloride and (v) Calcium carbonate

5. Give the names of the elements present in the following compounds:
   (a) Quick lime,
   (b) Hydrogen bromide,
   (c) Baking powder and
   (d) Potassium sulfate

6. Calculate the molar mass of the following substances.
   (a) Ethyne, C_2H_2
   (b) Sulfur molecule, S_8
   (c) Phosphorus molecule, P_4 (Atomic mass of phosphorus = 31)
   (d) Hydrochloric acid, HCl
   (e) Nitric acid, HNO_3

7. What is the mass of:
   (a) 1 mole of nitrogen atoms
   (b) 4 moles of aluminium atoms (atomic mass of aluminium = 27)
   (c) 10 moles of sodium sulfite (Na_2SO_3)

8. Convert into mole:
   (a) 12 g of oxygen gas
   (b) 20 g of water
   (c) 22 g of carbon dioxide

9. What is the mass of:
   (a) 0.2 moles of oxygen atoms
   (b) 0.5 moles of water molecules

10. Calculate the number of molecules of sulfur (S_8) present in 16 g of solid sulfur.

11. Calculate the number of aluminium ions present in 0.051 g of aluminium oxide.
    [Hint: The mass of an ion is the same as that of an atom of the same element. Atomic mass of Al = 27 u.]

Answers

In-Text Questions

1. Total mass of the reactants = Mass of sodium carbonate + Mass of ethanoic acid solution
   = 4.2 g + 10 g = 14.2 g
   Total mass of products = Mass of sodium ethanoate solution + Mass of carbon dioxide
   = 12 g + 2.2 g = 14.2 g
   Thus, Mass of products = Mass of the reactants. In other words, no change in mass takes place as a result of the given reaction. This shows that the given observations are in agreement with the law of conservation of mass.

2. According to the law of constant proportions, the composition of a compound is always fixed. By applying this
   As 1 g of hydrogen gas combines with oxygen
   = 8 g
   3 g of hydrogen gas will combine with oxygen
   = 8 × 3 = 24 g

3. The given postulate of Dalton's atomic theory is the result of the law of conservation of mass. 'Atoms are indivisible particles, which cannot be created or destroyed in a chemical reaction.'

4. The postulate of Dalton's atomic theory which says that "The number and kind of atoms in a given compound is fixed" can explain the law of definite proportions.

5. One atomic mass unit (amu) is a mass unit equal to exactly one twelfth (1/12th) the mass of one atom of carbon-12. The relative atomic masses of all the elements have been found with respect to an atom of carbon-12.

6. It is not possible to see it with the naked eye as an atom is extremely small in size. Generally, the radius of an atom is of the order of nanometres. For example, atomic radius of hydrogen atom is 10^{-10} m.

7. (i) Aluminium oxide:
       Symbol   Al   O
                     $= Al_2O_3$
       Charge valency 3+   2–
                        3   2
   (ii) Aluminium chloride:
       Symbol   Al   Cl
                     $= AlCl_3$
       Charge   3+   1–
   (iii) Hydrogen sulfide:
       Symbol   H   S
                     $= H_2S$
       Charge   1+   2–
   (iv) Calcium hydroxide:
       Symbol   Ca   (OH)
                     $= Ca(OH)_2$
       Charge   2+   1–

8. (i) $Al_2(SO_4)_3$ is Aluminium sulfate
   (ii) $MgCl_2$ is Magnesium chloride
   (iii) K_2SO_4 is Potassium sulfate
   (iv) KNO_3 is Potassium nitrate
   (v) $CaCO_3$ is Calcium carbonate

9. (a) Atomic mass of hydrogen = 1 u, oxygen = 16 u
       So the molecular mass of water, which contains two atoms of hydrogen and one atom of oxygen is = 2 × 1+ 1×16 = 18 u
   (b) Molecular mass of HNO_3 = Atomic mass of H + Atomic mass of N + 3 × Atomic mass of O = 1 + 14 + 48 = 63 u.

(c) Molecular mass of $H_3PO_4 = 3 \times$ Atomic mass of H + Atomic mass of P + $4 \times$ Atomic mass of O = 3 + 31 + 64 = 98 u.

10. Formula unit mass of $CaCl_2$ = Atomic mass of Ca + $(2 \times$ Atomic mass of Cl$)$ = $40 + 2 \times 35.5 = 40 + 71 = 111$ u.

11. (i) Formula unit mass of ZnO = 65 + 16 = 81 u
   (ii) Formula unit mass of $Na_2O = (23 \times 2) + (16 \times 1) = 46 + 16 = 62$ u
   (iii) Formula unit mass of K_2CO_3
   $= (39 \times 2) + (12 \times 1) + (16 \times 3) = 78 + 12 + 48 = 138$ u

12. (i) As 1 mole of He = Molar mass of He = 4 g
   4g of He = 1 mole of He
   Hence 52 g of He = $1/4 \times 52 = 13$ moles
   Number of moles (n)
   $= \dfrac{\text{Given mass (m)}}{\text{Molar mass (M)}} = \dfrac{52\,g}{4\,g} = 13$
   (ii) As 1 mole of He = Avogadro's number of atoms = 6.022×10^{23} atoms of He
   6.022×10^{23} atoms He = 1 mole of He
   12.044×10^{23} atoms of He
   $= \dfrac{1}{6.022 \times 10^{23}} \times 12.044 \times 10^{23} = 2$ moles

13. (i) 1 mole of N_2 – Molar mass of $N_2 = 28$ g
   $\therefore$ 0.5 mole of $N_2 = 28 \times 0.5 = 14$ g
   (ii) 1 mole of N atoms = Molar mass of N = 14 g
   $\therefore$ 0.5 mole of N atoms = $14 \times 0.5\,g = 7$ g
   (iii) As 1 mole of N atoms = 6.022×10^{23} atoms of N = Molar mass of N = 14 g
   6.022×10^{23} atoms have mass = 14 g
   Hence 3.011×10^{23} atoms will have mass
   $= \dfrac{14}{6.022 \times 10^{23}} \times 3.011 \times 10^{23}\,g = 7$ g
   (iv) As 1 mole of N_2 molecules = 6.022×10^{23} molecules = Molar mass of $N_2 = 28$ g
   6.022×10^{23} molecules of N_2 have mass = 28 g

14. (i) The number of atoms
   $= \dfrac{\text{Given mass}}{\text{Molar mass}} \times$ Avogadro's number
   $N = \dfrac{m}{M} \times N_A$
   $N = \dfrac{46}{23} \times 6.022 \times 10^{23}$

$N = 12.044 \times 10^{23}$

(ii) The number of molecules
$= \dfrac{\text{Given mass}}{\text{Molar mass}} \times$ Avogadro's number
$N = \dfrac{m}{M} \times N_A$

Atomic mass of oxygen = 8 u
So molar mass of O_2 molecules = $8 \times 2 = 16$ g

$N = \dfrac{8}{32} \times 6.022 \times 10^{23} = 1.51 \times 10^{23}$

(iii) The number of particles (atom)
$n \times N_A = 0.1 \times 6.022 \times 10^{23} = 6.0222 \times 10^{22}$

15. 1 mole carbon atom = 6.022×10^{23} atoms
Molar atomic mass = 12 g
6.022×10^{23} carbon atoms weigh = 12 g
1 carbon atom weighs
$= \dfrac{\text{Weight of carbon}}{\text{Number of carbon atoms}} = \dfrac{12}{6.022 \times 10^{23}}$
$= 1.99 \times 10^{-23}\,g$

16. Molar mass of sodium = 23 g
1 mole atom = 6.022×10^{23} atoms
23 g sodium contains = 6.022×10^{23} atoms
1 g sodium contains = 6.022×10^{23} atoms
100 g sodium contains = 6.022×10^{23} atoms = 2.618×10^{24} atoms
By the above method or by formula, we find the number of atoms in 100 g Fe;
Number of atoms of an element in a given mass
$= \dfrac{\text{Given mass}}{\text{Gram atomic}} \times$ Avogadro's number
$= \dfrac{100\,g}{56\,g} \times 6.022 \times 10^{23} = 1.075 \times 10^{24}$ atoms

It means 100 g of sodium has more atoms as compared to 100 g of iron.

Exercises

1. Mass of the compound = 0.24 g
Mass of boron = 0.096 g
Mass of oxygen = 0.144 g
Percentage of boron $= \dfrac{\text{Mass of boron}}{\text{Mass of the compound}} \times 100$
$= \dfrac{0.096\,g}{0.240\,g} \times 100$

$$\text{Percentage of oxygen} = \frac{\text{Mass of oxygen}}{\text{Mass of the compound}} \times 100$$

$$= \frac{0.144 \text{ g}}{0.240 \text{ g}} \times 100$$

Alternatively, % of oxygen can be found as follows:

% of oxygen = 100 % of boron = 100 − 40 = 60%

2. First, we find the proportion of mass of carbon and oxygen in carbon dioxide.

   In CO_2, C : O = 12 : 32 or 3 : 8

   In other words, it means

   12.00 g carbon reacts with oxygen = 32.00 g

   3.00 g carbon will react with oxygen $= \dfrac{32}{12} \times 3$

   = 8 g

   Hence 3.00 g of carbon will always react with 8.00 g of oxygen to give CO_2 (11 g), even if a large amount (50.00 g) of oxygen is present. This answer will be governed by 'the law of constant proportions'.

3. Polyatomic ions are groups of atoms which carry a fixed charge (either positive or negative) on them and behave as ions.

   For example, (i) Ammonium ion (NH_4^+),

   (ii) Sulfate ion (SO_4^{2-}),

   (iii) Carbonate ion (CO_3^{2-}).

4. (i) Magnesium chloride - $MgCl_2$
   (ii) Calcium oxide - CaO
   (iii) Copper nitrate - $Cu(NO_3)_2$
   (iv) Aluminium chloride - $AlCl_3$
   (v) Calcium carbonate - $CaCO_3$

5.

Compounds	Elements
(a) Quick lime – Calcium oxide – CaO	Calcium, oxygen
(b) Hydrogen bromide – HBr	Hydrogen, bromine
(c) Baking powder – Sodium hydrogen carbonate – $NaHCO_3$	Sodium, hydrogen, carbon, oxygen
(d) Potassium sulfate – K_2SO_4	Potassium, sulfur, oxygen

6. (a) Molar mass of C_2H_2 = (2 × Atomic mass of C) + (2 × Atomic mass of H)
   = (2 × 12) + (2 × 1) = 26 u
   (b) Molar mass of S_8 = (8 × Atomic mass of S)
   = 8 × 32 = 256 u

   (c) Molar mass of P_4 = 4 × Atomic mass of P = 4 × 31 = 124 u
   (d) Molar mass of HCl = Atomic mass of hydrogen + Atomic mass of Cl
   = 1 + 35.5 = 36.5 u
   (e) Molar mass of HNO_3
   = Atomic mass of H + Atomic mass of N + (3 × Atomic mass of O)
   = 1 + 14 + (3 × 16) = 15 + 48 = 63 u

7. (a) Molar mass of N atoms = Atomic mass of N.
   Mass of 1 mol of N atoms = 14 g
   (b) Mass of 1 mole Al atoms = 27 g
   Mass of 4 moles of Al atoms = 27 × 4 = 108 g
   (c) Mass of 1 mole of Na_2SO_3 = (23 ×2) + 32 + (16 × 3) = 46 + 32 + 48 = 126 g
   Mass of 10 moles of Na_2SO_3 = 126 × 10 = 1260 g

8. Use the general formula,

   $$\text{Mole} = \frac{\text{Given weight}}{\text{Molecular weight}}$$

   (a) Molar mass of oxygen (O_2) = 16 × 2 = 32 g
   32 g oxygen gas = 1 mol

   $$12 \text{ g oxygen gas} = \frac{1 \times 12 \text{ g}}{32 \text{ g}} = 0.375 \text{ mol}$$

   (b) Molar mass of water (H_2O) = 2 + 16 = 18g
   18 g water = 1 mol

   $$20 \text{ g water} = \frac{1 \times 20 \text{ g}}{18 \text{ g}} = 1.11 \text{ mol}$$

   (c) 22 g of carbon dioxide (CO_2)
   Molar mass of carbon dioxide (CO_2) = 12 + 32 = 44g
   44 g CO_2 = 1 mol

   $$22 \text{ } CO_2 = \frac{1 \times 22 \text{ g}}{44 \text{ g}} = 0.5 \text{ mol}$$

9. Use the relation Mass or weight = Mole × Molecular weight
   (a) Mass of 1 mole of O atoms = 16 g
   Mass of 0.2 mole of O atoms = 16 × 0.2 = 3.2 g
   (b) Mass of 1 mole of H_2O molecules = 18 g
   Mass of 0.5 mole of H_2O molecules = 18 × 0.5 = 9.0 g

10. Molar mass of sulfur (S_8) = 32 × 8 = 256 g

Number of S_8 molecules in 256 g of solid sulfur
$= 6.022 \times 10^{23}$

Number of S_8 molecules in 16 g of solid sulfur

$= \dfrac{6.022 \times 10^{23} \times 16\,g}{256\,g} = 3.76 \times 10^{23}$

11. Molar mass of
$Al_2O_3 = (27 \times 2) + (16 \times 3) = 54 + 48 = 102\,g$

$$\underset{\text{1 mol (102 g)}}{Al_2O_3} \rightleftharpoons \underset{\text{2 mol}}{2Al^{3+} + 3O^{2-}}$$

102 g Al_2O_3 contains Al^{3+} ions

$= 2 \times 6.022 \times 10^{23}$

Therefore, 0.051 g Al_2O_3 will contain Al^{3+} ions $= \dfrac{2 \times 6.022 \times 10^{23}}{102} \times 0.051$

$= 6.022 \times 10^{20}\ Al^{3+}$ ions

CHAPTER 4

Structure of the Atom

Learning Objectives

After studying this unit, you will be able to:

- Explain cathode rays and the discovery of electrons
- Explain anode rays and the discovery of protons
- List the characteristics and distinguishing properties of electrons, protons and neutrons
- Describe the structure of the atom in terms of Thomson's model, Rutherford's model and Bohr's model
- Define atomic number and mass number
- Write the electronic configuration of elements
- Explain valence electron and valency
- Describe isotopes and isobars

4.1 INTRODUCTION

Atoms and molecules are the building blocks of matter. The existence of different types of matter is due to the different types of atoms and molecules present in the matter. For a long time it was believed that atoms were indivisible, without any inner structure. However, now we know that atoms are divisible and they have an inner structure (Fig. 4.1). Atoms have smaller particles known as **subatomic particles**.

PROTONS
Charge: Positive
Mass: 1 amu
Location: Nucleus

NEUTRONS
Charge: None
Mass: 1 amu
Location: Nucleus

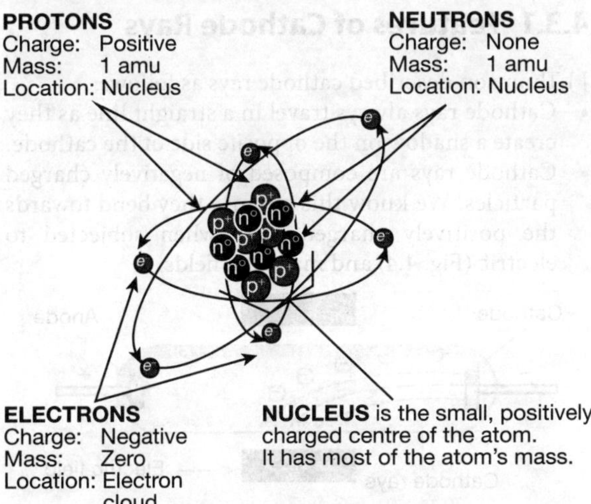

ELECTRONS
Charge: Negative
Mass: Zero
Location: Electron cloud

NUCLEUS is the small, positively charged centre of the atom. It has most of the atom's mass.

Fig. 4.1 The structure of an atom

Atoms are made up of three types of subatomic particles: electrons, protons and neutrons. **Electrons** have a negative charge, **protons** have a positive charge, while **neutrons** have no charge or are neutral. Protons and neutrons are present in the small nucleus at the centre of the atom. Electrons are present outside the nucleus (in the post-nucleus region). They revolve around the nucleus in circular orbits or energy shells. As an atom is neutral, it means it contains an equal number of electrons and protons.

The atoms of different elements differ in the number of electrons, protons and neutrons they contain. In this unit, we shall discuss these subatomic particles and the various atomic models.

4.2 CHARGED PARTICLES IN MATTER

You may have noticed that when you run a comb through dry hair (Fig. 4.2), the comb attracts small pieces of paper. Similarly, when you rub a glass rod with a piece of silk cloth and bring it near an inflated balloon, the glass rod attracts the balloon. We know that an electrically charged object can attract an uncharged object. This means that on rubbing with dry hair, a comb acquires an electric charge, and on rubbing with silk cloth, a glass rod also acquires an electric charge. Now, the question that arises is where does this electric charge come from? The obvious answer is from within the atoms present inside the comb and glass rod. Such simple experiments tell us that some charged particles are present in the atoms of matter. Hence, the atom is divisible.

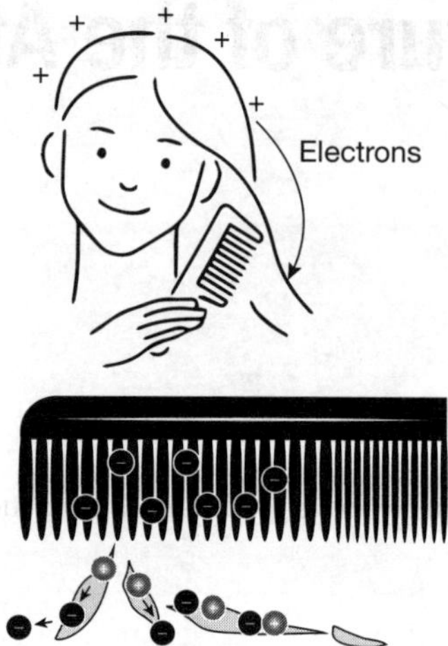

Fig. 4.2 The electric charge produced on the comb (on rubbing in hair) comes from the atoms present in the comb. This shows that some charged particles are present in the atoms of the comb (and hence of other matter).

4.3 Cathode Rays and the Discovery of the Electron

Sir William Crookes designed a discharge tube (called Crookes discharge tube), which is a long glass tube fitted with metal electrodes on either end across which high voltage can be applied (Fig. 4.3). The electrode which is connected to the negative terminal of the power source is known as the cathode, while the electrode which is connected to the positive terminal is known as the anode.

J J Thomson modified it as follows. He connected the discharge tube to a vacuum pump for controlling the pressure of the gas inside the tube. When the pressure of the gas is reduced to nearly 10^{-2} atm and a potential difference of nearly 10000 volts is applied to the electrodes, an electric current flows and at the same time, light is emitted by the gas. The fluorescence (green coloured) was caused due to the bombardment of the walls of the tube by rays emanating from the cathode. These rays are known as cathode rays. Actually, these rays are beams or streams of negatively charged small particles.

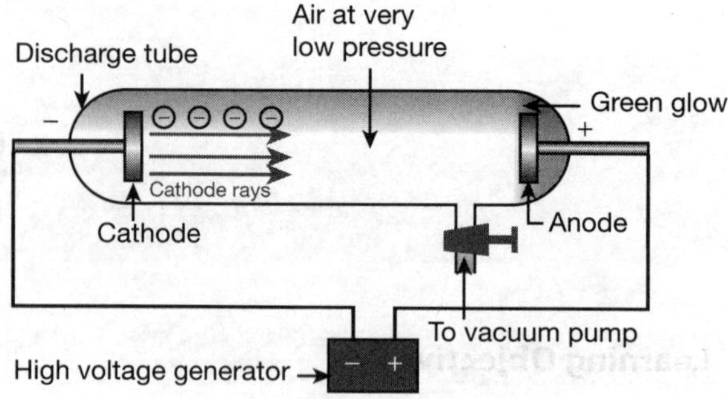

Fig. 4.3 Cathode ray discharge tube

Competition Edge

When the atmospheric pressure inside the tube is normal, no current flows through the air between the electrode as at normal pressure, air is a poor conductor of electricity. If the pressure is less than 0.001 mm, there is no emission and the tube appears dark. However, the glass glows with a greenish-yellow light (fluorescence).

The colour emitted depends upon the nature of the gas taken in the discharge tube; for example, in the case of neon gas, a reddish, orange light is emitted. The television picture tube is a cathode ray tube in which a picture is produced on the screen coated with a suitable material.

4.3.1 Features of Cathode Rays

J J Thomson described cathode rays as follows:
- Cathode rays always travel in a straight line as they create a shadow on the opposite side of the cathode.
- Cathode rays are composed of negatively charged particles. We know this because they bend towards the positively charged plate when subjected to electric (Fig. 4.4) and magnetic fields.

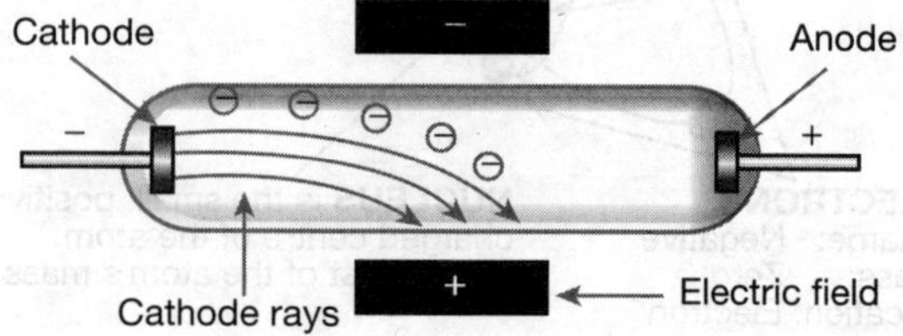

Fig. 4.4 Effect of electric field on cathode rays

- Cathode rays are composed of material particles and possess energy, so they can produce mechanical effects. For example, they can rotate a light paddle wheel placed in their way (Fig. 4.5).

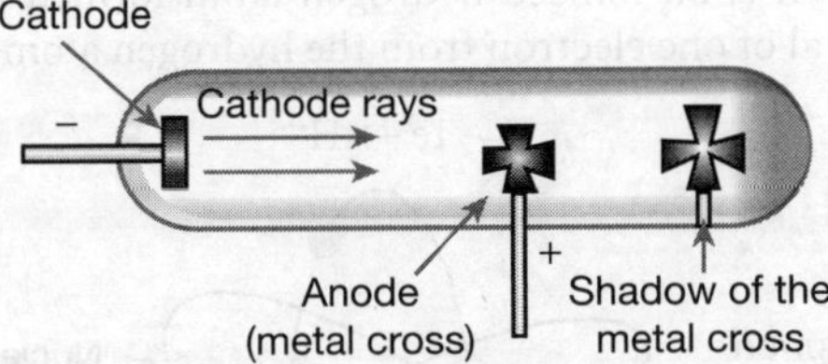

Fig. 4.5 Cathode rays cast a shadow of the objects placed in their path

- Cathode rays can heat up the object on which they fall. When they strike an object, a part of the kinetic energy is transferred to the object, as a result of which there is a rise in temperature.
- Cathode rays can penetrate through thin metallic sheets.
- Cathode rays can ionise the gas through which they travel.
- Cathode rays can produce a green fluorescence on a glass surface or on certain substances like zinc sulfide.
- When cathode rays fall on certain hard metals like copper and tungsten, X-rays are formed. These cannot be deflected by any electrical or magnetic fields but they can pass through opaque material and be stopped by solid objects such as bones.

4.3.2 Electron

The electron was discovered by J J Thomson during a study of cathode rays. Cathode rays consist of small, negatively charged particles known as electrons. An electron is a negatively charged particle found in the atoms of all elements (hydrogen atom has only 1 electron). Electrons are located outside the nucleus, in the extra or post-nucleus region of an atom. It is usually represented by the symbol e^- or $_{-1}^{0}e$.

Features of an Electron

- **Mass of an electron:** The mass of an electron is nearly 1/1837 of the mass of a hydrogen atom. Since the mass of a hydrogen atom is 1 u (lowest), the absolute mass of an electron is 0.0005488 amu or 9.1×10^{-28} grams or 9.1×10^{-31} kg. It is important to note that due to its almost negligible mass, the electron is ignored during the calculation of the atomic mass of an atom.

- **Charge of an electron:** It was confirmed by Milliken's oil-drop experiment and its absolute value is 1.6×10^{-19} Q or -4.8×10^{-10} esu. As it is the smallest possible negative charge occupied by any particle, it is taken as the unit of negative charge. It means the electron has 1 unit of negative charge (−1).

4.4 Anode Rays and the Discovery of the Proton

In the previous section, the formation of cathode rays showed us that all atoms contain negatively charged particles or electrons. Now, as an atom is electrically neutral, it must also have some equal positively charged particles to balance the negative charge of electrons. It has been found by experiments that all atoms have positively charged particles known as protons.

The existence of the proton was proved by Goldstein when he discovered anode rays.

Goldstein conducted the cathode ray experiment by using a perforated cathode and discovered anode rays which are composed of positively charged particles (protons). When the pressure in the tube is decreased, it was observed that in addition to cathode rays, a new kind of ray was also found which came through the perforations (holes) of the cathode. These rays travelled in the opposite direction of the cathode rays and passed through the holes of the cathode and struck the other end of the discharge tube (Fig. 4.6). When these radiations struck the end of the tube, fluorescent radiations were seen. Such rays were called **canal rays** as they passed through the holes or canals in the cathode. These rays were also called anode rays as they move from the anode toward the cathode. It was observed that anode rays consist of positively charged particles. So, they were also called positive rays.

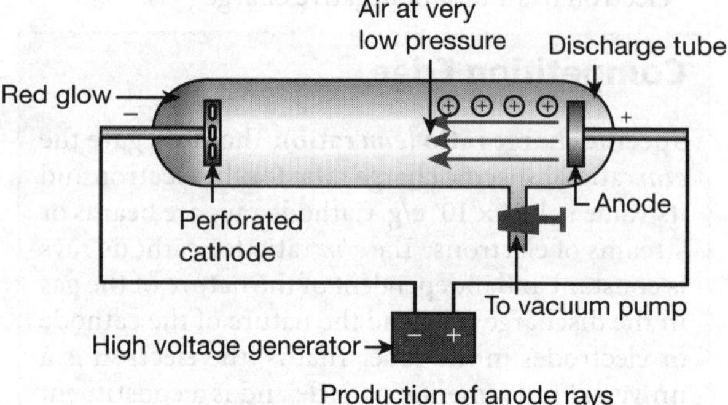

Fig. 4.6 Anode ray discharge tube

4.4.1 Features of Anode Rays

Goldstein described anode rays as follows:

- Anode rays travel in straight lines.
- Anode rays consist of material particles.
- Anode rays are deflected by electric and magnetic fields towards the negatively charged plate which indicates that they are positively charged or composed of positively charged particles.
- The charge to mass ratio of the particles in the anode rays was determined by W Wien using Thomson's technique. This ratio depends upon the nature and mass of the gas taken in the discharged tube. The mass of the particles was the same as the atomic mass of the gas taken in the discharge tube.

As hydrogen is the lightest gas and the hydrogen atom is the lightest atom, the positive particles obtained from hydrogen are the lightest and have the smallest charge. The anode rays obtained from hydrogen are made up of the same type of positive particles. These particles are known as protons (named by Rutherford). Hence, the anode rays obtained from hydrogen gas consist of protons. A proton is formed by the removal of an electron from a hydrogen atom.

4.4.2 Proton

The proton was discovered by Goldstein and described by Rutherford. A proton is a positively charged particle found in the atoms of all elements. Protons are located

in the nucleus (Fig. 4.7). The hydrogen atom contains only one proton in its nucleus; the atoms of all other elements contain more than one proton. A proton is usually represented by the symbol p^+ or ^1_1p. In fact, a proton is an ionised hydrogen atom formed by the removal of one electron from the hydrogen atom.

$$H - 1e \rightarrow H^+$$

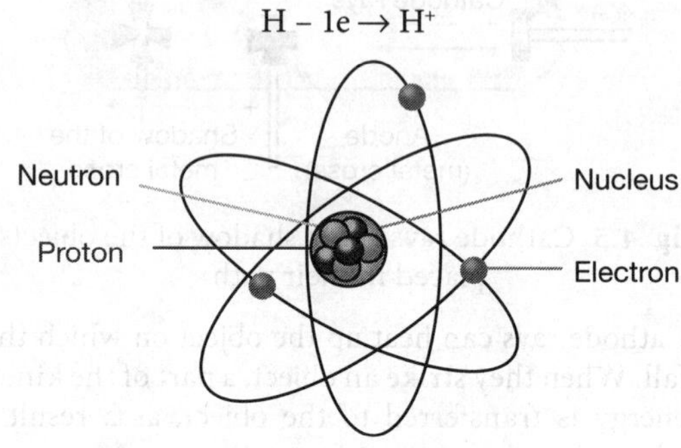

Fig. 4.7 Protons

Features of a Proton

- **Mass of proton:** The mass of a proton has been found to be equal to 1.673×10^{-27} kg. This is almost equal to that of an atom of hydrogen. Since the mass of a hydrogen atom is 1 amu, the relative mass of a proton is nearly 1.0072 amu. The mass of a proton is nearly 1837 times that of an electron.
- **Charge of proton:** The charge present on a proton is equal and opposite to the charge present on an electron. The value of the charge on a proton is 1.602×10^{-19} coulomb or $+4.8 \times 10^{-10}$ esu of positive charge. As 1.6×10^{-19} coulomb is the smallest positive charge carried by any particle, this value is taken as the unit of positive charge. It means that a proton carries 1 unit positive charge or **the relative charge of a proton is +1.**

4.5 Neutron

After the discovery of proton and electron, it was observed that all the mass of an atom could not be accounted for on the basis of only the protons and electrons present in it. For example, an oxygen atom has 8 protons and 8 electrons. The mass of electrons is very small, so the atomic mass of oxygen should be only 8 u, which is the mass of 8 protons. However, the actual mass of one oxygen atom is 16 u. Now how can anyone explain this extra mass of 8 units? This was solved by James Chadwick in 1932 with the discovery of another subatomic particle, the neutron. The neutron is a neutral particle found in the nucleus of an atom. It is represented by the symbol $_0^1 n$. It was discovered by bombarding Be atoms with a-particles.

$$_4Be^9 + {}_2He^4 \rightarrow {}_6C^{12} + {}_0n^1$$

The subatomic particle not present in a hydrogen atom is neutron. A hydrogen atom contains only one proton and one electron.

We can now explain why the atomic mass of oxygen is 16 u and not 8. An oxygen atom has 8 protons and 8 neutrons, each having a mass of 1 u.

Atomic mass of oxygen = Mass of 8 protons + Mass of 8 neutrons

$= 8 \times 1 + 8 \times 1$

$= 16$ u

Now it is clear that the atomic mass is equal to the sum of the masses of protons and neutrons present in the nucleus of an atom. The mass of electrons present in an atom is very small and can therefore be ignored.

4.5.1 Features of a Neutron

- **Mass of a neutron:** The mass of a neutron is slightly higher than the mass of a proton. The relative mass of a neutron is 1 u. The absolute mass of a neutron is 1.674×10^{-27} kg or 1.0086 amu.
- **Charge of a neutron:** A neutron has no charge and it is electrically neutral.

Table 4.1 provides a comparison between proton, neutron and electron

Table 4.1 Comparison between proton, neutron and electron

	Electron	**Proton**	**Neutron**
Symbol	e/e^-	p/p^+	n
Nature	Negatively charged	Positively charged	Neutral
Relative mass	1/1840 of a H atom	Equal to H atom	Equal to H atom
Actual mass	9.1×10^{-28} g	1.673×10^{-24} g	1.674×10^{-24} g
Charge	(-1) $(1.602 \times 10^{-19}$ C$)$	$(+1)$ $(1.602 \times 10^{-19}$ C$)$	No charge
Location	Outside the nucleus	Inside the nucleus	Inside the nucleus

4.6 STRUCTURE OF THE ATOM AND DEVELOPMENT OF THE ATOMIC MODEL

After the discovery of the electron and proton, scientists began thinking about the arrangement of these particles in an atom. This led to the development of the concept of atomic models. An atomic model depicts the arrangement of fundamental particles in an atom. J J Thomson was the first scientist to propose a model for the structure of the atom.

There are three atomic models: Thomson's, Rutherford's and Bohr's.

4.6.1 Thomson's Atomic Model

Thomson was the first to propose a detailed model of the atom just after the discovery of electrons. He proposed that an atom consists of a uniform sphere of positive electricity in which the electrons are distributed or embedded more or less uniformly. The negative and the positive charges are equal in magnitude. Thus, the atom as a whole is electrically neutral. This model of the atom is known as the plum pudding model or watermelon model (Fig. 4.8).

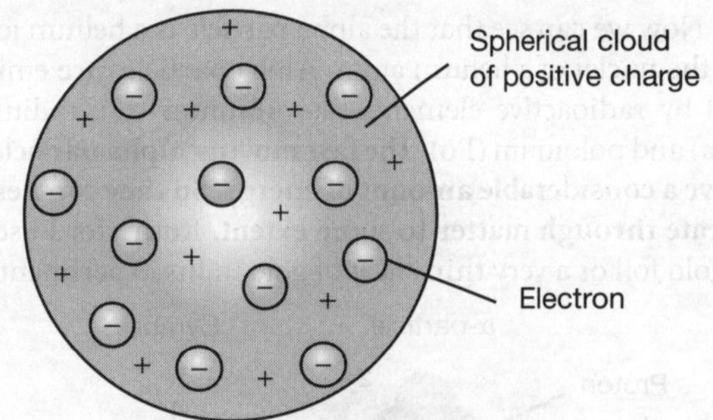

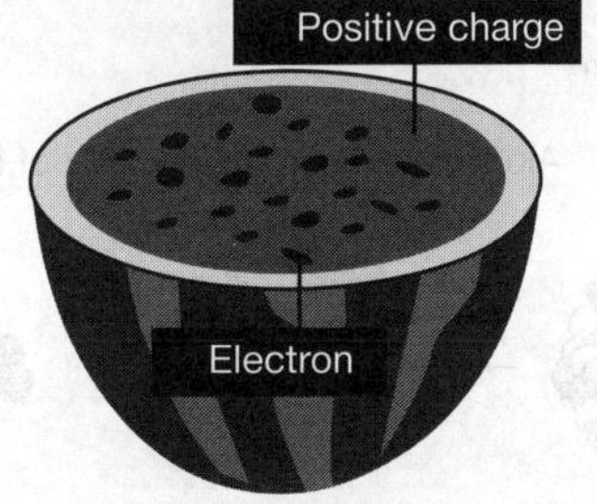

Fig. 4.8 Thomson's model of the atom

The salient features of Thomson's atomic model are:

- An atom consists of a positively charged sphere and the negatively charged electrons are uniformly embedded in it.
- The negative and positive charges are equal in magnitude. So, the atom as a whole is electrically neutral (overall no charge).

Although Thomson's model could explain why atoms are electrically neutral, it could not explain how the positively charged particles are shielded from the negatively charged electrons without getting neutralised. The results of experiments carried out by other scientists (Rutherford and others) could not be explained by this model, so it was rejected.

4.6.2 Rutherford's Atomic Model

Before we discuss Rutherford's experiment, let us first examine the alpha particle. An alpha particle is a positively charged particle having 2 units of positive charge and 4 units of mass (Fig. 4.9). It is a helium ion (He^{2+}), which is formed when a helium atom loses electrons.

$$^{4}_{2}He \xrightarrow{\ -2e\ } He^{2+} \text{ or } ^{4}_{2}\alpha$$

Now we can see that the alpha particle is a helium ion or the nuclei of a helium atom. Alpha particles are emitted by radioactive elements like uranium (U), radium (Ra) and polonium (Po). **The fast moving alpha particles have a considerable amount of energy. So they can penetrate through matter to some extent.** Rutherford used a gold foil or a very thin sheet of gold in his experiment.

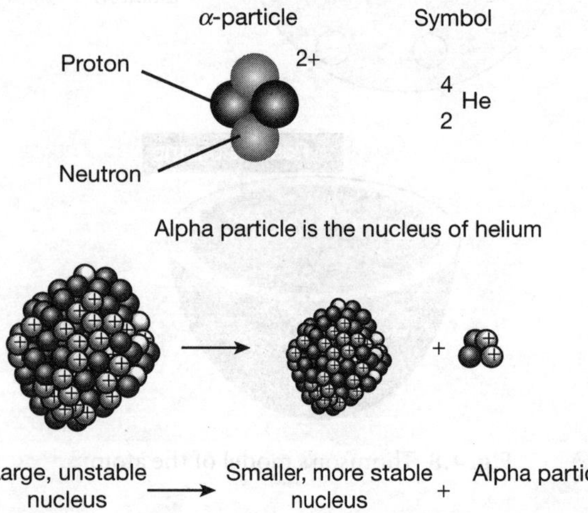

Fig. 4.9 Representation and formation of alpha particles

Rutherford's Experiment

Thus far, we have studied the discovery and properties of electrons, protons and neutrons and it is quite clear that an atom is composed of these three subatomic particles. Many experiments were carried out to find out the arrangement of these subatomic particles in an atom. It was Rutherford's alpha particle scattering experiment which led to the discovery of a small positively charged centrally located space, that is, the nucleus in an atom which contains the protons and neutrons (combined called nucleons).

Rutherford and his students (Hans Geiger and Ernest Marsden) bombarded a very thin gold foil with α-particles (helium nuclei). This famous α-particle scattering experiment is represented in Fig. 4.10 given below.

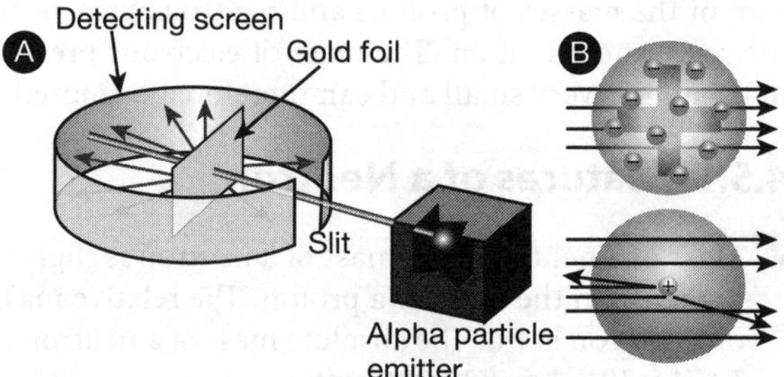

Fig. 4.10 Rutherford's experiment

A stream of high-energy α-particles from a radioactive source was directed at a thin foil (thickness < 100 nm) of gold metal. The thin gold foil had a circular fluorescent zinc sulfide screen around it. Whenever α-particles struck the screen, a tiny flash of light was produced at that point. The results of the scattering experiment were quite unexpected. According to Thomson's model of the atom, the mass of each gold atom in the foil should have been spread evenly over the entire atom, and α-particles had enough energy to pass directly through such a uniform distribution of mass. It was expected that the particles would slow down and change direction only by small angles as they passed through the foil. Instead, it was observed that:

- Most of the α-particles passed through the gold foil undeflected
- A small fraction of the α-particles was deflected by small angles
- Very few α-particles (~1 in 20,000) bounced back, that is, were deflected by nearly 180°

Based on these observations, Rutherford introduced his atomic model and drew the following conclusions:

(i) Most of the atom is empty as most of the α-particles passed straight through without any deflection. This empty space is known as the post- or extra-nuclear region.

(ii) Electrons occupy positions in this empty space (extra-nuclear region).

(iii) As very few α-particles were deflected at various angles up to a maximum of 180°, there is a solid, compact, positively charged space at the centre of the atom. This central part has all the positive charge and nearly the whole mass of the atom and is called the **nucleus**. The particles present in the nucleus are known as **nucleons**. The discovery of the nucleus was the main outcome of this experiment, for which Rutherford received the Nobel Prize.

The radius of an atom is 10^5 times more than the radius of a nucleus. We can imagine this like a football placed in the middle of a ground.

(iv) The radius of an atom is 10^{-10} m or 10^{-8} cm. The radius of the nucleus $= 10^{-15}$ m or 10^{-13} cm. Atomic radius > nucleus radius by 10^5 times.

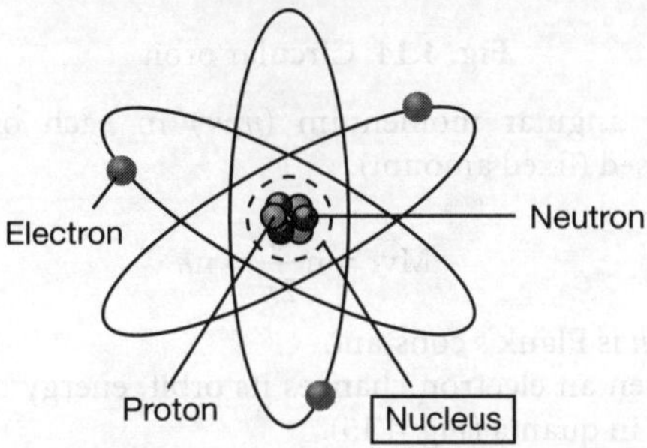

Fig. 4.11 Structure of the nucleus

(v) A centrifugal force develops between the electrons and the nucleus (Fig. 4.11). So the electrons revolve around the nucleus just like stars move around the sun.

As at the time that Rutherford's model was introduced, the neutron had not yet been discovered, later on this model was further improved by including neutrons.

Merits of Rutherford's Model

- It explains the discovery of the nucleus and position of protons and electrons within an atom.
- It also explains circulatory rotation of the electrons around the nucleus (planetary rotation).
- According to Rutherford, the simplest atom hydrogen has a small nucleus which contains one proton and

one electron revolving around it. Hydrogen is the only atom which does not have any neutrons (Fig. 4.12).

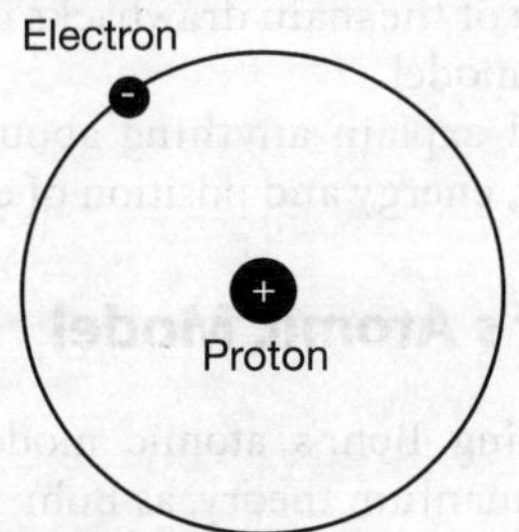

Fig. 4.12 A hydrogen atom

Demerits of Rutherford's Model

(i) This model was unable to explain the stability of an atom. According to Rutherford's postulate, electrons revolve at a very high speed around the nucleus of an atom in a fixed orbit. However, Maxwell explained that accelerated charged particles release electromagnetic radiations. Hence, electrons revolving around the nucleus will release electromagnetic radiation (Fig. 4.13). The electromagnetic radiation will possess energy from the electronic motion, as a result of which the orbits will gradually shrink. Finally, the orbits will shrink and collapse into the nucleus of the atom. According to the calculations, if Maxwell's explanation is followed, Rutherford's model would collapse within 10^{-8} seconds. Therefore, Rutherford's atomic model did not follow Maxwell's theory and it was unable to explain an atom's stability.

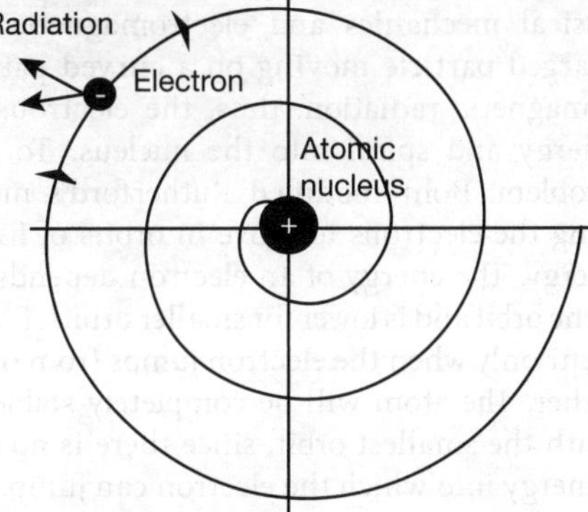

Fig. 4.13 How energy is lost and electrons move closer to the nucleus

(ii) Rutherford's theory was incomplete because it did not mention anything about the arrangement of electrons in the extra-nuclear region, which was later explained by Bohr. This was one of the main drawbacks of Rutherford's atomic model.

(iii) It could explain anything about the number, velocity, energy and position of electrons.

4.6.3 Bohr's Atomic Model

Before discussing Bohr's atomic model, let us first take a look at quantum theory, as Bohr was the first to use this theory to explain the atomic structure. It was introduced by Max Planck. It states that a hot vibrating body does not emit or absorb energy continuously but emits or absorbs discontinuously in the form of small energy packets or bundles known as quanta (photon in the case of light energy). The energy of radiation (E) is directly proportional to the frequency of radiation (v).

$$E \propto v \quad E = h v$$

Here, h is Planck's constant and its value is 6.6253×10^{-34} J s or kg m² s⁻¹.

Absorption or emission in the form of multiples of quanta is known as quantisation of energy.

$$E = n h v$$

where n = number of photons absorbed or emitted per second.

Bohr's Model

In 1913, Bohr proposed his quantised shell model of the atom to explain how electrons can have stable orbits around the nucleus. The motion of the electrons in the Rutherford model was unstable because, according to classical mechanics and electromagnetic theory, any charged particle moving on a curved path emits electromagnetic radiation; thus, the electrons would lose energy and spiral into the nucleus. To remedy this problem, Bohr modified Rutherford's model by requiring the electrons to move in orbits of fixed size and energy. The energy of an electron depends on the size of the orbit and is lower for smaller orbits. Radiation can occur only when the electron jumps from one orbit to another. The atom will be completely stable in the state with the smallest orbit, since there is no orbit of lower energy into which the electron can jump.

Features of Bohr's Atomic Model

Bohr introduced the circular orbit concept (Fig. 4.14) based on Planck's quantum theory. Around the nucleus, there are circular regions: orbits or shells.

K   L   M   N   O...
n = 1   2   3   4   5...
→ Energy and distance from nucleus increase

Each orbit has a fixed amount (quantised amount) of energy, so it is called energy level. An electron revolves round the nucleus in a particular orbit having definite energy without any change of energy or any radiation of energy. That is why these orbits are called stationary states.

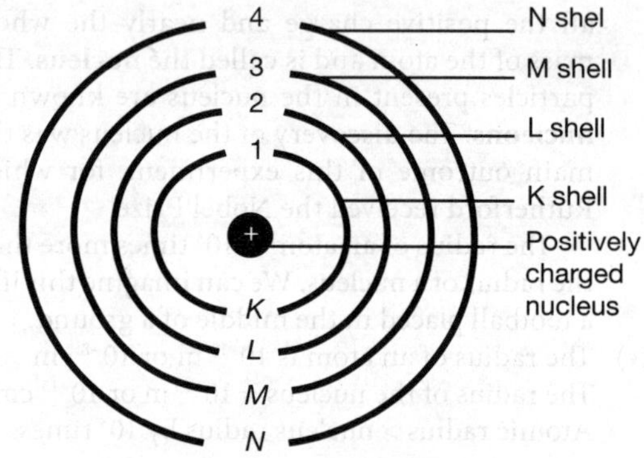

Fig. 4.14 Circular orbit

The angular momentum (mvr) in each orbit is quantised (fixed amount).

$$Mvr = n\frac{h}{2\pi} = n\hbar$$

where h is Planck's constant.

When an electron changes its orbit, energy change occurs in quanta (Fig. 4.15).

$$\Delta E = E_2 - E_1 = hv \text{ or } = \frac{hc}{\lambda} \text{ quanta}$$

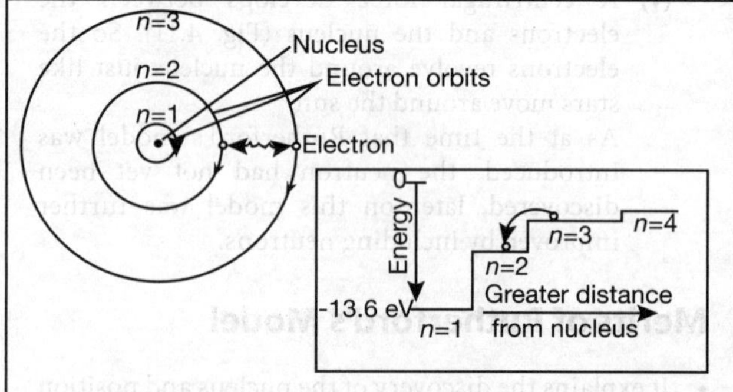

Fig. 4.15 Bohr's model

- **Excited state:** Here, an electron jumps from a lower to a higher orbit or energy level, by absorbing energy in quanta.

- **De-excited state**: Here, an electron jumps from a higher to a lower energy level by releasing energy in quanta.

Merits of Bohr's Model

Bohr's atomic model not only overcomes all the drawbacks of Rutherford's atomic model, but it also has many other advantages over it:

(i) It explains that the centripetal force developed between the nucleus and the electrons due to the rotation of the electrons around the nucleus overcomes the force of attraction between them; as a result, the electrons do not fall into the nucleus. Hence, it explains the stability of the atom.

(ii) According to this theory, both energy and angular momentum are quantised in an orbit.

(iii) By using the concept of quantisation of energy, Bohr's model can be used to find the radius of any orbit, velocity and energy of an electron in any orbit.

Drawbacks of Bohr's Model

- It is not applicable to species having more than one electron, such as Li, He, and so on.
- It cannot explain why atoms undergo chemical combination.
- It can only explain the particle nature of electrons. There is no explanation for the wave nature, that is, it does not follow de Broglie and Heisenberg's theory.

Competition Edge

With the help of Bohr's model, the value of the radius of orbit, energy of electron and velocity of electron can be found out for any orbit.

Radius of any orbit $(r_n) = 0.529 \times n^2/Z$ Å

Velocity of electron in any orbit $(V_n) = 2.18 \times 10^6 \times Z/n$ m/sec

Energy of an electron in any orbit $(E_n) = -2.18 \times 10^{-19} \times Z^2/n^2$ J/mole

Here, n = number of orbits, Z = atomic number.

Quick Review

- Atoms are made up of three subatomic particles: electrons, protons and neutrons.
- Cathode rays are beams or streams of negatively charged small particles.
- Anode rays consist of positively charged particles. So, they are called positive rays.
- Electrons are present outside the nucleus while protons and neutrons are present inside the nucleus.
- An electron is a negatively charged particle found in the atoms of all elements. Electrons are located outside the nucleus in an atom. An electron is usually represented by the symbol e^- or $_{-1}^{0}e$.
- The absolute mass of an electron is 0.0005488 amu or 9.1×10^{-28} g or 9.1×10^{-31} kg.
- Charge on an electron (e^-) was confirmed by Milliken's oil-drop experiment and its value is 1.6×10^{-19} Q or -4.8×10^{-10} esu.
- A proton ($_1^1p$) is a positively charged particle found in the atoms of all elements. Protons are located in the nucleus of an atom.
- The value of the charge on a proton is 1.602×10^{-19} coulomb or $+4.8 \times 10^{-10}$ esu of positive charge.
- The neutron ($_0^1n$) is a neutral particle found in the nucleus of an atom.
- Thomson's atomic model: An atom consists of a uniform sphere of positive electricity in which the electrons are distributed or embedded more or less uniformly. The negative and the positive charge are equal in magnitude. Thus, the atom as a whole is electrically neutral.
- Rutherford's atomic model: Rutherford bombarded a very thin gold foil with α-particles (helium nuclei) called Rutherford's α-particle scattering experiment.
- The centrally located small, solid, compact part having all the positive charge and nearly the whole mass is called the nucleus.
- Radius of atom is = 10^{-10} m or 10^{-8} cm. Radius of nucleus = 10^{-15} m or 10^{-13} cm. Atomic radius > nucleus radius by 10^5 times.
- Rutherford's model cannot explain stability of atoms, number and velocity of electrons and the linear nature of spectrum.
- Bohr modified and removed all the demerits of Rutherford's model and introduced the circular orbit concept in which both energy and angular momentum are quantised (fixed values).

Exercise 4.1

Questions marked * are practical based.

Section A: Multiple Choice Questions
(1 Mark)

1. Which subatomic particle has a negative charge?
   (a) Neutron
   (b) Nucleus
   (c) Proton
   (d) Electron

2. Which subatomic particle has a positive charge?
   (a) Neutron
   (b) Nucleus
   (c) Proton
   (d) Electron

3. Nucleon refers to:
   (a) Electrons and neutrons
   (b) Electrons only
   (c) Protons and neutrons
   (d) Protons and electrons

4. Which subatomic particle has no charge?
   (a) Neutron
   (b) Nucleus
   (c) Proton
   (d) Electron

5. Which of the following statements accurately describes the masses of subatomic particles?
   (a) Mass of 1 neutron equals the mass of 1 proton
   (b) Mass of 1 electron equals the mass of 1840 neutron
   (c) Mass of 1 proton equals the mass of 1 electron
   (d) Mass of 1 neutron equals the mass of 1840 proton

6. Which of the following is used to identify an element?
   (a) Number of nuclei
   (b) Number of protons
   (c) Number of electrons
   (d) Number of nucleons

7. In which of the following pairs are both species charged?
   (a) e, n
   (b) p, n
   (c) e, p
   (d) α, n

8. Thomson's atomic model is known as:
   (a) Watermelon model
   (b) Plum pudding model
   (c) Planetary model
   (d) Both (a) and (b)

9. Select the incorrect statement regarding Rutherford's atomic model:
   (a) Nucleus contains the proton
   (b) Electrons are present in the extra-nulcear region
   (c) Electrons revolve around the nucleus
   (d) Electrons revolve in orbits

10. Who used Planck's theory to give the atomic model?
    (a) Thomson
    (b) Rutherford
    (c) Bohr
    (d) Crookes

11. Circular regions in which electrons revolve are known as:
    (a) Orbits
    (b) Energy levels
    (c) Shells
    (d) All of these

12. Radius of nucleus is:
    (a) 10^{-10} m
    (b) 1 Å
    (c) 100 picometer
    (d) All of these

13. The correct decreasing order of e/m ratio is:
    (a) $n > p > e$
    (b) $e > p > n$
    (c) $p > n > e$
    (d) $e = p > n$

14. Select the set showing correct statements:
    (i) For every orbit, 'n' has a specific value
    (ii) When 'n' is 1, it is the K shell.
    (iii) As 'n' increases, the value of energy decreases
    (iv) As 'n' increases, the size or radius of the atom increases
    (a) (i), (ii)
    (b) (ii), (iii)
    (c) (i), (ii), (iv)
    (d) (i), (ii), (iii), (iv)

15.

Column I (Particle)	Column II (Discover)
1. Electron	(i) Chadwick
2. Proton	(ii) JJ Thomson
3. Neutron	(iii) Rutherford

Select the option showing the correct match:
(a) 1-(i), 2-(ii), 3-(iii)
(b) 1-(iii), 2-(ii), 3-(i)
(c) 1-(ii), 2-(iii), 3-(i)
(d) 1-(ii), 2-(i), 3-(iii)

Assertion-Reason Questions

Direction: In the following question two statements (Assertion) A and Reason (R) are given Mark.
(a) if A and R both are correct and R is the correct explanation of A;
(b) if A and R both are correct but R is not the correct explanation of A;
(c) A is true but R is false;
(d) A is false but R is true

Assertion	Reason
1. Cathode rays are made up of negatively charged particles (electrons).	1. In an electric field, they are deflected towards the positive plate.
2. Anode rays originate from the anode.	2. Anode rays are made up of positively charged particles (protons).
3. Electron is a fundamental particle.	3. Specific charge ratio for cathode rays does not depend upon the nature of the gas taken in the tube.
4. Charge on electron was confirmed by Milliken's oil-drop experiment.	4. Electron is almost mass-less.
5. Thomson's model explained neutrality of an atom.	5. According to Thomson, electrons are uniformly embedded in the positively charged sphere.
6. The radius of the nucleus is 1 lakh times smaller than the radius of the atom.	6. The nucleus contains all the positive charge and almost all the mass of the atom.
7. Bohr introduced circular orbit concept.	7. It explained the wave nature of electrons.

Section B: Very Short Answer Questions   (2 Marks)

1. Which experiment was used to find the charge on electrons?

2. Which subatomic particle has the highest specific charge ratio?

3. What is the location of origin of anode rays?

4. Define protons in terms of the hydrogen atom.

5. Why was the neutron discovered later?

6. Name the element which does not have any neutron.

7. What is the main outcome of Thomson's atomic model regarding an atom?

8. Name the particles which are known as nucleons.

9. What is the radius of nucleus of an atom in Å?

10. Who introduce circular orbit concept?

11. The value of energy and angular momentum in an orbit are fixed. True or false?

12. Why is an orbit called energy level or shell?

Section C: Short Answer Questions   (3 Marks)

1. How can you say that cathode rays are made up of material particles?

2. Who discovered the electron and what is the value of charge on the electron in coulomb and is its mass in kg?

3. Write any two differences between cathode and anode rays.

4. What will be the effect on current if we kept the pressure in the discharge tube below 1 mm?

5. What is e/m or specific charge ratio? Write the decreasing order of specific charge ratio for electron, proton and neutron.

6. In Rutherford's experiment:
   (i) Most of the alpha particles passed straight through. What conclusion can you draw from it?
   (ii) Very few alpha particles were deflected back. What conclusion can you draw from it?

7. Why does Rutherford's model not explain the stability of an atom?

8. Who discovered the neutron and how?

9. Define excited and de-excited state according to Bohr's model.

10. How was Bohr's model able to remove the drawbacks of Rutherford's atomic model?

Section D: Long Answer Questions
(5 Marks)

1. *(i) What happens in the discharge tube if high voltage is applied at a gas pressure of one atmosphere?
   (ii) What will happen if pressure is reduced to 1 mm?
   (iii) What will happen if pressure is further reduced to 0.001 mm?

2. (i) How were cathode rays discovered?
   (ii) Write four properties.
   (iii) Write any commercial applications of cathode ray tubes.

3. (i) What are the main outcomes of Rutherford's atomic model?
   (ii) Write any two drawbacks of it.
   (iii) On which model is it based?

4. (i) Which theory did Bohr use to introduce his atomic model?
   (ii) Write any three major outcomes of Bohr's atomic model.
   (iii) Write any two drawbacks of Bohr's atomic model.

Section E: Case Study or Passage-Based Questions
(4 Marks)

1. In the cathode ray discovery, a discharge tube with low pressure and high voltage was used. Cathode rays showed fluorescence on the walls of the discharge tube as well as on some metal sulfide. They also showed heating effect and produced X-rays on striking against the surface of hard metals.

   (i) What is value of pressure in mm and voltage in volts in the discharge tube used for cathode ray discovery?

   (ii) Why are discharge tubes known as Crookes tubes?

   (iii) A television picture tube is an example of which ray tube?

   (iv) Which rays are formed when cathode rays are made to strike on a hard metal?

   (v) *Name a substance whose screen is used for fluorescence?

2. In order to explain the atomic structure, many scientists like Thomson, Rutherford and Bohr introduced their atomic models. Every model had some merits and some demerits. Today, these models do not have much significance, but still we obtained valuable correct information regarding the atomic structure from these models.

   (i) Thomson's model was able to confirm that:
   (a) Atom has a nucleus
   (b) Atom is neutral
   (c) Electrons revolve around the nucleus
   (d) All of these

   (ii) The main contribution of Rutherford's atomic model was:
   (a) Discovery of nucleus
   (b) Orbit concept
   (c) Energy is quantised in orbit
   (d) Both (a) and (b)

   (iii) Select the correct statement regarding Bohr's orbit:
   (a) Energy is quantised but not angular momentum in an orbit
   (b) Angular momentum is quantised but not energy in an orbit
   (c) Both energy and angular momentum are quantised in an orbit
   (d) Quantisation of energy in any orbit is not fixed

   (iv) Select the set showing correct statements:
   (i) Thomson's model is known as the plum pudding model.
   (ii) Rutherford used alpha particles.
   (iii) Bohr introduced the orbit concept.
   (iv) Bohr's model can explain the wave nature of electrons.
   (a) (i), (ii) (b) (ii), (iii)
   (c) (i), (ii), (iii) (d) (i), (ii), (iii), (iv)

High Order Thinking Skills (HOTS) Questions

1. How can you say that the electron is a universal fundamental particle?

2. What experimental evidence did Rutherford have that led to the conclusion that the nucleus exists?

3. (i) In Rutherford's scattering experiment, what would happen to the alpha rays

if instead of gold, the foil of some light metal was used?

(ii) Electrons carry a negative charge whereas the nucleus carries a positive charge. Explain why the electrons do not fall into the nucleus.

4. How are the different colours of light emitted by elements under certain conditions explained by the Bohr model?

5. (i) What basic premise of the Bohr model was discounted by modern theory?

(ii) What is wrong with this statement: "In the quantum mechanical model, the electron is pictured as a particle in an orbit around the nucleus, much like a planet is in an orbit around the sun."

Answers

Section A: Multiple Multiple Choice Questions

1. (d)	**2.** (c)	**3.** (c)	**4.** (a)
5. (a)	**6.** (b)	**7.** (c)	**8.** (d)
9. (d)	**10.** (c)	**11.** (d)	**12.** (d)
13. (b)	**14.** (c)	**15.** (c)	

Assertion–Reason Questions

1. (a)	**2.** (d)	**3.** (a)	**4.** (b)
5. (a)	**6.** (b)	**7.** (c)	

Section B: Very Short Answer Questions

1. Milliken's oil-drop experiment.

2. Electron.

3. Anode rays originate in the region between the anode and cathode.

4. Refer to Section 4.4.2, Proton.

5. Neutron was discovered late as it was chargeless.

6. Hydrogen atom.

7. It explains the neutrality of the atom.

8. Proton and Neutron

9. Radius of nucleus of an atom is 1 Å or 10^{-10} m.

10. Bohr.

11. True.

12. Refer to Section 4.6.3, Bohr's Atomic Model.

Section C: Short Answer Questions

1. It can be proved by the fact that when a light paddle wheel made of mica is placed in the path of the cathode rays so that the rays strike the blades of the upper half, it begins to rotate.

2. Refer to Section 4.3.2, Electron.

3. (i) Cathode rays originate from the cathode, while anode rays originate from a region between the cathode and anode.

(ii) Cathode rays are made up of negatively charged particles (electron) while anode rays are made up of positively charged particles (proton).

4. Refer to Section 4.3.1, Features of Cathode Rays.

5. Refer to Section 4.3.1, Features of Cathode Rays.

6. (i) Most of an atom is empty.
   (ii) Discovery of the nucleus.

7. Refer to Section 4.6.2, Rutherford's Atomic Model.

8. Refer to Section 4.5, Neutron.

9. Refer to Section 4.6.3, Bohr's Atomic Model.

10. Refer to Section 4.6.3, Bohr's Atomic Model.

Section D: Long Answer Questions

1. (i) No current flows between the electrodes as gases at normal pressure are poor conductors of electricity.

(ii) When the pressure of the gas is reduced to 1 mm, the current begins to flow between the electrodes and the entire discharge tube begins to glow. The colour of the glow depends upon the nature of the gas taken in the tube.

(iii) The glow disappears and the tube appears completely dark. However, the wall of the glass tube opposite the cathode glows, showing that some invisible rays are coming from the cathode.

2. Refer to Section 4.3, Cathode Rays and the Discovery of the Electron.

3. Refer to Section 4.6.2, Rutherford's Atomic Model

4. Refer to Section 4.6.3, Bohr's Atomic Model.

Section E: Case Study or Passage-Based Questions

1. (i) 0.001 mm and 10000 V.
    (ii) Crookes studied different gases using the discharge tube; so they are called Crookes tubes.
    (iii) Cathode ray tube.
    (iv) X-rays.
    (v) Zinc sulfide.

2. (i) (b)   (ii) (a)   (iii) (c)   (iv) (c)

High Order Thinking Skills (HOTS) Questions

1. Refer to Section 4.3.2, Electron.

2. A small percentage of the alpha particles were deflected from their path, and a very small percentage reflected back from the foil. A very high percentage of the alpha particles came straight through the foil. This supported the idea that most of an atom's mass and positive charge were concentrated in a small core, the nucleus.

3. (i) Alpha particles are heavy particles with 4 units of mass. If a light metal is used, the nucleus will be light and the alpha-particles may not be deflected back but may be pushed forward.
    (ii) Electrons keep on revolving around the nucleus as the centrifugal force acting outwards balances the force of attraction; so the electrons do not fall into the nucleus.

4. Different colours of light represent different energies of light. According to the Bohr model, the different electron orbits also represent different energies. Bohr theorised that electrons can gain or lose energy and move from one orbit to another. The energies that electrons can lose when jumping from a higher orbit to a lower orbit represent the different colours of light.

5. (i) That the electron is a particle in orbit around the nucleus was discounted with the modern theory.
    (ii) It is a different model, the obsolete "Bohr model," not the quantum mechanical model, that describes the electrons that way.

4.7 COMPOSITION OF THE NUCLEUS

Moseley introduced the **atomic number**, denoted by 'Z'. The number of unit positive charges carried by the nucleus of an atom is called the atomic number of the element; that is,

Atomic number (Z) = Number of protons (p) = Number of electrons (e)

For example, in the case of the carbon atom

Atomic number = Number of protons = Number of electrons = 6

Every element has a fixed and natural number value of atomic number.

For example, the atomic number of carbon, nitrogen and oxygen are 6, 7 and 8, respectively.

Competition Edge

In neutral atoms, as the number of protons and electrons is the same, the atomic number will be equal in both. However, in an ion, as the number of electrons changes, the atomic number will be equal to the number of protons only. In short, we can say that

$Z = p = e$ (from atoms), $Z = p$ (for ions)

For sodium atom (Na) $Z = p = e = 11$

For sodium ion (Na^+) $Z = p = 11$

$e = 10$

An atom has protons, neutrons and electrons, but the mass of electrons is almost negligible. So the real mass of an atom can be determined by using the mass of protons and neutrons. **Mass number** is equal to the sum of protons and neutrons in an atom.

Mass number (A) = No. of protons (p) + No. of neutrons (n)

In short, A = p + n,   as p = Z so A = Z + n

Mass number is equal to atomic mass and nearly equal to atomic weight.

For example, in one magnesium atom, there are 12 protons and 12 neutrons, which means the mass number of magnesium is 24.

Mass number of Mg = Number of protons + Number of neutrons

$$= 12 + 12 = 24$$

Representation of an Element or Atom

Generally, an atom is represented by its symbol. Atomic number is written on the lower side of the symbol and the mass number is written on the upper side (Fig. 4.16).

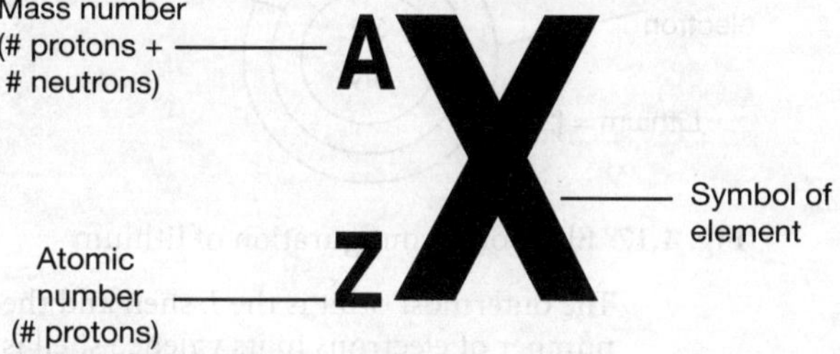

Fig. 4.16 Representation of atomic number and mass number

A = mass number, Z = atomic number, X = symbol of element.

For example, $^{7}_{3}\text{Li}$ indicates that lithium has an atomic number equal to 3 and a mass number equal to 7. An atom of carbon whose atomic number is 6 and mass number 12 is represented as $^{12}_{6}\text{C}$.

Relationship between Mass Number and Atomic Number

We can relate between the mass number and atomic number of an element easily as the number of protons in an atom is equal to the atomic number of the element. So we can rewrite the relation by putting "Atomic number" in place of "No. of protons" as follows:

Mass number = No. of protons + No. of neutrons
Mass number = Atomic number + No. of neutrons

TEST YOUR KNOWLEDGE

1. The number of electrons in an atom is 9 and the number of protons is also 9.
    (a) What is the atomic number of the atom?
    (b) What is the charge on the atom?

Solution: (a) Atomic number is equal to the number of protons in one atom. Here, this atom contains 9 protons, so the atomic number is 9.

(b) This atom contains an equal number of positively charged protons and negatively charged electrons (9 each), so it has no overall charge. That is, the charge on this atom is 0 (zero).

2. The atomic nucleus of an element has mass number 23 and number of neutrons 12. What is the atomic number of the element?

Solution: We know that

Mass number = No. of protons + No. of neutrons

23 = No. of protons + 12

∴ No. of protons = 23 – 12 = 11

Now, Atomic number = No. of protons = 11

3. A helium atom has an atomic mass of 4 u and two protons in its nucleus. How many electrons and neutrons does it have?

Solution: The number of protons in the helium nucleus has been given to be 2 and the number of electrons is equal to the number of protons, so it has 2 electrons.

Mass number = No. of protons + No. of neutrons

4 = 2 + No. of neutrons

No. of neutrons = 4 – 2 = 2

Hence, the helium atom has 2 electrons and 2 neutrons.

4. Find the number of:
    (i) electrons   (ii) protons
    (iii) neutrons   (iv) nucleons
    In an atom with mass number = 39, atomic number = 19.

Solution: (i) We know that

Atomic number = No. of protons = No. of electrons

No. of electrons = 19

> (ii) No. of protons = 19
> (iii) Mass number = No. of neutrons + No. of electrons
> No. of neutrons = Mass number − No. of protons
> = 39 − 19 = 20
> (iv) Nucleons = No. of protons + No. of neutrons
> = 19 + 20 = 39

4.8 Electron Distribution in Orbits or Electronic Configuration of Elements

The arrangement of electrons in various energy levels or orbits of an atom is known as the electronic configuration of the atom. The distribution of electrons in different orbits or shells is governed by a scheme known as the **Bohr–Bury scheme**.
According to this scheme:

(i) In general, the maximum number of electrons that can be present in any orbit is given by $2n^2$, where n is the number of the energy shells or orbits.

- The first or innermost energy shell (K or $n = 1$) can have only two electrons.
- The second shell (L or $n = 2$) can have up to 8 electrons.
- From the third shell (M or $n = 3$) onwards, the shells become larger and the third shell can have a maximum of 18 electrons (Table 4.2).

Table 4.2 Maximum number of electrons in various orbits

Orbit	Value of n	Maximum number of electrons in the orbit
K	1	$2 \times 1^2 = 2$
L	2	$2 \times 2^2 = 8$
M	3	$2 \times 3^2 = 18$
N	4	$2 \times 4^2 = 32$

(ii) The electrons are arranged around the nucleus in different energy levels or energy shells. The electrons first occupy the shell with the lowest energy, that is, closest to the nucleus (lower n value).

(iii) The outermost shell or orbit of an atom cannot have more than 8 electrons, even if it has the capacity to have more. Similarly, the shell next to the outermost shell cannot have more than 18 electrons, even if it has the capacity to have more. This arrangement gives a stable electronic configuration.

(iv) **Valence shell and valence electrons:** The outermost orbit of an atom is known as its valence shell and the electrons present in the outermost orbit are known as valence electrons. For example,

(a) The atomic number of lithium is 3 and the electronic configuration of lithium is 2, 1 or it may be represented as:

K L
2 1

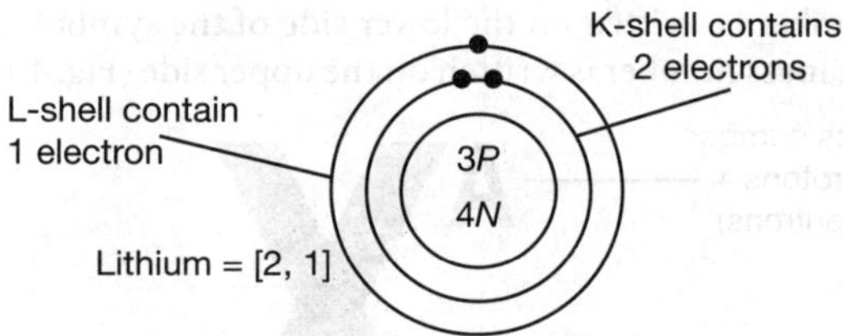

Fig. 4.17 Electronic configuration of lithium

The outermost orbit is the L shell and the number of electrons in its valence shell is one (Fig. 4.17).

(b) The atomic number of beryllium is 4 and the electronic configuration of beryllium is 2, 2 or it may be represented as:

K L
2 2

The outermost orbit is the L shell and the number of electrons in its valence shell is two.

(c) The atomic number of sodium is 11 and the electronic configuration of sodium is 2, 8, 1 or it may be represented as:

K L M
2 8 1

The outermost orbit is the M shell and the number of electrons in its valence shell is one (Fig. 4.18).

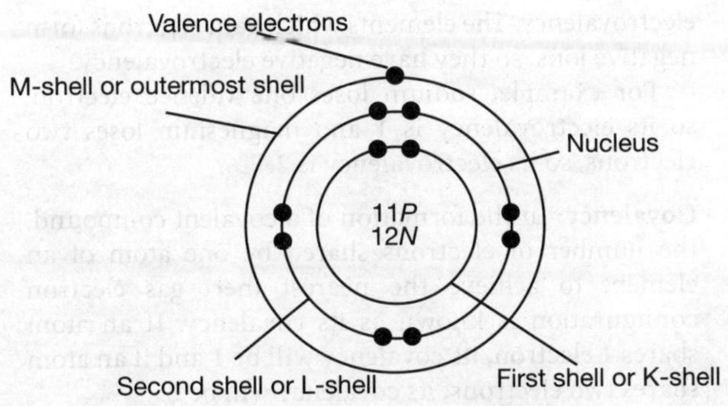

Fig. 4.18 Sodium atom

(d) The atomic number of magnesium is 12 and the electronic configuration of magnesium is 2, 8, 2 or it may be represented as:

K	L	M
2	8	2

The outermost orbit is the M shell and the number of electrons in its valence shell is two.

- **It is important to note that in case of sodium, we cannot write 2, 9 and in case of magnesium 2, 10 as the outer orbit cannot have more than 8 electrons and the maximum capacity of the M shell is 8.**

4.8.1 Electronic Configurations of the First Twenty Elements

The electronic configurations of the first twenty elements having atomic numbers from 1 to 20 are given in Table 4.3.

Table 4.3 Electronic configurations of the first twenty elements

Element	Symbol	Number of protons or atomic number	K shell $n = 1$	L shell $n = 2$	M shell $n = 3$	N shell $n = 4$	Electronic configuration	Valency V = Valence electron or 8 – Valence electrons
Hydrogen	H	1	1				1	1
Helium	He	2	2				2	0
Lithium	Li	3	2	1			2, 1	1
Beryllium	Be	4	2	2			2, 2	2
Boron	B	5	2	3			2, 3	3
Carbon	C	6	2	4			2, 4	4
Nitrogen	N	7	2	5			2, 5	8 – 5 = 3
Oxygen	O	8	2	6			2, 6	8 – 6 = 2
Fluorine	F	9	2	7			2, 7	8 – 7 = 1
Neon	Ne	10	2	8			2, 8	8 – 8 = 0
Sodium	Na	11	2	8	1		2, 8, 1	1
Magnesium	Mg	12	2	8	2		2, 8, 2	2
Aluminium	Al	13	2	8	3		2, 8, 3	3
Silicon	Si	14	2	8	4		2, 8, 4	4
Phosphorus	P	15	2	8	5		2, 8, 5	8 – 5 = 3
Sulfur	S	16	2	8	6		2, 8, 6	8 – 6 = 2
Chlorine	Cl	17	2	8	7		2, 8, 7	8 – 7 = 1
Argon	Ar	18	2	8	8		2, 8, 8	8 – 8 = 0
Potassium	K	19	2	8	8	1	2, 8, 8, 1	1
Calcium	Ca	20	2	8	8	2	2, 8, 8, 2	2

4.8.2 Valency

The valency of an element is the combining capacity of the atoms of the element with atoms of the same or different elements. The valency of an element depends on the number of valence electrons or outermost orbit electrons in its atom as only the valence electrons can take part in chemical bonding. **The valency of an element is either equal to the number of valence electrons in its atom or equal to the number of electrons required to complete eight electrons in the valence shell (octate state).**

For metals: Valency of a metal = No. of valence electrons in its atom

For non-metals:Valency of a non-metal = 8 – No. of valence electrons in its atom

As metals have 1, 2, 3 electrons, they lose electrons; on the other hand, most non-metals have 4, 5, 6, 7 electrons, so they accept electrons to get an 8-valence electron configuration (octet state). This is why for non-metals, we use valency equals to 8 – Number of valence electrons.

(i) Let us discuss a few metals

Element	Symbol	Atomic number	K	L	M	Valency
Sodium	Na	11	2	8	1	1
Magnesium	Mg	12	2	8	2	2
Aluminium	Al	13	2	8	3	3

(ii) Let us discuss a few non-metals

Element	Symbol	Atomic number	K	L	Valency
Nitrogen	N	7	2	5	8 – 5 = 3
Oxygen	O	8	2	6	8 – 6 = 2
Fluorine	F	9	2	7	8 – 7 = 1

There is only one exception to this rule and that is the valency of hydrogen. The valency of hydrogen is equal to the number of valence electrons, that is 1.

Types of Valency

There are two types of valency: electrovalency and covalency. If an element combines by the loss or gain of electrons to form electrovalent compounds (or ionic compounds), its valency is known as **electrovalency**, and if an element combines by the sharing of electrons to form covalent compounds (or molecular compounds), its valency is known as **covalency**.

Electrovalency: In the formation of an electrovalent compound (or ionic compound), the number of electrons lost or gained by one atom of an element to achieve the nearest inert gas electron configuration is known as its electrovalency. The elements which lose electrons form positive ions, so they have positive electrovalency. The elements which gain electrons form negative ions, so they have negative electrovalency.

For example, sodium loses one valence electron, so its electrovalency is 1 and magnesium loses two electrons, so its electrovalency is 2.

Covalency: In the formation of a covalent compound, the number of electrons shared by one atom of an element to achieve the nearest inert gas electron configuration is known as its covalency. If an atom shares 1 electron, its covalency will be 1 and if an atom shares two electrons, its covalency will be 2.

For example, hydrogen shares one electron, so its covalency is 1; oxygen shares two electrons, so its covalency is 2; and nitrogen shares three electrons, so its covalency is 3.

Competition Edge

Variable Valency: Some elements like phosphorus and sulfur show more than one valency (variable valences). For example, P (15) = 2, 8, 5
Phosphorus has 2 valences 5 and 3
Valency = Valence electron = 5
Valency = 8 – Valence electron = 3

Inertness of Noble Gases

There are some elements which do not react or combine with other elements to form compounds or they are chemically inert. Such compounds are known as noble gases or inert gases (Table 4.4). For example, helium, neon, argon, krypton, xenon and radon. We know that only the outermost electrons (valence electrons) of an atom take part in a chemical reaction. As the noble gases are chemically unreactive, we must conclude that the electronic arrangements in their atoms are very highly stable and do not allow the outermost electrons to take part in chemical reactions.

Table 4.4 Electronic configurations of noble gases (or inert gases)

Name	Symbol	Atomic number	Electronic configuration	No. of electrons in the outermost shell
Helium	He	2	2	2
Neon	Ne	10	2,8	8

(Continued)

Table 4.4   (Continued)

Name	Symbol	Atomic number	Electronic configuration	No. of electrons in the outermost shell
Argon	Ar	18	2,8,8	8
Krypton	Kr	36	2,8,18,8	8
Xenon	Xe	54	2,8,18,18,8	8
Radon	Ra	86	2,8,18,32,18,8	8

From Table 4.4, it is quite clear that only one helium has 2 electrons in its outermost shell and the rest have 8 electrons in the outermost shells of their atoms. We know that the outermost shell of an atom can accommodate a maximum of 8 electrons (except the K shell which can hold only a maximum of 2 electrons). This means that **all these noble gases have completely filled outermost shells (fulfilled configuration)**. To have '8 electrons' in the outermost shell of an atom is known as 'octet' of electrons and atoms having 8 electrons (or octet of electrons) in their outermost shell are highly stable and are chemically unreactive too. Having 2 electrons in the outermost shell (K shell) is known as a 'duplet' of electrons and it is a stable arrangement of electrons. Helium is the only inert gas having a duplet of electrons in its outermost shell.

As the atoms of inert gases are highly stable and unreactive, they can exist in the free state as individual atoms; that is, inert gases exist as single atoms (monoatomic).

For example, helium, neon, argon, all exist in the form of monoatomic molecules He, Ne, Ar, Kr, respectively.

Cause of Chemical Combination

Every species in this world wants to become more and more stable. In the case of atoms, stability means having the electronic arrangement of an inert gas (octet state). The atoms combine with one another to achieve the inert gas electronic configuration and to become more stable. When atoms combine to form chemical compounds, they do so in such a way that every atom gets 8 electrons in its outermost shell (octet) or 2 electrons in the outermost K shell (duplet).

An atom can achieve the inert gas (octet state) electronic arrangement in three ways:

(i) by **losing** one or more electrons to another atom

(ii) by **gaining** one or more electrons from another atom

(iii) by **sharing** one or more electrons with another atom

It is important to note that if an element has 1, 2 or 3 electrons in the outermost shell of its atom, then it loses these electrons to obtain the next inert gas electronic arrangement of eight valence electrons and forms a positively charged ion (cation). It is not possible to add 5, 6 or 7 electrons to an atom due to high energy considerations (practically not feasible).

For example (Figs 4.19 and 4.20),

$$Na \xrightarrow{-1e} Na^+$$
$$2, 8, 1 \qquad\quad 2, 8$$

$$Mg \xrightarrow{-2e} Mg^{2+}$$
$$2, 8, 2 \qquad\quad 2, 8$$

$$Al \xrightarrow{-3e} Al^{3+}$$
$$2, 8, 3 \qquad\quad 2, 8$$

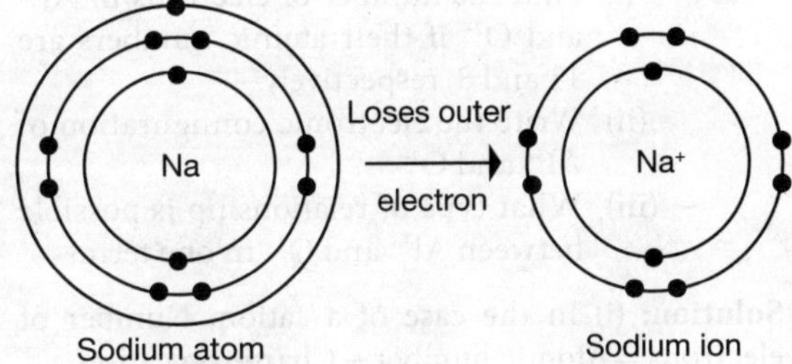

Fig. 4.19   Formation of Na^+ from Na

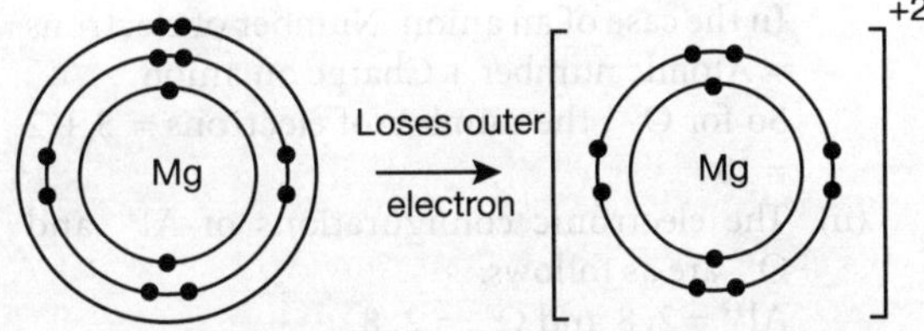

Fig. 4.20   Formation of Mg^{2+} from Mg

If an element has 5, 6 or 7 electrons in its the outermost shell of its atom, then it accepts or gains 3, 2, 1 electrons, respectively, to obtain the stable next inert gas configuration of eight valence electrons, and forms a negatively charged ion (anion). It is not possible to remove 5, 6 or 7 electrons from an atom due to very high energy requirement (practically not feasible).

For example,

$$N \xrightarrow{3e} N^{3-}$$
$$2, 5 \qquad 2, 8$$

$$O \xrightarrow{2e} O^{2-}$$
$$2, 6 \qquad 2, 8$$

$$F \xrightarrow{1e} F^-$$
$$2, 7 \qquad 2, 8$$

If an element has 4 electrons in the outermost shell of its atom, it can neither lose 4 electrons nor gain 4 electrons due to energy considerations. Such an element can obtain the inert gas electronic arrangement of eight valence electrons only by sharing its 4 outermost electrons with the 4 electrons of the other atoms.

For example, carbon has 4 electrons and it completes its octet state by sharing these 4 electrons with 4 electrons of other atoms. For example, in methane (CH_4).

*An atom that loses or gains an electron forms an ion and the charge present on the ion represents its valency. For example, the valences of Na^+, Mg^{2+} and Al^{3+} are 1, 2, 3, respectively. Similarly the valences of F^-, O^{2-} and N^{3-} are 1, 2, 3, respectively. Valency is simply a number; so negative or positive sign is not used before it.

TEST YOUR KNOWLEDGE

1.   (i)   Find the number of electrons in Al^{3+} and O^{2-} if their atomic numbers are 13 and 8, respectively.
     (ii) Write the electronic configuration of Al^{3+} and O^{2-}.
     (iii) What type of relationship is possible between Al^{3+} and O^{2-} in one term?

Solution: (i) In the case of a cation, Number of electrons = Atomic number − Charge on cation

So for Al^{3+}, the number of electrons = 13 − 3 = 10

In the case of an anion, Number of electrons = Atomic number + Charge on anion

So for O^{2-}, the number of electrons = 8 + 2 = 10

(ii) The electronic configurations of Al^{3+} and O^{2-} are as follows:

Al^{3+} = 2, 8 and O^{2-} = 2, 8

(iii) As both Al^{3+} and O^{2-} have the same number of electrons, they are iso-electronic.

4.9 CONCEPT OF ISOTOPES AND ISOBARS

4.9.1 Isotopes

In nature, most elements have a number of atoms with the same atomic number (Z), but different mass number (A). Such atoms of an element are known as 'isotopes' ('isos' means 'equal' and 'topos' means 'the same place' with reference to the periodic table). They

were discovered by Soddy. The isotopes of an element have the same atomic number as they have the same number of protons and electrons. Isotopes of an element differ in mass number as they contain different number of neutrons.

Examples:

(i) *Isotopes of hydrogen*: It has three isotopes (Fig. 4.21). These three isotopes are commonly known as hydrogen (H) or protium, deuterium (D) and tritium (T). As the atomic number is the same for all the three, they all have one electron and one proton but different numbers of neutrons (Table 4.5).

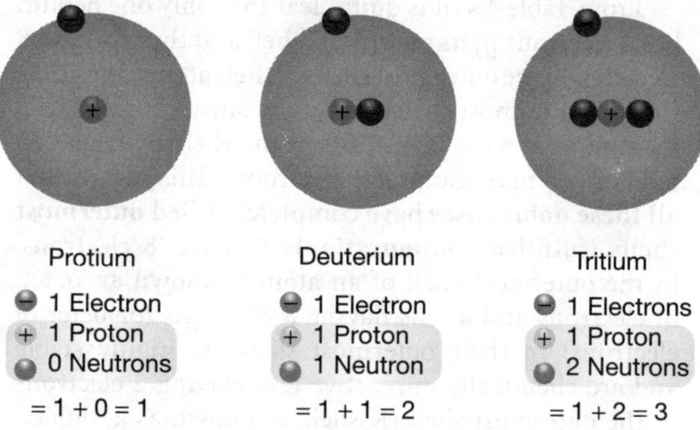

Fig. 4.21 Isotopes of hydrogen

Table 4.5 Isotopes of hydrogen

	Protium	Deuterium	Tritium
Atomic number, Z	1	1	1
Mass number, A	1	2	3
Number of protons	1	1	1
Number of electrons	1	1	1
Number of neutrons	0	1	2
Electronic configuration	K	K	K
	1	1	1

(ii) *Isotopes of carbon*: Carbon has the following three isotopes (Fig. 4.22):

$^{12}_{6}C$	$^{13}_{6}C$	$^{14}_{6}C$
$Z = 6$	6	6
$p = 6$	6	6
$n = 6$	7	8
$A = 12$	13	14

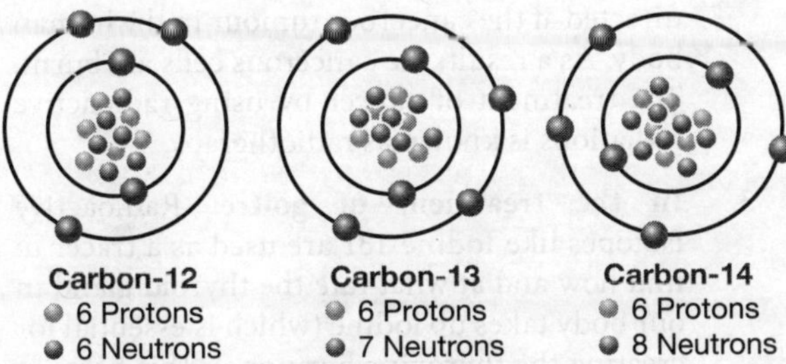

Fig. 4.22 Isotopes of carbon

(iii) *Isotopes of chlorine*: Chlorine has the following two isotopes (Fig. 4.23):

$$^{35}_{17}Cl \qquad ^{37}_{17}Cl$$

$$Z = 17 \qquad\qquad 17$$
$$p = 17 \qquad\qquad 17$$
$$n = 18 \qquad\qquad 20$$
$$A = 35 \qquad\qquad 37$$

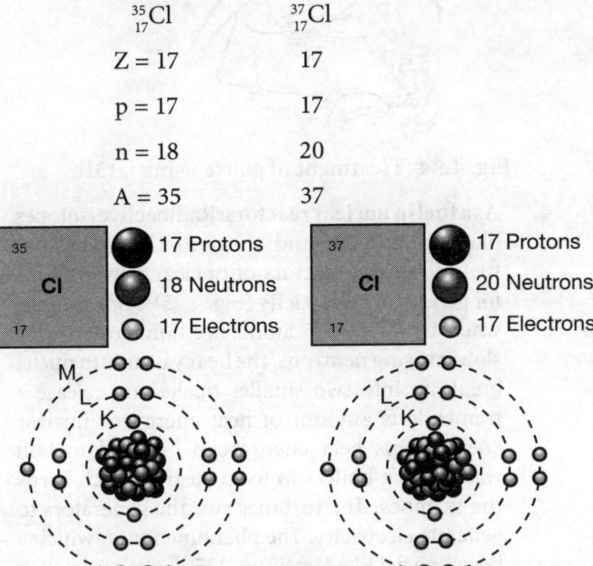

Fig. 4.23 Isotopes of chlorine

Features of Isotopes

(i) Isotopes have the same atomic number, same number of protons and electrons but differ in the number of neutrons. So in the periodic table, their position is the same.

For example, $^{12}_{6}C$ and $^{14}_{6}C$ have atomic number 6, the number of protons and electrons is 6 but the number of neutrons is 6 and 8, respectively.

(ii) **Isotopes have the same chemical properties:** The chemical properties of an atom of the element depend on the atomic number or the number of protons and electrons and not on the number of neutrons. As all the isotopes of an element have the same atomic number, their chemical properties are the same. The identical chemical properties can be explained on the basis of their electronic configurations as follows:

For example, the two isotopes of chlorine, $^{35}_{17}Cl$ and $^{37}_{17}Cl$, have the same number of electrons (17), so both of them have the same electronic configuration of 2, 8, 7 with 7 valence electrons; so they show identical chemical properties.

(iii) **Isotopes have different physical properties:** We know that the physical properties of an element depend on the mass of the atoms or atomic mass. As the masses of the isotopes of an element are slightly different due to the presence of different number of neutrons, the physical properties of the isotopes of an element are slightly different. Mass-dependent physical properties like densities, melting points and boiling points differ in isotopes. For example, the three isotopes of carbon, $^{12}_{6}C$ and $^{13}_{6}C$ and $^{14}_{6}C$ have slightly different physical properties as they have slightly different atomic masses of 12 u, 13 u and 14 u, respectively.

(iv) **Fractional atomic weight:** The atomic masses of many elements may be in fractions and not whole numbers. For example, the atomic mass of chlorine is 35.5 u whereas that of copper is 63.5 u. The fractional atomic mass of elements is due to the existence of their isotopes with different masses. As the atomic mass of an element is the average relative mass of all the natural isotopes of that element, most of the elements have fractional atomic masses.

Average atomic mass of an element can be found out by using the atomic weight and percentage of different atoms in the formula here.

Average atomic mass

$$= \left(\text{Atomic weight} \times \frac{\text{Percentage}}{100} \right)_1$$
$$+ \left(\text{Atomic weight} \times \frac{\text{Percentage}}{100} \right)_2$$

For example, natural chlorine has two types of atoms (Cl-35 and Cl-37), one having a mass of 35 u and the other having a mass of 37 u in the proportion of 75% and 25%, respectively. So the average mass of a chlorine atom will be 75% of 35 and 25% of 37, which is 35.5 u. This value is taken as the atomic mass of chlorine.

Average atomic mass of chlorine

$$= 35 \times \frac{75}{100} + 37 \times \frac{25}{100}$$

$$= \frac{2625}{100} + \frac{925}{100}$$

$$= 26.25 + 9.25 = 35.5 \, \text{u}$$

Radioactive Isotopes

There are two kinds of isotopes: those which are stable and those which are unstable. The isotopes which are unstable due to the presence of extra neutrons in their nuclei and emit various types of radiations like alpha, beta and gamma are known as radioactive isotopes. Mostly, the heaviest isotope is radioactive in nature. For example, Carbon-14, Sodium-24, Cobalt-60, Arsenic-74, Iodine-131 and Uranium-235. As the high energy radiations emitted by such isotopes are quite harmful to human beings and other creatures, these isotopes must be used with care by taking all suitable precautions and at proper concentrations to avoid damage.

Applications of Radioactive Isotopes

Radioactive isotopes are used in various fields such as medicine, agriculture, biology, chemistry, engineering and industry. Radioactive isotopes are used as fuels in nuclear reactors of nuclear power plants for generating electricity. Here are some of their important applications:

1. **As tracers to detect tumours and blood clots:** Some radioactive isotopes can be used as 'tracers' in medicine to detect the presence of tumours and blood clots in the human body. In this process, a small amount of the low-activity radioactive compound or tracer is either injected into the body of or given orally. This radioactive compound moves through the body and accumulates in the area of the tumour or blood clot. The exact position of the accumulated radioactive tracer can be found out by using a device known as Geiger counter. This gives the exact position of the tumour or blood clot and is of great help to doctors in deciding further treatment. For example, Sodium-24 tracer is used to detect the presence of blood clots and Arsenic-74 tracer is used to detect the presence of tumours.

2. **In the treatment of cancer:** Some radioactive isotopes like Cobalt-60 are used in the treatment of cancer. Here, the high-energy gamma radiations emitted by Cobalt-60 are directed at the cancerous tumour in the human body. As a result, the cancerous cells are burnt. The treatment of cancer by using radioactive radiations is known as radiotherapy.

3. **In the treatment of goitre:** Radioactive isotopes like Iodine-131 are used as a tracer to find how and at what rate the thyroid gland in our body takes up iodine (which is essential for creating the thyroxine hormone). This helps in the treatment of diseases like goitre (Fig. 4.24).

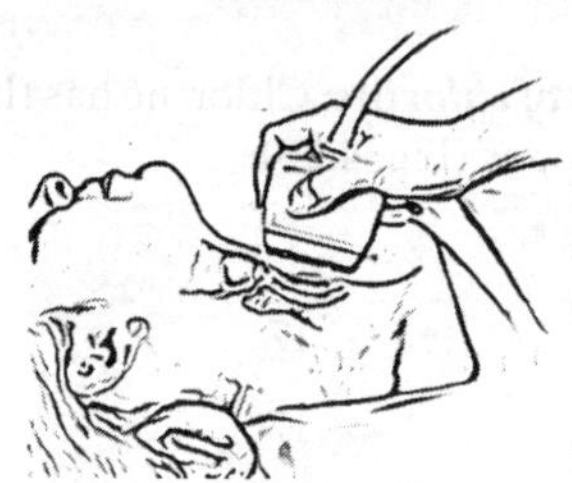

Fig. 4.24 Treatment of goitre using I-131

4. **As a fuel in nuclear reactors:** Radioactive isotopes like Uranium-235 and Thorium-232 are used as fuel in nuclear reactors of nuclear power plants for generating electricity (Fig. 4.25). For example, when Uranium-235 atoms are bombarded with slow-moving neutrons, the heavy uranium nuclei break up into two smaller nuclei and release a tremendous amount of heat energy or nuclear energy. This heat energy can be used to boil water in big boilers to form steam, which turns the turbines. The turbines run the generators to generate electricity. The phenomenon in which a heavy nuclei like Uranium-235 nuclei are broken into smaller nuclei to obtain energy is known as nuclear fission and this energy is known as nuclear energy. It is a controlled process and if the process is uncontrolled, it can be used to create the atom bomb.

Fig. 4.25 Nuclear power plant

Source: Tihange Nuclear Power Station, Huy, Belgium, Hullie, 2009, https://commons.wikimedia.org/wiki/File:Tihange_-_nuclear_power_plant.JPG. Licensed under: CC-BY-SA-3.0

5. **To detect leaks in underground pipes:** Some radioactive isotopes are used in industry to detect leaks in underground pipelines of gas, oil and water. In order to check the leakage in a metal pipeline (Fig. 4.26), a solution of the radioactive substance is introduced in the pipeline. At the place of crack or leakage in the pipeline, the radioactive solution will leak out, and the radioactive detector or Geiger counter will indicate a higher level of radiations. In this way, the leak can be found.

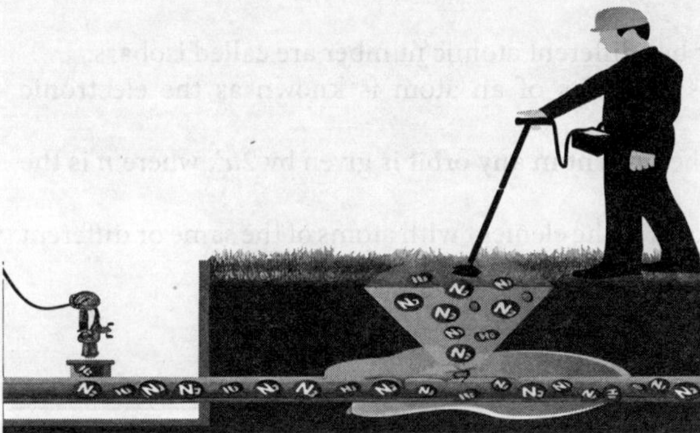

Fig. 4.26 Radioactive isotopes used for detecting leakage

4.9.2 Isobars

Atoms of different elements with the same mass number (A) but different atomic number (Z) are called isobars.

In nature, some atoms of different elements having different atomic numbers are found to have the same mass number; such atoms are known as isobars. Isobars have different number of protons in their nuclei but the total number of nucleons or the sum of protons and neutrons is the same.

For example, argon, $_{18}^{40}$Ar, potassium $_{19}^{40}$K and calcium $_{20}^{40}$Ca are isobars as argon, potassium and calcium are atoms of different elements having different atomic numbers of 18, 19 and 20, respectively, but the same mass number or sum of protons and neutrons of 40.

Argon ($_{18}^{40}$Ar)	Potassium ($_{19}^{40}$K)	Calcium ($_{20}^{40}$Ca)
Atomic number = 18	Atomic number = 19	Atomic number = 20
Mass number = 40	Mass number = 40	Mass number = 40
No. of electrons = 18	No. of electrons = 19	No. of electrons = 20
No. of protons = 18	No. of protons = 19	No. of protons = 20
No. of neutrons = 22	No. of neutrons = 21	No. of neutrons = 20

$_6C^{14}$ and $_7N^{14}$ are also isobars as they have different atomic number 6 and 7, respectively, but the same mass number 14.

Features of Isobars

- Isobars have different number of protons and neutrons, but the sum of the protons and neutrons is the same.
- Isobars have different chemical properties as they have different atomic numbers which determine chemical properties, but they have the same physical properties related to mass like density and boiling point.

Competition Edge

Isoelectric: Atoms, ions or molecules having the same number of electrons are said to be isoelectronic. For example, CH_4, Ne, Na^+, NH_4^+, Mg^{2+}, F^-, O^- have 10 electrons, so they are isoelectronic.

Isotones: Species having different atomic number (Z) and mass number (A) but the same number of neutrons are called isotones. For example, carbon-14, nitrogen-15 and oxygen-16 are isotones.

	$_6C^{14}$	$_7N^{15}$	$_8O^{16}$
n:	14 − 6	15 − 7	16 − 8
	= 8	= 8	= 8

Quick Review

- The number of unit positive charges carried by the nucleus of an atom is called the atomic number of the element. $Z = p = e$ (from atoms), $Z = p$ (for ions).
- Mass number (A) is equal to the sum of protons and neutrons, or atomic number and neutrons. Mass number is nearly equal to atomic weight.
  Mass number = Atomic number + No. of neutrons
  Or $(A = p + n = Z + n)$
  Number of neutrons = Mass number − Atomic number or Number of protons
- Generally, an atom is represented by its chemical symbol. Atomic number is written on the lower side of the symbol and the mass number is written on the upper side.
- Atoms of the same element with the same atomic number (Z) but different mass number (A) are called isotopes.
- Atoms of different elements with the same mass number but different atomic number are called isobars.
- The arrangement of electrons in various energy levels or orbits of an atom is known as the electronic configuration of the atom.
- In general, the maximum number of electrons that can be present in any orbit is given by $2n^2$, where n is the number of the energy shell.
- Valency of an element is the combining capacity of the atoms of the element with atoms of the same or different elements. Valency is equal to number of valence electrons.
- Valency = Number of valence electron or 8 − Valence electrons.
- Number of electrons in cation = Atomic number − Charge on cation.
- Number of electron in anion = Atomic number + Charge on anion.

Exercise 4.2

Questions marked * are practical based.

Section A: Multiple Choice Questions
(1 Mark)

1. The atomic number is the same as:
   (a) Proton number
   (b) Electron number
   (c) Neutron number
   (d) Nuclei number

2. The electronic configuration of calcium is:
   (a) 2,8,8,2
   (b) 2,8,7
   (c) 2,8,8
   (d) 2,8,2

3. Electrons present in the outermost shell are called:
   (a) Duplet electrons
   (b) Valence electrons
   (c) Orbit electrons
   (d) Octet electrons

4. Which of following can have a duplet of valence electrons?
   (a) He
   (b) Li
   (c) H⁻
   (d) Both (a) and (c)

5. Which of the following pairs is the odd one out?
   (a) C-12, C-14
   (b) H-2, H-3
   (c) Ar-40, Ca-40
   (d) N-14, N-15

6. How many of the following elements can show a valency of 2?

   He, Mg, K, Ca, S, O, N

   (a) 2 only
   (b) 3 only
   (c) 4 only
   (d) 1 only

7. Which of following isotopes is used by the Archeological Department to find the age of fossils?
   (a) P-32
   (b) O-18
   (c) C-14
   (d) U-238

8. Which of following pairs contains the same number of valence electrons?
   (a) Na, Na⁺
   (b) K⁺, Ca²⁺
   (c) P, S
   (d) Cl⁻, K

9. Which of following pairs can show variable valences?
   (a) O, F
   (b) S, P
   (c) Na, Mg
   (d) Ne, N

10. Which pair have 8 electrons in their L-shell (n = 2)?
    (a) N, O   (b) Na, F
    (c) Mg, Ne   (d) O, S

11. Which of the following pairs has 8 valence electrons?
    (a) P, S   (b) Ca^{2+}, K^+
    (c) F, Ne   (d) Ne, Na

12. Which pair can show a valency of 1?
    (a) Na, Mg   (b) O, S
    (c) Na, F   (d) N, Mg

13. Which of the following options represents a correct statement set?
    (i) Phosphorus and sulfur can show variable valences.
    (ii) K-40 and Ca-40 have the same sum of n + p
    (iii) N-14 and C-14 have the same number of neutrons
    (iv) S^{2-} and K^+ have the same electronic configuration
    (a) (i), (ii)   (b) (i), (ii), (iii)
    (c) (i), (ii), (iv)   (d) (i), (ii), (iii), (iv)

14. Select the option showing the correct match:

Column I	Column II
1. C-13, C-14	(i) Isobar
2. Na^+, F^-	(ii) Inert gas
3. Ar-40, Ca-40	(iii) Isotope
4. Ne, Ar	(iv) Isoelectric

    (a) 1-(i), 2-(ii), 3-(iii), 4-(iv)
    (b) 1-(iii), 2-(iv), 3-(i), 4-(ii)
    (c) 1-(iv), 2-(iii), 3-(i), 4-(ii)
    (d) 1-(iii), 2-(iv), 3-(ii), 4-(i)

15. Select the option showing the correct match:

Column I	Column II
1. Ne	(i) 7 valence electrons
2. Na	(ii) 8 valence electrons
3. Mg	(iii) 1 valence electron
4. F	(iv) 2 valence electrons

    (a) 1-(i), 2-(ii), 3-(iii), 4-(iv)
    (b) 1-(iii), 2-(ii), 3-(iv), 4-(i)
    (c) 1-(ii), 2-(iii), 3-(iv), 4-(i)
    (d) 1-(ii), 2-(i), 3-(iv), 4-(iii)

Assertion–Reason Questions

Direction: In the following question two statements (Assertion) A and Reason (R) are given Mark.
(a) if A and R both are correct and R is the correct explanation of A;
(b) if A and R both are correct but R is not the correct explanation of A;
(c) A is true but R is false;
(d) A is false but R is true

Assertion	Reason
1. Atomic number is always a natural number.	1. Atomic number is equal to the number of protons which cannot be zero or in fraction.
2. Atomic weight can be in fraction.	2. Due to isotopes, the average atomic weight is taken.
3. F^- and Na^+ have the same number of electrons.	3. Both these have the same number of neutrons also.
4. Helium and neon have zero valency.	4. The number of valence electrons in both are the same.
5. $_1H^1$, $_1H^2$, $_1H^3$ are isotopes.	5. They differ in the number of neutrons.
6. The electronic configuration of an element with atomic number 20 is 2, 8, 10.	6. The maximum number of electrons in any atom is equal to its atomic number.
7. The electronic configuration of aluminum is 2, 8, 3.	7. Aluminum is a metal.

Section B: Very Short Answer Questions   (2 Marks)

1. If you are given an element E with atomic number Z and mass number A, how will you represent this element with atomic number and mass number.

2. If an element has atomic number 8 and it also contains 8 neutrons, find its mass number.

3. Which two valences are possible for phosphorus (atomic number = 15)?

4. Write any two elements with zero valency.

5. $^{40}_{18}$Ar and $^{40}_{20}$Ca are examples of _________.

6. The number of valence electrons in neon with atomic number 10 is _________.

7. Iodine-131 is used in the treatment of _________ .

8. Isotopes have the same chemical properties. Write true or false.

9. Isobars have the same chemical properties. Write true or false.

10. Carbon-14 is used to find the age of fossils. Write true or false.

11. O^{2-} and Mg^{2+} have the same number of electrons. Write true or false.

12. How does one isotope of an element differ from another isotope of the same element?

13. How do $_1H^1$ and $_1H^2$ differ in terms of subatomic particles?

14. An atom with a nucleus of eight protons and nine neutrons is an isotope of oxygen. Is an atom with nine protons and eight neutrons another isotope of oxygen? Explain your answer.

15. The two most common isotopes of chlorine are chlorine-35 and chlorine-37, but the atomic weight that is given in the periodic table for chlorine is 35.453. Explain why this atomic weight is not the same as the weight of either of these isotopes.

Section C: Short Answer Questions (3 Marks)

1. Define atomic number and mass number. Write the relation between these.

2. What are isotopes and how do they differ? Give an example.

3. What are isobars? Explain with an example.

4. Isotopes of the same element have different physical properties but the same chemical properties. Why is this?

5. Write the applications of C-14, Co-60, I-131 briefly.

6. What do you mean by the term valence electrons and valency? Give an example.

7. Differentiate between covalency and electro valency with examples.

8. An atom has atomic number 13 and mass number 27. How many protons, neutrons and electrons does it contain? Identify this element and write its symbol.

9. How many protons, electrons and neutrons are present in a sulfide ion, S^{2-}? Sulfur has atomic number 16 and mass number 32.

10. How many protons, neutrons and electrons are present in a potassium ion, K^+? Potassium has atomic number 19 and mass number 39.

11. Silver has two isotopes: 51.35% of the atoms are silver-107 and 48.65% of the atoms are silver-109. Calculate the relative atomic mass of silver.

12. How many electrons may be accommodated in the first three energy levels?

13. What is the same about the electron structures of:
    (i) Lithium, sodium and potassium
    (ii) Beryllium, magnesium and calcium

14. What is the mass number of an isotope of magnesium if it has 13 neutrons?

15. Name the element described by the following: mass number = 19, number of neutrons = 10.

Section D: Long Answer Questions (5 Marks)

1. In five words or less, state what each of the following scientists is famous for as far as atomic theories are concerned:
    (i) Dalton   (ii) Crookes
    (iii) Thomson   (iv) Milliken
    (v) Rutherford   (vi) Chadwick
    (vii) Bohr

2. Fill in the blanks.
    (i) The maximum number of electrons that can be found in the n = 3 principal level is _________.
    (ii) The total number of orbitals in the n = 3 principal level are _________.

3. Define each of the following terms:
    (i)   Mass
    (ii) Atomic mass unit
    (iii) Atomic weight
    (iv) Atomic number

4. Give the atomic number, mass number and the number of neutrons for each of the following:
    (i)   Carbon-14   (ii) Iron-58
    (iii) Bismuth-209   (iv) Nitrogen-15
    (v)   Chlorine-37   (vi) Sulfur-36

5. Draw the electrons in their shells for each of the following elements:
    (i)   Nitrogen (atomic number 7)
    (ii) Magnesium (atomic number 12)
    (iii) Chloride ion, Cl– (atomic number 17)
    (iv) Calcium ion, Ca^{2+} (atomic number 20).

Section E: Case Study or Passage-Based Questions   (4 Marks)

1. In the case of a neutral atom, the atomic number (Z) is equal to the number of protons (p) or electrons (e). However, in the case of ions, the number of electrons is not equal to the atomic number or number of protons as the ion is formed by the loss or gain of electrons. Loss of electron gives a cation while gain of electron gives an anion. The number of electrons for cation is equal to the atomic number minus electron loss and in the case of anion, the number of electrons is equal to the atomic number plus electron gained.
    (i)   Write the electronic configuration of an atom (F) and F^- (atomic number 9).
    (ii) Write the number of protons and electrons in Al^{3+}. (Atomic number = 13)
    (iii) Write the number of protons and electrons in N^{3-}. (Atomic number = 7)
    (iv) Write the electronic configuration of O^{2-} (atomic number = 8).
    (v)   How many of the following ions have the same number of electrons?
          Na^+, K^+, O^{2-}, N^{3-}, Mg^{2+}, Ne, Ca
          (Atomic number of Na = 11, K = 19, O = 8, N = 7, Mg = 12, Ne = 10, Ca = 20)

2. If atoms of the same element have different masses, they are called isotopes. The difference in masses arises because the nucleus contains different number of neutrons but the same number of protons. The isotopes have the same electronic configuration and the same chemical properties. Because of the mass differences between the isotopes, they show much greater difference in physical properties.
    (i)   In the isotope of oxygen O^{18}, the number of neutrons in the nucleus would be:
          (a) 8   (b) 10
          (c) 18   (d) 2
    (ii) The atomic weight of the isotope of hydrogen which contains 2 neutrons in the nucleus would be:
          (a) 2   (b) 3
          (c) 1   (d) 4
    (iii) Hydrogen and deuterium differ in:
          (a) Reactivity with oxygen
          (b) Reactivity with chlorine
          (c) Melting point
          (d) Reducing action
    (iv) The atomic weight of chlorine is 35.5. Chlorine has two isotopes with atomic weights 35 and 37. Natural chlorine will have the two isotopes in the numerical ratio of
          (a) 3 : 1   (b) 1 : 2
          (c) 2 : 1   (d) 3 : 4
    (v)   Which of the following options has a correct set of statements?
    (I)   Hydrogen atom has only 2 isotopes
    (II) Isotopes differ in number of neutrons
    (III) Isotopes differ in physical properties related to mass
    (IV) Isotopes have different positions in the periodic table
          (a) (I), (II)   (b) (II), (III)
          (c) (I), (II), (IV)   (d) (II), (III), (IV)

High Order Thinking Skills (HOTS) Questions

1. Find the number of total electrons which will together weigh 10 g.

2. An ion has 1 neutron more than the number of protons but it contains 1 electron less than the number of protons. Write any two possible ions which satisfy these conditions.

3. The nucleus of an atom has a total mass almost equal to 23.436×10^{-24} g and a total charge of 11.214×10^{-19} coulomb.
    (i)   Find the atomic number,

(ii) mass number,
(iii) identify this element and
(iv) write its electronic configuration.

4. The diagram below shows the electronic configuration of a particular element X.

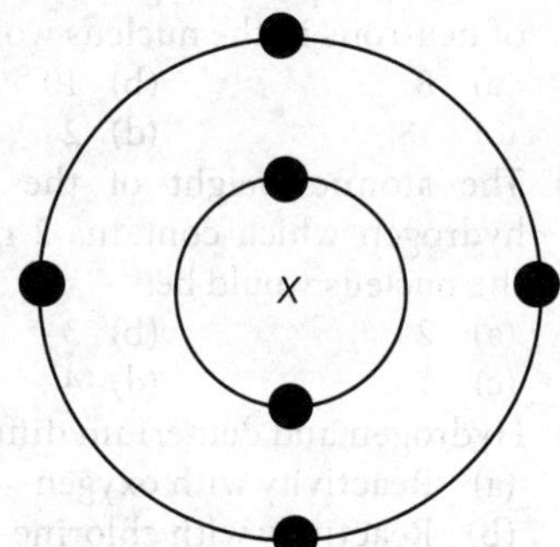

(i) How many valence electrons does element X have?
(ii) Write its electronic configuration.
(iii) If element X is a neutral atom and has 6 neutrons, what is its atomic number and mass number.
(v) What element is X?

5. Chlorine has atomic number 17. Its atomic weight is 35.5. The mass number of naturally occurring isotopes of chlorine are 35 and 37. Chlorine is a member of seventh group of the periodic table and is highly electronegative.

On the basis of this, answer the following questions:
(i) Which of the following statements are true?
  (a) All chlorine atoms have the same nuclear charge
  (b) Almost all of the mass of the chlorine atom is concentrated in its nucleus
  (c) Some chlorine nuclei contain 18 protons
  (d) Some chlorine nuclei contain 20 neutrons
(ii) Which of the following statements are false?
  (a) The two isotopes Cl-35 and Cl-37 have similar chemical properties
  (b) Cl-35 is two times more abundant than Cl-37
  (c) 6×10^{23} molecules of chlorine weigh 71 grams
  (d) The chlorine nucleus contains 17 protons and 18.5 neutrons
(iii) Which of the following statements are true?
  (a) Chlorine is the first member of the seventh group
  (b) The chlorine atom has seven electrons in outermost orbit
  (c) Chlorine shows a valency of 7
  (d) Chlorine is an active oxidising agent

Answers

Section A: Multiple Choice Questions

1. (a)	**2.** (a)	**3.** (b)	**4.** (d)
5. (c)	**6.** (c)	**7.** (c)	**8.** (b)
9. (b)	**10.** (c)	**11.** (b)	**12.** (c)
13. (c)	**14.** (b)	**15.** (c)	

Assertion–Reason Questions

1. (a)	**2.** (a)	**3.** (c)	**4.** (c)
5. (b)	**6.** (d)	**7.** (b)	

Section B: Very Short Answer Questions

1. $^{A}_{Z}E$

2. Mass number = Atomic number + Number of neutron = 8 + 8 = 16

3. 3 and 5

4. Helium (He), Neon (Ne)

5. Isobar

6. 8

7. Thyroids

8. True

9. False

10. True

11. True

12. An isotope of an element is an atom of that element that has a particular number of neutrons in its nucleus. One isotope of a given element differs from another in this number of neutrons.

13. They differ in number of neutrons as $_1H^1$ and $_1H^2$ have zero and one neutron, respectively.

14. No. The different number of protons changes the identity of the element. An atom with nine protons is an atom of fluorine.

15. Atomic weight is the average weight of all the isotopes of an element, not the weight of just one of the atoms.

Section C: Short Answer Questions

1. Refer to Section 4.2, Charged Particles in Matter.

2. Refer to Section 4.9.1, Isotope.

3. Refer to Section 4.9.2, Isobar.

4. Refer to Section 4.9.1, Isotope.

5. Refer to Section 4.9.1, Isotope.

6. Refer to Section 4.8.2, Valency.

7. Refer to Section 4.8.2, Valency.

8. The atomic number is the number of protons, which is 13 here.
   Since an atom is neutral so the number of electrons = Atomic number = 13.
   As the number of protons plus the number of neutrons is 27 (mass number).
   So, the number of neutrons = Mass number – Atomic number = 27 – 13 = 14.
   The element is aluminium and it is usually written as $_{13}^{27}Al$.

9. Sulfur has atomic number 16, so a sulfur atom contains 16 protons. Since the electrical charge of a sulfide ion is '2–', it contains two more electrons than it does protons.
   Therefore, the number of electrons = Atomic number + negative charge = 16 + 2 = 18.
   As the atomic mass = 32.
   The number of neutrons = Mass number – Number of protons = 32 – 16 = 16.
   As sulfide ion contains 16 protons, 16 neutrons and 18 electrons so it is written as $_{16}^{32}S^{2-}$.

10. Potassium has atomic number 19, so a potassium atom contains 19 protons. Since the electrical charge of a potassium ion is '1+', it contains one more proton than it does electrons. Therefore, the number of electrons is 18.
    As the atomic mass is 39, the number of neutrons = 39 – 19 = 20.
    Hence, a potassium ion contains 19 protons, 20 neutrons and 18 electrons and it is written as $_{19}^{39}K^+$.

11. The fraction that is $^{107}Ag^+$ is 51.35/100; the fraction that is $^{109}Ag^+$ is 48.65/100.
    Weighted mean relative atomic mass =
    $$= \frac{51.35}{100} \times 107 + \frac{48.65}{100} \times 109 = 107.97$$
    Or 108.0 to 1 decimal place.

12. In the first energy level ($n = 1$), total number of electrons = $2n^2 = 2 \times 1^2 = 2$.
    In the second energy level ($n = 2$), total number of electrons = $2n^2 = 2 \times 2^2 = 8$.
    In the third energy level ($n = 3$), total number of electrons = $2n^2 = 2 \times 3^2 = 18$.

13. (i) Lithium, sodium and potassium contain 1 valence electron.
    (ii) Beryllium, magnesium and calcium contain two valence electrons.

14. The number of neutrons given = 13.
    The number of protons = Atomic number of magnesium = 12.
    Therefore, Mass number = Number of protons + Number of neutrons = 12 + 13 = 25

15. The atomic number (the number of protons) determines an element's identity.
    Since the mass number and the number of neutrons is given, the number of protons can be calculated as follows:
    Number of protons = Mass number – Number of neutrons = 19 – 10 = 9

Section D: Long Answer Questions

1. (i) Dalton—first comprehensive atomic theory
   (ii) Crookes—cathode rays
   (iii) Thomson—plum pudding model
   (iv) Millikan—value of charge on electron
   (v) Rutherford—existence of nucleus

(vi) Chadwick—neutron

(vii) Bohr—solar system model

2. (i) $2n^2$ = maximum number of electrons in principal level n. In this case, $2n^2 = 18$.

(ii) n^2 = the number of orbitals in principal level n. In this case, $n^2 = 9$

3. (i) Mass is the amount of material as measured by measuring its weight on the earth's surface.

(ii) The atomic mass unit is one-twelfth the mass of an atom of the carbon-12 isotope.

(iii) The atomic weight of an element is the average weight of all atoms that exist for that element.

(iv) The atomic number of an element is the number of protons in the nucleus of its atom. It is the number that identifies or characterises a given element.

4. (i) Atomic number is 6; mass number is 14; number of neutrons is 8.

(ii) Atomic number is 26; mass number is 58; number of neutrons is 32.

(iii) Atomic number is 83; mass number is 209; number of neutrons is 126.

(iv) Atomic number is 7; mass number is 15; number of neutrons is 8.

(v) Atomic number is 17; mass number is 37; number of neutrons is 20.

(vi) Atomic number is 16; mass number is 36; number of neutrons is 20.

5. (i) Nitrogen (7) = 2, 5

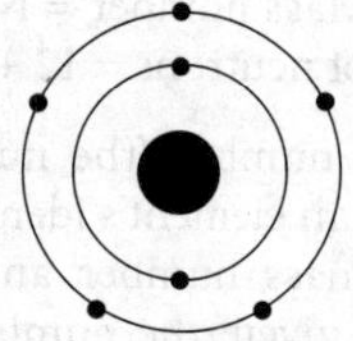

(ii) Magnesium (12) = 2, 8, 2

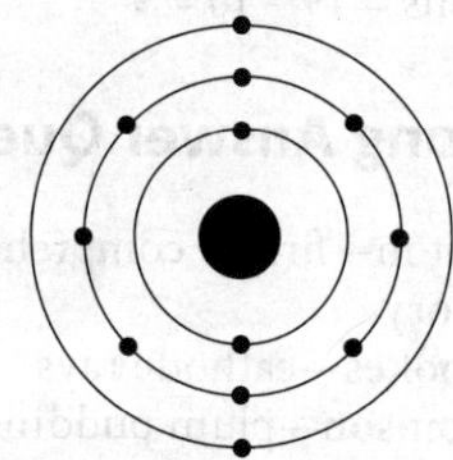

(iii) Chloride ion, Cl– (17)

Total electrons = 17 + 1 = 18
Electronic configuration = 2, 8, 8

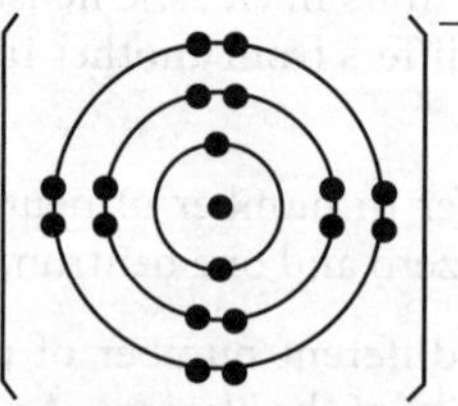

(iv) Calcium ion, Ca^{2+} (20)
Total electrons = 20 – 2 = 18
Electronic configuration = 2, 8, 8

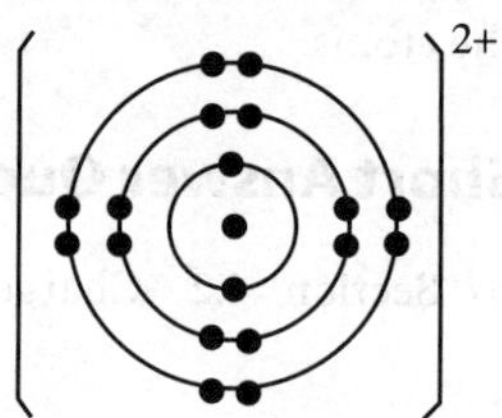

Section E: Case Study or Passage-Based Questions

1. (i) F (9) = 2, 7
    F^- (9 + 1 = 10) = 2, 8

(ii) Number of protons = Atomic number = 13
Number of electron in cation = Atomic number – Charge on cation = 13 – 3 = 10

(iii) Number of protons = Atomic number = 7
Number of electron in anion = Atomic number + Charge on anion = 7 + 3 = 10

(iv) Number of electron in anion = Atomic number + Charge on anion = 8 + 2 = 10
O^{2-} (10 e) = 2, 8

(v) Na+, O^{2-}, N^{3-}, Mg^{2+}, Ne have 10 electrons; so they are isoelectronic.

2. (i) (b)   (ii) (b)   (iii) (c)   (iv) (a)   (v) (b)

High Order Thinking Skills (HOTS) Questions

1. As mass of 1 electron = 9.11×10^{-28} g
    A mass of 9.11×10^{-28} g = 1 electron
    Mass of 10 g will have number of electrons

    $$= \frac{\text{Given mass}}{\text{Mass of 1 e}} = \frac{10}{9.11 \times 10^{-28}} = 1.098 \times 10^{28}$$

2. (i) Sodium atom has 11 protons ($Z = 11$) and mass number 23. Here, the ion has 1 electron less, it is Na^+. In Na^+
Number of protons = 11
Number of electron = Atomic number – Positive charge = 11 – 1 = 10
Number of neutrons = Mass number – Number of protons = 23 – 11 = 12

(ii) Potassium atom has 19 protons ($Z = 19$) and mass number 39. The ion has 1 electron less, so it is K^+. In K^+
Number of protons = 19
Number of electrons = Atomic number – Positive charge = 19 – 1 = 18
Number of neutrons = Mass number – Number of protons = 39 – 19 = 20
Hence, the two possible ions are Na^+ and K^+ with 1 electron less and 1 extra neutron than the number of protons in them.

3. (i) Charge on proton = 1.602×10^{-19} Coulomb
Total charge on nucleus = 11.214×10^{-19} Coulomb

$$\text{Number of protons} = \frac{\text{Total charge}}{\text{Charge on nucleon}}$$

$$= \frac{11.214 \times 10^{-19}}{1.602 \times 10^{-19}} = 7$$

As Atomic number = Number of protons = 7

(ii) Mass of nucleon (proton) = 1.674×10^{-24} g

Total mass of nucleus = 23.436×10^{-24} g

$$\text{Now number of nucleon} = \frac{\text{Total mass}}{\text{Mass of nucleon}}$$

$$= \frac{23.436 \times 10^{-24}}{1.674 \times 10^{-24}} = 14$$

Mass number = Number of nucleon = 14

(iii) Atomic number 7 and mass number 14 is of nitrogen ($^{14}_{7}N$)

(iv) Electronic configuration of nitrogen (7) = 2, 5.

4. (i) 4 valence electrons

(ii) 2, 4

(iii) Atomic number = 6. (Since this element is neutral, its number of electrons equals the number of protons)
Mass number = Number of protons + Number of neutrons = 6 + 6 = 12

(iv) As the atomic number is 6 and mass number is 12, the element is carbon ($^{12}_{6}C$)

5. (i) Here a, b and d are correct. Only c is incorrect as all Cl-atoms will have 17 protons.

(ii) Here b and d are incorrect as Cl-35 is three times more abundant than Cl-37 and the number of neutrons in them are 18 and 20, respectively.

(iii) Here a, b and d are correct and only c is wrong as the major valency of chlorine is 1.

Hints

Section A: Multiple Choice Questions

5. Ar-40, Ca-40 are isobars while the other pairs are isotopes.

6. Mg, Ca, S, O can show a valency of 2.

13. N-14 and C-14 have different number of neutrons; that is 7 and 8, respectively; so this statement is incorrect while the rest are correct.

Activities with Discussion and Conclusion

Activity 4.1 (To show the existence of charged particles in nature)

(i) Comb dry hair. Does the comb then attract small pieces of paper?

(ii) Rub a glass rod with a silk cloth and bring the rod near an inflated balloon. Observe what happens.

Discussion:

(i) Small pieces of paper are attracted towards the comb due to the development of attractive forces between the pieces of paper and the comb. When we pass a comb through dry hair, it gets electrified. Now when this comb is brought near small pieces of paper, its atoms get polarised where charge is induced on the pieces of paper with polarity opposite to that of comb. This is why small pieces of paper are attracted towards the comb.

(ii) When a glass rod is rubbed with a silk cloth and it is brought near an inflated balloon, the balloon is attracted towards the rod due to the development of attractive forces between the rod and balloon. This is due to the development of opposite charges on them. When you rub the glass rod with a silk cloth, electrons are stripped away from the atoms in the glass and transferred to the silk cloth. This leaves the glass rod with more positive than negative charge. Glass happens to lose electrons easily, and silk grabs them away from the glass atoms, so after rubbing, the glass becomes positively charged and the silk becomes negatively charged. When you rub a balloon on wool, this causes the electrons to move from the wool to the balloon's surface. The rubbed part of the balloon now has a negative charge.

Conclusion: From these observations we can say that on rubbing two objects together, they become electrically charged. Now the question arises from where does the charge come when two objects are rubbed together? The charge produced here confirms that an atom is divisible and has charged particles (electrons and protons).

Activity 4.2 (To make static atomic models that display the electronic configuration of elements 1–18)

In order to make a static atomic model displaying the electronic configuration of any atom, let us take 1 red ball to represent the nucleus. Now take wires of different lengths so that they can be converted into circular shapes by bending to represent circular orbits or cells with different radii. Now put blue balls on this wire to represent electrons in a particular orbit.

Keep the first circle small (K shell) followed by a larger second circle (L shell) followed by a slightly larger third circle (M shell) and place the blue balls on these circles as per the rules of electronic configuration; that is, maximum 2 on the first circle, 8 on the second and 8 on the third circle. It is important to note that the placement of blue balls must be according to the requirements to show electronic configuration.

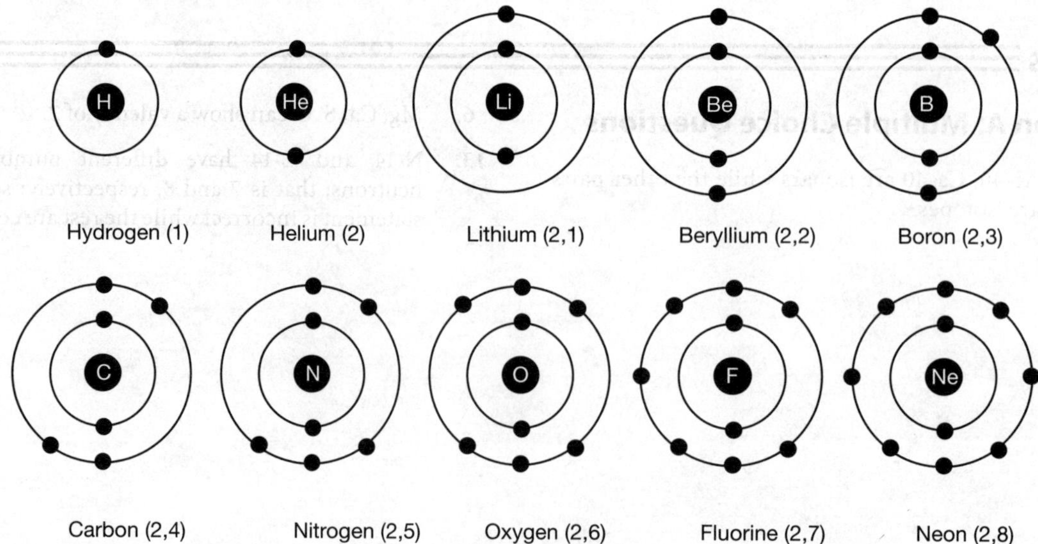

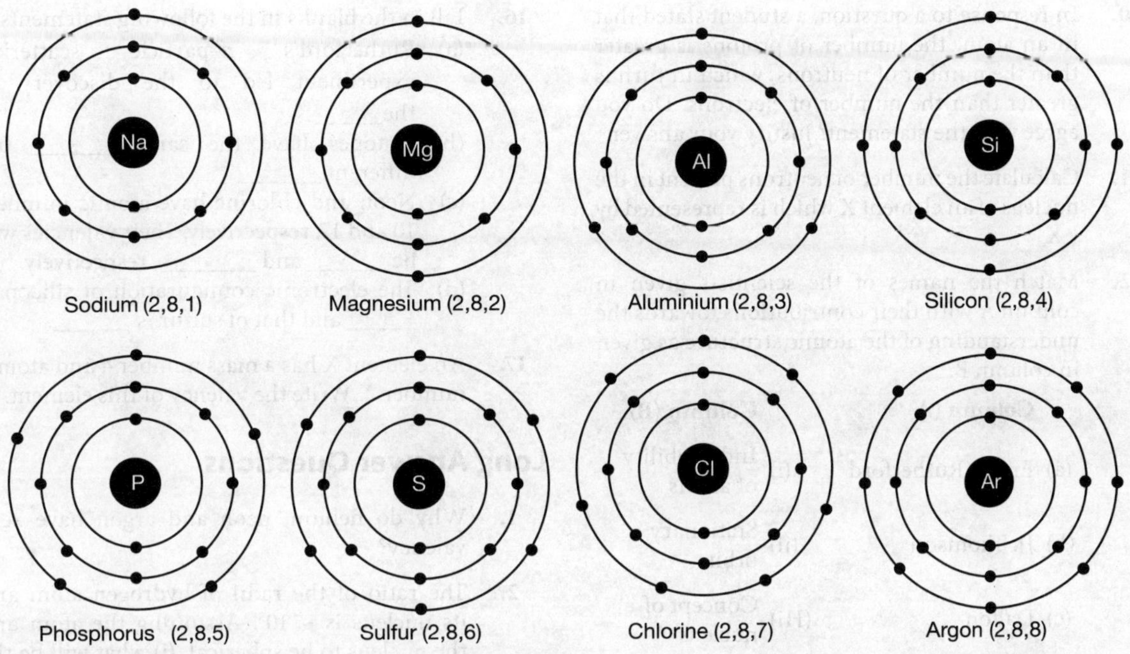

Sodium (2,8,1) Magnesium (2,8,2) Aluminium (2,8,3) Silicon (2,8,4)

Phosphorus (2,8,5) Sulfur (2,8,6) Chlorine (2,8,7) Argon (2,8,8)

Exemplar Problems

Short Answer Questions

1. Is it possible for the atom of an element to have one electron, one proton and no neutron. If so, name the element.

2. Write any two observations which support the fact that atoms are divisible.

3. Will ^{35}CI and ^{37}CI have different valences? Justify your answer.

4. Why did Rutherford select a gold foil for his α-ray scattering experiment?

5. Find out the valency of the atoms represented by the figures (a) and (b)

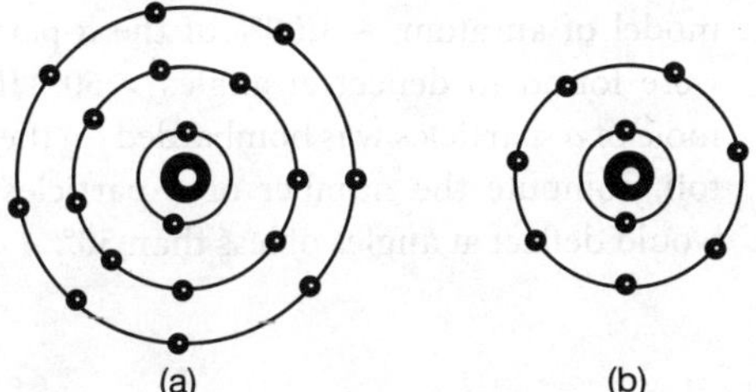

(a) (b)

6. One electron is present in the outermost shell of the atom of an element X. What would be the nature and value of charge on the ion formed if this electron is removed from the outermost shell?

7. Write down the electron distribution of a chlorine atom. How many electrons are there in the L shell? (Atomic number of chlorine is 17).

8. In the atom of an element X, 6 electrons are present in the outermost shell. If it acquires noble gas configuration by accepting the requisite number of electrons, what would be the charge on the ion so formed?

9. What information do you get from the figure about the atomic number, mass number and valency of atoms X, Y and Z? Give your answer in tabular form.

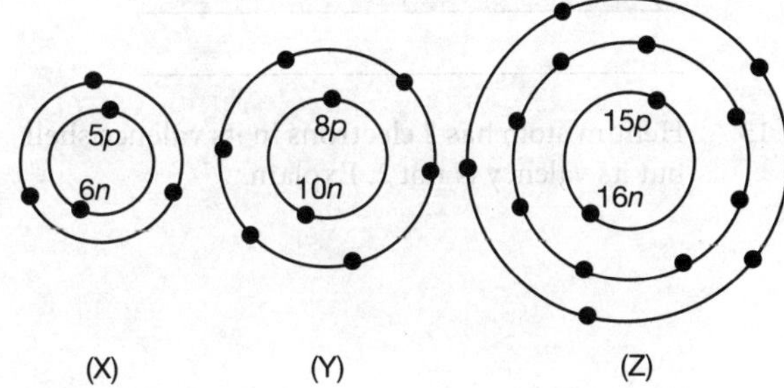

10. In response to a question, a student stated that in an atom, the number of protons is greater than the number of neutrons, which in turn is greater than the number of electrons. Do you agree with the statement? Justify your answer.

11. Calculate the number of neutrons present in the nucleus of an element X which is represented by $_{15}^{31}X$

12. Match the names of the scientists given in column A with their contributions towards the understanding of the atomic structure as given in column B:

Column (A)		Column (B)
(a) Ernest Rutherford	(i)	Indivisibility of atoms
(b) JJ. Thomson	(ii)	Stationary orbits
(c) Dalton	(Hi)	Concept of nucleus
(d) Niels Bohr	(iv)	Discovery of electrons
(e) James Chadwick	(v)	Atomic number
(f) E. Goldstein	(vii)	Neutron
(g) Mosley	(vii)	Canal rays

13. The atomic number of calcium and argon are 20 and 18, respectively, but the mass number of both these elements is 40. What is the name given to such a pair of elements?

14. Complete the table on the basis of the information available in the symbols given below.

    (a) $_{17}^{35}Cl$ (b) $_{6}^{12}C$ (C) $_{35}^{81}Br$

Element	n_p	n_n

15. Helium atom has 2 electrons in its valence shell but its valency is not 2. Explain.

16. Fill in the blanks in the following statements.
    (a) Rutherford's α-particle scattering experiment led to the discovery of the______.
    (b) Isotopes have the same _______ but different _______.
    (c) Neon and chlorine have atomic numbers 10 and 17, respectively. Their valencies will be_______ and _______, respectively.
    (d) The electronic configuration of silicon is ______ and that of sulfur is_______.

17. An element X has a mass number 4 and atomic number 2. Write the valency of this element.

Long Answer Questions

1. Why do helium, neon and argon have zero valency?

2. The ratio of the radii of hydrogen atom and its nucleus is $\sim 10^5$. Assuming the atom and the nucleus to be spherical, (i) what will be the ratio of their sizes? If the atom is represented by planet earth $R_e = 6.4 \times 10^6$m, estimate the size of the nucleus.

3. Enlist the conclusions drawn by Rutherford from his α-ray scattering experiment.

4. In what way is Rutherford's atomic model different from that of Thomson's?

5. What were the drawbacks of Rutherford's model of an atom?

6. What are the postulates of Bohr's model of an atom?

7. Show diagrammatically the electron distributions in a sodium atom and a sodium ion and also give their atomic number.

8. In the gold foil experiment of Geiger and Marsden, that paved the way for Rutherford's model of an atom, $\sim 1.00\%$ of the α-particles were found to deflect at angles $> 50°$. If one mole of α-particles was bombarded on the gold foil, compute the number of α-particles that would deflect at angles of less than 50°.

Answers

Short Answer Questions

1. Yes, it is possible. For example, hydrogen has one electron, one proton and no neutron. It is represented as $_1^1H$.

2. Atoms are divisible. With the discovery of electrons and protons, it was established that atoms are divisible and are made up of negatively charged electrons and positively charged protons along with some neutral particles called neutrons.

3. No, both ^{35}Cl and ^{37}Cl will have the same valences, as ^{35}Cl and ^{37}Cl are isotopes. As isotopes have the same number of electrons and protons and they differ only in the number of neutrons, their electron distribution (2, 8, 7) will be the same.

4. A light metal cannot be used because on being hit by fast moving α-particles, the atom of light metal will be pushed forward and no scattering can occur. So Rutherford selected gold foil as gold is a heavy metal with high mass number, highly malleable and can be beaten to get very thin foils also.

5. (a) It has electronic configuration 2,8,8. Its outermost shell has a complete octet. So, its valency is 0.
    (b) It has electronic configuration 2, 7. It can easily gain one electron to complete its outermost octet. So, its valency is –1. (Negative sign indicates tendency to gain electron)

6. $X - 1e^- \rightarrow X^+$.
    Here the ion formed by X by the loss of one electron will have a positive nature and one positive (+1) charge on the cation formed (X^+).

7. The electron distribution of the chlorine atom is as follows:

Cl	K	L	M
$17 \rightarrow$	2	8	7

    Hence, the L shell will have 8 electrons.

8. $X - 2e^- \rightarrow X^{2-}$
    When an atom (X) has 6 electrons in its outermost shell and it accepts 2 electrons, it gets two negative charges (X^{2-}).

9.

Element	Atomic number	Mass number	Valency
X	5	11	3
Y	8	18	2
Z	15	31	3,5

10. Here, the given statement is not correct as the number of protons can never be greater than the number of neutrons. The number of neutrons can be equal to or greater than the number of protons but the number of protons is equal to the number of electrons for an atom as it is neutral.

11. In $_{15}^{31}X$, Number of protons = Number of electrons = 15
     Number of neutrons = Mass number (A) – No. of protons (p) = 31 – 15 = 16

12. (a) (iii), (b) (iv), (c) (i), (d) (ii), (e) (vi), (f) (vii) and (g) (v).

13. A pair of elements in which the elements have the same mass number but different atomic numbers are called isobars. Hence $_{20}^{40}Ca$ and is $_{18}^{40}Ar$ are isobars.

14.

Element	n_p	n_n
$_{17}^{35}Cl$	17	18
$_6^{12}C$	6	6
$_{35}^{81}Br$	35	46

15. Helium has only one shell (K) and the maximum number of electrons in a K shell can be 2, so it cannot lose or gain electrons; hence, its valency is zero.

16. (a) Nucleus, (b) Atomic number, mass number, (c) 0 and 1 and (d) Silicon – 2, 8, 4, sulfur – 2, 8, 6.

17. As the element X has atomic number as 2, an atom of X contains two electrons. These 2 electrons fill the K shell completely. So, the valency of X is zero.

Long Answer Questions

1. Helium has only a K shell and it is completely filled with 2 electrons. Argon and neon have 8 electrons in their outermost shell which is the maximum number of electrons that can be accommodated in the outermost shell. Hence, their valency is zero as they do not accept or lose any electrons.

2. (i) Volume of the sphere $= \dfrac{4}{3}\pi r^3$

   Let R be the radius of the atom and r be that of the nucleus $= R = 10^3\, r$

   Volume of the atom $= \dfrac{4}{3}\pi R^3 = \dfrac{4}{3}\pi (10^5 r)^3$

   $= \dfrac{4}{3}\pi r^3 \times 10^{15}$

   Volume of the nucleus $= \dfrac{4}{3}\pi r^3$

   Ratio of the size of the atom to that of the

   nucleus $= \dfrac{\dfrac{2}{3} \times 10^{15} \times \pi r^3}{\dfrac{4}{3}\pi r^3} = 10^{15}$

   (ii) If the atom is represented by the planet earth ($R_e = 6.4 \times 10^6$ m), then the radius of the nucleus would be

   $r_n = \dfrac{R_e}{10^5}$

   $r_n = \dfrac{6.4 \times 10^6\, m}{10^5} = 6.4 \times 10\, m = 64\, m$

3. Refer to Section 4.6.2, Rutherford's Atomic Model.

4. Refer to Section 4.6.2, Rutherford's Atomic Model.

5. Refer to Section 4.6.3, Bohr's Atomic Model.

6. Refer to Section 4.6.3, Bohr's Atomic Model.

7. $_{11}$Na has 2,8,1 (11 electrons)
   Na $- 1e^- \rightarrow$ Na$^+$ (10 electrons) electronic configuration 2, 8.

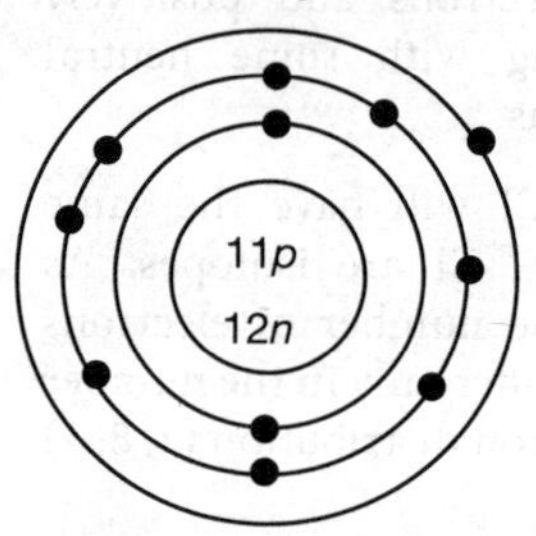

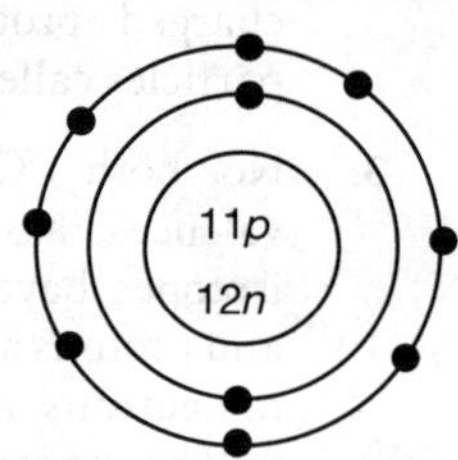

   The atomic number of an element is equal to the number of protons in its atom. As sodium atom and sodium ion contain the same number of protons, the atomic number of both is 11.

8. % of α-particles deflected more than 50° = 1%
   % of α-particles deflected less than 50° = 100 − 1 = 99%
   Number of α-particles bombarded = 1 mole = 6.022×10^{23} particles
   Number of particles deflected at an angle less than 50° = 99/100 × 6.022 × 10²³ 100

   $= \dfrac{99}{100} \times 6.022 \times 10^{23} = \dfrac{596.178}{100} \times 10^{23} = 5.96 \times 10^{23}$

NCERT Corner

In-Text Questions

1. What are canal rays?

2. If an atom contains one electron and one proton, will it carry any charge or not?

3. On the basis of Thomson's model of an atom, explain how the atom is neutral as a whole.

4. On the basis of Rutherford's model of an atom, which subatomic particle is present in the nucleus of an atom?

5. Draw a sketch of Bohr's model of an atom with three shells.

6. What do you think would be the observation if the alpha-particle scattering experiment is

carried out using a foil of a metal other than gold?

7. Name the three subatomic particles of an atom.

8. Helium atom has an atomic mass of 4 u and two protons in its nucleus. How many neutrons does it have?

9. Write the distribution of electrons in carbon and sodium atoms.

10. If the K and L shells of an atom are full, what would be the total number of electrons in the atom?

11. How will you find the valency of chlorine, sulfur and magnesium?

12. If the number of electrons in an atom is 8 and the number of protons is also 8, (i) what is the atomic number of the atom and (ii) what is the charge on the atom?

13. With the help of Table 4.1, find out the mass number of oxygen and sulfur.

14. For the symbols H, D and T, tabulate three subatomic particles found in each of them.

15. Write the electronic configuration of any one pair of isotopes and isobars.

Exercises

1. Compare the properties of electrons, protons and neutrons.

2. What are the limitations of J.J. Thomson's model of the atom?

3. What are the limitations of Rutherford's model of the atom?

4. Describe Bohr's model of the atom.

5. Compare all the proposed models of an atom given in this chapter.

6. Summarise the rules for writing the distribution of electrons in various shells for the first 18 elements.

7. Define valency by taking examples of silicon and oxygen.

8. Explain with examples (i) Atomic number, (ii) Mass number, (iii) Isotopes and (iv) Isobars. Give any two uses of isotopes.

9. Na^+ has completely filled K and L shells. Explain.

10. If bromine atom is available in the form of, say, two isotopes 79/35Br (49.7%) and 81/35Br (50.3%), calculate the average atomic mass of the bromine atom.

11. The average atomic mass of a sample of an element X is 16.2 u. What are the percentages of isotopes 16/8 X and 18/8 X in the sample?

12. If Z = 3, what would be the valency of the element? Also, name the element.

13. The composition of the nuclei of two atomic species X and Y are given as under

	X	Y
Protons =	6	6
Neutrons =	6	8

Give the mass numbers of X and Y. What is the relation between the two species?

14. For the following statements, write T for true and F for false:
    (a) J.J. Thomson proposed that the nucleus of an atom contains only nucleons.
    (b) A neutron is formed by an electron and a proton combined together. Therefore, it is neutral.
    (c) The mass of an electron is about 2000 times that of a proton.
    (d) A radioactive isotope of iodine is used for making iodine tincture, which is used as a medicine.

15. Rutherford's alpha-particle scattering experiment was responsible for the discovery of:
    (a) Atomic nucleus   (b) Electron
    (c) Proton   (d) Neutron

16. Isotopes of an element have:
    (a) The same physical properties
    (b) Different chemical properties
    (c) Different number of neutrons
    (d) Different atomic numbers

17. Number of valence electrons in a Cl^- ion are:
    (a) 16   (b) 8
    (c) 17   (d) 18

18. Which of the following is a correct electronic configuration of sodium?
    (a) 2, 8   (b) 8, 2, 1
    (c) 2, 1, 8   (d) 2, 8, 1

19. Complete the following table.

Atomic number	Mass number	Number of neutrons	Number of protons	Number of electrons	Name of the atomic species
9	–	10	–	–	–
16	32	–	–	–	Sulfur
–	24	–	12	–	–
–	2	–	1	–	–
–	1	0	1	0	–

Answers

In-Text Questions

1. Canal rays are positively charged radiations that can pass through perforated cathode plates and these rays are made up of positively charged particles known as protons.

2. As an electron is a negatively charged particle, whereas a proton is a positively charged particle and the magnitude of their charges is equal, an atom containing one electron and one proton will not carry any charge. Thus, it will be a neutral atom.

3. As per Thomson's model of the atom, an atom consists both negative and positive charges which are equal in number and magnitude. Hence, they balance each other as a result of which the atom as a whole is electrically neutral in nature.

4. On the basis of Rutherford's model of an atom, protons are present in the nucleus of an atom.

5. Refer to Section 4.6.3, Bohr's Atomic Model.

6. If the α-particle scattering experiment is carried out using a foil of any metal as the thin as gold foil used by Rutherford, there would be no change in observations. But as other metals are not as malleable as gold, such a thin foil would be difficult to obtain. If we use a thick foil, more α-particles would bounce back and no idea about the location of positive mass in the atom would be available with such accuracy.

7. The three subatomic particles of an atom are:
    (i) Protons, (ii) Electrons and (iii) Neutrons.

8. Number of neutrons = Atomic mass (A) − Number of protons (p)
    So, the number of neutrons in the atom = 4 − 2 = 2

9. (i) The total number of electrons in a carbon atom is 6. The distribution of electrons in a carbon atom is as follows:
    First orbit or K-shell = 2 electrons
    Second orbit or L-shell = 4 electrons
    That is, the distribution of electrons in a carbon atom is 2, 4.

    (ii) The total number of electrons in a sodium atom is 11. The distribution of electrons in a sodium atom is as follows:
    First orbit or K-shell = 2 electrons
    Second orbit or L-shell = 8 electrons
    Third orbit or M-shell = 1 electron
    That is, the distribution of electrons in a sodium atom is 2, 8, 1.

10. The maximum capacity of the K shell is 2 electrons and that of the L shell is 8 electrons. So, there will be a maximum of 10 electrons in the atom.

11. If the number of electrons in the outermost shell (valence electron) of the atom of an element is less than or equal to 4, the valency of the element is equal to the number of electrons in the outermost shell. On the other hand, if the number of electrons in the outermost shell of the atom of an element is greater than 4, the valency of that element is determined by subtracting the number of electrons in the outermost shell from 8.
    As the distribution of electrons in chlorine, sulfur and magnesium atoms are 2, 8, 7; 2, 8,

6 and 2, 8, 2, respectively, so, the number of electrons in the outermost shell of chlorine, sulfur and magnesium atoms are 7, 6 and 2, respectively.

Valency of chlorine = 8 − 7 = 1

Valency of sulfur = 8 − 6 = 2

The valency of magnesium = 2

12. (i) As the atomic number is equal to the number of protons, the atomic number of the atom is 8.

(ii) As the number of both electrons and protons is equal, the charge on the atom is 0.

13. Mass number of oxygen = Number of protons + Number of neutrons = 8 + 8 = 16

Mass number of sulfur = Number of protons + Number of neutrons = 16 +16 = 32

14.

Symbol	Proton	Neutron	Electron
H	1	0	1
D	1	1	1
T	1	2	1

15. $^{12}_{6}C$ and $^{14}_{6}C$ are isotopes of carbon and they have the same electronic configuration as (2, 4)

$^{22}_{10}Ne$ and $^{22}_{11}Na$ are isobars. They have different electronic configuration as given below:

$^{22}_{10}Ne$ = 2, 8, $\qquad$ $^{22}_{11}Na$ = 2, 8, 1

Exercise

1. Refer to Table 4.1, Comparison between proton, neutron and electron.

2. The limitations of J.J. Thomson's model of the atom are as follows:

It could not explain the result of scattering experiment performed by Rutherford.

It did not have any experimental support.

3. The limitations of Rutherford's model of the atom are:

It failed to explain the stability of an atom.

It does not explain the spectrum of hydrogen and other atoms.

4. Refer to Section 4.6.3, Bohr's Atomic Model.

5.

Thomson's model	Rutherford's model	Bohr's model
An atom consists of a positively charged sphere and the electrons are embedded in it.	An atom consists of a positively charged centre called the nucleus. The mass of the atom is contributed mainly by the nucleus.	Bohr proved almost all the points stated by Rutherford, except the one regarding the revolution of electrons for which he added that there are certain orbits known as discrete orbits inside the atom in which electrons revolve around the nucleus.
The negative and positive charges are equal in magnitude. As a result, the atom is electrically neutral.	The nucleus is very small (less than 100000 times as compared to the size of the atom).	While revolving in their discrete orbits, the electrons do not radiate any form of energy.

6. Refer to Section 4.7, Composition of the Nucleus.

7. The valency of an element is the combining capacity of that element. The valency of an element is determined by the number of valence electrons present in the atom of that element.

Valency of Silicon: It has electronic configuration 2,8,4; hence, the valency of silicon is 4 as these electrons can be shared with others to complete the octet.

Valency of Oxygen: It has electronic configuration 2,6; hence, the valency of oxygen is 2 as it will gain 2 electrons to complete its octet.

8. Refer to Section 4.9, Concept of Isotopes and Isobars.

9. The atomic number of sodium is 11. So, a neutral sodium atom has 11 electrons and its electronic configuration is 2, 8, 1. But Na^+ is formed by losing 1 electron, so it has 10 electrons. Out of 10, K-shell contains 2 and L-shell 8 electrons, respectively, and hence, Na^+ has completely filled K and L shells.

10. It is given that two isotopes of bromine are 79/35Br (49.7%) and 81/35Br (50.3%). Then, the average atomic mass of a bromine atom is given by:

$$79 \times \frac{49.7}{100} + 81 \times \frac{50.3}{100}$$

$$= \frac{3926.3}{100} + \frac{4074.3}{100} = \frac{8000.6}{100} = 80.006 \text{ u}$$

11. It is given that the average atomic mass of the sample of element X is 16.2 u. Let the percentage of isotope 18/8 X be y%. Thus, the percentage of isotope 16/8 X will be $(100 - y)$%. Therefore,

$$18 \times \frac{y}{100} + 16 \times \frac{(100 - y)}{100} = 16.2$$

$$\frac{18y}{100} + \frac{16(100 - y)}{100} = 16.2$$

$$\frac{18y + 1600 - 16y}{100} = 16.2$$

$$18y + 1600 - 16y = 1620$$
$$2y + 1600 = 1620$$
$$2y = 1620 - 1600$$
$$y = 10$$

Therefore, the percentage of isotope 18/8 X is 10%.

And, the percentage of isotope 16/8 X is $(100 - 10)$ % = 90%.

12. Z = 3 means that the atomic number of the element is 3. As its electronic configuration is 2, 1, the valency of the element is 1 (as the outermost shell has only one electron).
Hence, the element with Z = 3 is lithium (Li).

13. Mass number of X = Number of protons + Number of neutrons = 6 + 6 = 12
Mass number of Y = Number of protons + Number of neutrons = 6 + 8 = 14
As these two atomic species X and Y have the same atomic number, but different mass numbers, they are isotopes.

14. (a) F   (b) F   (c) T   (d) F

15. (a) Atomic nucleus

16. (c) Different number of neutrons.

17. (b) 8

18. (d) 2, 8, 1

19.

Atomic number	Mass number	Number of neutrons	Number of protons	Number of electrons	Name of the atomic species
9	19	10	9	9	Fluorine
16	32	16	16	16	Sulfur
12	24	12	12	12	Magnesium
1	2	1	1	1	Deuterium
1	1	0	1	0	Hydrogen ion